The Application of Hydraulic and Sediment Transport Models in Fluvial Geomorphology

The Application of Hydraulic and Sediment Transport Models in Fluvial Geomorphology

Special Issue Editors

Artur Radecki-Pawlik

Tomáš Galia

MDPI • Basel • Beijing • Wuhan • Barcelona • Belgrade • Manchester • Tokyo • Cluj • Tianjin

Special Issue Editors
Artur Radecki-Pawlik
Cracow University of Technology
Poland

Tomáš Galia
University of Ostrava
Czech Republic

Editorial Office
MDPI
St. Alban-Anlage 66
4052 Basel, Switzerland

This is a reprint of articles from the Special Issue published online in the open access journal *Water* (ISSN 2073-4441) (available at: https://www.mdpi.com/journal/water/special_issues/Fluvial_Geomorphology).

For citation purposes, cite each article independently as indicated on the article page online and as indicated below:

LastName, A.A.; LastName, B.B.; LastName, C.C. Article Title. *Journal Name* **Year**, *Article Number*, Page Range.

ISBN 978-3-03936-451-0 (Hbk)
ISBN 978-3-03936-452-7 (PDF)

Contents

About the Special Issue Editors

Artur Radecki-Pawlik (Prof., Ph.D., PE) is a river engineer who works at the Division of Structural Mechanics and Material Mechanics, Faculty of Civil Engineering, Cracow University of Technology. His research is related to river and mountain stream hydraulics, low-head hydraulic structures, river engineering, river morphology, hydrology, and sediment transport. Artur is an author or co-author of more than 70 scientific articles published in respectable journals (e.g., *River Research and Applications, Science of the Total Environment, Geomorphology, Water, Journal of Mountain Science,* and others). He is also an author and a co-editor of 10 books, including 2 international books: *Rivers – Physical, Fluvial and Environmental Processes* (Springer, 2015) and *Open Channel Hydraulics, River Hydraulic Structures and Fluvial Geomorphology* (CRC—Taylor and Francis, 2018).

Tomáš Galia (Ph.D.) is a fluvial geomorphologist who works at the Dpt. of Physical Geography and Geoecology, University of Ostrava, Czech Republic. His research is related to geomorphology and sediment transport in steep mountain streams, which also includes the impacts of human disturbances on sediment connectivity. Moreover, he has been working with processes related to instream wood in Mediterranean and Central European landscapes over the past few years. Tomáš is an author or co-author of more than 35 scientific articles published in respectable journals (e.g., *Water Resources Research, Science of the Total Environment, Catena,* and others).

Preface to "The Application of Hydraulic and Sediment Transport Models in Fluvial Geomorphology"

To study and understand fluvial geomorphology processes, it is important to look into the role of different kinds of models in hydraulic and sediment transport. In this Special Issue, we present 14 papers in the area of this research subject. These papers serve as an example of the present understanding of the application of hydraulic and sediment transport modeling of fluvial processes and their effect on streams and rivers.

The presented collection of papers is not exhaustive, but it highlights some priorities related to this subject and aims to fulfill knowledge gaps in this field of research.

Artur Radecki-Pawlik, Tomáš Galia
Special Issue Editors

Article

Comparison of Hydrodynamics Simulated by 1D, 2D and 3D Models Focusing on Bed Shear Stresses

Kurt Glock [1,*], Michael Tritthart [1], Helmut Habersack [2] and Christoph Hauer [1]

1 Christian Doppler Laboratory for Sediment Research and Management, Institute of Hydraulic Engineering and River Research, Department of Water-Atmosphere-Environment, University of Natural Resources and Life Sciences, Muthgasse 107, A-1190 Vienna, Austria; michael.tritthart@boku.ac.at (M.T.); christoph.hauer@boku.ac.at (C.H.)

2 Institute of Hydraulic Engineering and River Research, Department of Water-Atmosphere-Environment, University of Natural Resources and Life Sciences, Muthgasse 107, A-1190 Vienna, Austria; helmut.habersack@boku.ac.at

* Correspondence: kurt.glock@boku.ac.at; Tel.: +43-1-47654-81916

Received: 20 December 2018; Accepted: 26 January 2019; Published: 29 January 2019

Abstract: For centuries, scientists have been attempting to map complex hydraulic processes to empirical formulas using different flow resistance definitions, which are further applied in numerical models. Now questions arise as to how consistent the simulated results are between the model dimensions and what influence different morphologies and flow conditions have. For this reason, 1D, 2D and 3D simulations were performed and compared with each other in three study areas with up to three different discharges. A standardized, relative comparison of the models shows that after successful calibration at measured water levels, the associated 2D/1D and 3D/1D ratios are almost unity, while bed shear stresses in the 3D models are only about 62–86% of the simulated 1D values and 90–100% in the case of 2D/1D. Reasons for this can be found in different roughness definitions, in simplified geometries, in different calculation approaches, as well as in influences of the turbulence closure. Moreover, decreasing 3D/1D ratios of shear stresses were found with increasing discharges and with increasing slopes, while the equivalent 2D/1D ratios remain almost unchanged. The findings of this study should be taken into account, particularly in subsequent sediment transport simulations, as these calculations are often based on shear stresses.

Keywords: hydraulic models; numerical simulations; flow resistance; bed shear stress

---

1. Introduction

The interaction between gravity and flow resistance is the mechanism underlying both fundamental hydraulic processes and river morphological changes [1,2].

Following Powell [2], flow resistance is composed of the boundary resistance, vegetation resistance [3–7], channel resistance [8–12], spill resistance [13–18] and sediment transport resistance [19–25] and affects bed shear stress and energy dissipation. The challenge in river hydraulics is to achieve the complexity of flow resistance through physical relationships and engineering approaches to gain knowledge about flow velocities, water depths, bed shear stresses and energy losses in channels [2].

Classical approaches for the calculation of flow variables under the influence of flow resistance are the equations of Darcy–Weisbach, Chezy and Manning–Strickler [26–28]. These are empirical methods, which were developed from observations in channels in which the flow resistance was determined almost exclusively from the boundary of the flow—much like a block sliding down a plane [29]. Attempts for modifications of these classical approaches are particularly found for relatively rough beds in numerous subsequent research works [7,25,30–37]. However, the Manning–Strickler equation and the related bed roughness values of Manning's n and Strickler's k_{St} are still used in engineering practice and particularly in numerical modelling.

For the determination of the bed roughness value, k_{St}, Strickler [38] developed an equation considering the grain size distribution of the bed material. On the basis of this investigation Meyer-Peter and Müller [39] as well as Jäggi [40] proposed improved approaches by separately considering the surface layer of the bed material or the variable water depths in the channel, respectively. In addition, guidelines for the determination of k_{St}-values [41–43] have been developed for different channel types. However, the Strickler equation was developed for simplified channel geometries without considering lateral heterogeneities. To address these differences in compound and composite cross sections different approaches were elaborated in various research works [44–47]. Of particular interest in this context was vegetation in riverine landscape, which has a significant impact on flow resistance [48–50].

Despite some uncertainties, including the water depth dependencies of k_{St}-values, related to the use of Strickler's value, this approach is employed in a substantial percentage of all numerical simulation studies for the definition of the flow resistance. Most well-known 1D numerical models based on the Saint Venant equations (e.g., HEC-RAS, MIKE 11, etc.) and also 2D numerical models employing the depth-averaged Reynolds-equations (e.g., Hydro_AS-2D, MIKE21, TELEMAC-2D, etc.) define the roughness by Strickler's or Manning's value. In contrast, most 3D numerical models (e.g., Delft3D, Fluent, RSim-3D, SSIIM) solving the fully three-dimensional Reynolds-equations with different turbulence closures, typically permit the application of the equivalent sand roughness k_s [51], which is based on the sediment size and also allows accounting for the average height of bed forms. To date, numerous studies [52–61] have been published based on the mentioned numerical models to calculate hydrodynamic properties and their impacts on rivers.

Moreover, the modelled hydrodynamics are often used for predicting the sediment transport as well as morphological changes in rivers. In this context, the bed shear stress is of particular importance, due to the fact that well-known bed load transport equations [39,62–65] are based on this variable and are used both in analytical calculations [66,67] as well as numerical simulations [68–71].

These issues show the paramount importance of roughness values in the calculation of bed shear stresses in the context of numerical simulations. Independent of the dimensionality of the models this roughness value usually represents the main opportunity for parameterizing flow resistance in numerical simulations. Moreover, it is the main variable parameter of choice to achieve successful model calibration implying reasonable difference between measured and modelled water surface elevations as well as flow velocities.

Thus, the central research question of this paper is whether valid calculations of water surface elevations in different model dimensions (1D, 2D and 3D) imply a consistent depiction of flow velocities and near-bed flow forces. Of particular interest is a comparison of simulated bed shear stresses of different models, due to their substantial influence on the simulation of bed load transport. The influence of certain characteristic morphologies including artificial bathymetries, pool-riffle sections as well as river bends is investigated on the basis of three different case studies.

2. Materials and Methods

2.1. Case Studies

The locations of the different case studies are depicted in Figure 1a. Case study 1 (C.1) "S-Shaped Trapezoidal Channel" (Figure 1b) represents controlled conditions within a laboratory experiment and was selected to focus on the influence of river bends without any morphological structures. C.2 "Ybbs" (Figure 1c) was chosen to investigate the effects of a pool-riffle section in a stretched natural river site [72]. The complex morphology of the second natural river site C.3 "Sulm" (Figure 1d) combines the different characteristics representing a river meander with pool-riffle sections [72].

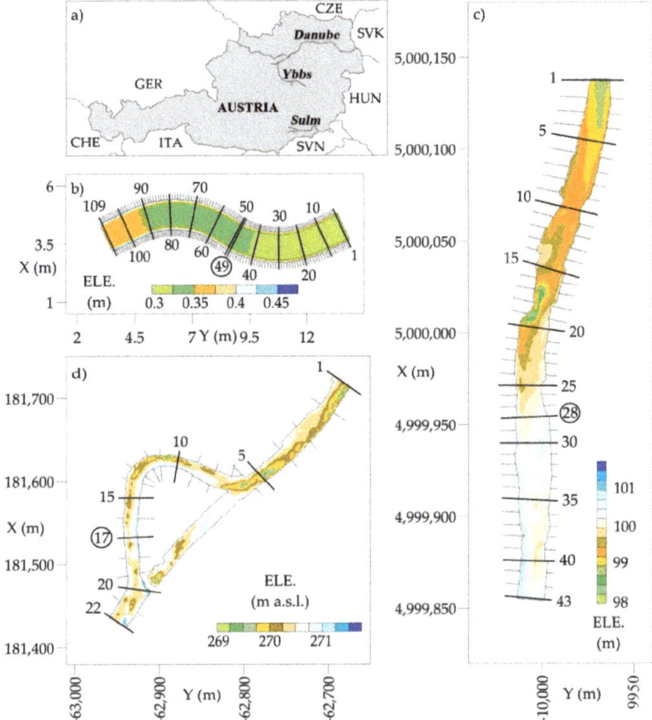

Figure 1. Location of case studies in Austria (**a**); Overview of case studies including cross-sections used for a generalized comparison (grey and black lines) and the indicated cross-section for detailed analysis (encircled): Design of laboratory experiment representing C.1 (**b**), bed elevations of natural river site "Ybbs" representing C.2 (**c**) and bed elevations of natural river site "Sulm" representing C.3 (**d**).

2.1.1. S-Shaped Trapezoidal Channel (C.1)

The laboratory experiment was developed at the Hydraulic Engineering Institute of the University of Innsbruck, consisting of a physical laboratory channel 3.0 m in length followed by two consecutive bends with a mean radius of 4.0 m and apex angle of 60 each, and finally an exit channel being 2.8 m long with a bed slope of 0.005. The banks are sloped by a ratio of 2:3. The bed sediments are characterized by a mean diameter d_m of around 5 mm. One set-up of the physical experiment, with a discharge of 0.09 $m^3\ s^{-1}$, was part of the validation study of the numerical model RSim-3D [73], which is also used within this research. Measurements of the water surface elevation gained during the experiment form the basis for calibrating the numerical models.

2.1.2. Ybbs (C.2)

The river site C.2 "Ybbs" (Figure 1c) has a total length of around 300 m with an average bed slope of 0.0055 and is classified as a straight river due to natural morphological constraints [74]. The bed sediments are characterized by a mean diameter d_m of around 70 mm [72], excluding the riffle section (located between cross-section 21 and 31) with a d_m of around 93 mm. The hydrology is defined by conditions during low flow ($Q_L = 1.6\ m^3\ s^{-1}$), mean flow ($Q_M = 4.5\ m^3\ s^{-1}$), and a flood with a return period of one year ($HQ_1 = 64.0\ m^3\ s^{-1}$). In the course of a monitoring campaign in 2006, the bathymetry as well as the water surface elevation were measured at low flow conditions in a local reference system and are the basis for calibrating the numerical models within this study. The calibrated models are further used in the course of the comparison study including the simulation of Q_M and HQ_1. Due to the fact that the valley margins of C.2 are close to the river bed, the inundation areas are greatly reduced, thus this river site is particularly suitable for a sensitivity analysis of flow resistances during floods.

2.1.3. Sulm (C.3)

The river site C.3 "Sulm" (Figure 1d) has a total length of around 500 m with an average bed slope of 0.0010 and is classified as a meandering river [74] with a mean diameter d_m of the surface layer of around 42 mm [72]. The hydrology is characterized by conditions during low flow ($Q_L = 2.7$ m^3 s^{-1}), mean flow ($Q_M = 9.0$ m^3 s^{-1}) and a flood with a return period of one year ($HQ_1 = 95.0$ m^3 s^{-1}). During a monitoring campaign in 2002, the bathymetry as well as the water surface elevation were measured at low flow conditions and are the basis for calibrating numerical models within this study. Additionally, Q_M is simulated with the calibrated models in the course of the comparison study.

2.2. Basic Principles

2.2.1. Empirical Relationship between the Equivalent Sand Roughness k_s and the Strickler Value k_{st}

Strickler [38] established the eponymous roughness value k_{St} considering the grain size distribution of the bed material, which is given by

$$k_{St} = \frac{a}{\sqrt[6]{k_S}} \tag{1}$$

where a is an empirical parameter and k_S is the equivalent sand roughness, which is a characteristic value of the grain size distribution. Following this approach Meyer-Peter and Müller [39] as well as USACE [43] adjusted a and k_S (Table 1). According to Table 1, k_S ranges between grain size diameters of d_{50} and d_{90}, for which 50% and 90% of the cumulative mass are finer, respectively. Jäggi [40] retained the fundamental concept of Strickler, whereby he developed a new approach for calculating a

$$a = (2.5\, g^{0.5}\, \ln{(6.1\, Z)}/Z^{0.16})(1 - e^{-0.02Z/I^{0.5}})^{0.5} \tag{2}$$

which depends on the relative roughness $Z = h/k_S$—including k_S and the water depth h—and the bed slope I, resulting in a range of possible a values (Table 1).

Table 1. Overview of parameters, a and k_S.

Author	a	k_s	Restriction
Strickler [38]	21.1	d_{90}	base layer
Meyer-Peter and Müller [39]	26	d_{90}	armor layer
USACE [43]	29.4	d_{50}	no limitation
Jäggi [40]	2.0–23.3	d_{90}	no limitation

Independent of the different approaches, the equations are applicable only to medium-range values for the Strickler parameter. Naudascher [42] specified the applicability of these equations by defining a range of typical Reynolds numbers in rivers between $20 < 4R/k_S < 1000$ where R denotes the hydraulic radius [73].

2.2.2. Calculation of Bed Shear Stress

Before going into the details of any results, the numerical principles of bed shear stress calculations in different models are demonstrated. The classical approach in hydraulic engineering for determining the average bed shear stress τ in a cross-section is known as the Du Boys equation and is given by

$$\tau = \rho g R I_f \tag{3}$$

where ρ is the density of water, g is the gravitational acceleration, R is the hydraulic radius in a cross-section and I_f is the friction (or energy) slope. Due to the fact that in very broad channels, characterized by a channel width B higher than 30 times the water depth h, the hydraulic radius R can be replaced by the h, the Du Boys equation is simplified to

$$\tau = \rho g h I_f, \text{ for } B \geq 30h \tag{4}$$

1D numerical models usually employ the approach of Du Boys (Equation (3)).

This equation, however, is also applied in the case of 2D numerical models for determining bed shear stresses. Nevertheless, the calculation is improved due to the fact that hydraulic information (water depth h, flow velocity v) is available in each computational node leading to a higher resolution of results. The friction slope I_f is expressed by the Strickler-equation

$$I_f = \frac{v^2}{k_{St}{}^2 R^{\frac{4}{3}}} \tag{5}$$

Considering the simplification of replacing the hydraulic radius R by the water depth h and taking into account Equation (5), the approach for calculating bed shear stresses in the case of 2D numerical models is given by

$$\tau = \frac{\rho g v^2}{k_{St}{}^2 h^{\frac{1}{3}}} \tag{6}$$

A different approach to 1D and 2D modelling based on the law of the wall is usually employed in the case of 3D numerical models. A well-established method is given by [75]

$$\tau = \rho C_\mu{}^{\frac{1}{4}} k_P{}^{\frac{1}{2}} \frac{v_{p,t}}{v^+} \tag{7}$$

where C_μ is an empirical constant of the k$-\varepsilon$ model, k_P denotes the turbulent kinetic energy (which can also be interpreted as turbulent flow velocity $v_{k_P} = \sqrt{k_P}$ at the near-wall node), $v_{p,t}$ is the near-wall tangential velocity and v^+ is representative of the velocity profile at turbulent flow conditions, according to the law of the wall [76] given by

$$v^+ = \frac{1}{\kappa} \ln\left(\frac{y}{y_k}\right) \tag{8}$$

where κ is the Von Karman constant which takes the value of 0.41, y is the wall distance and y_k is a parameter dependent of the equivalent sand roughness of Nikuradse (k_s):

$$y_k = k_s{}^{-8.0\kappa} \tag{9}$$

Equations (8) and (9) describe the dependence of v^+ on the geometry as well as on the roughness.

Considering that the square root of k_P results in a velocity depending on the turbulence and $v_{p,t}$ as near-wall tangential velocity, the similarity between Equation (6) and (7) becomes visible. However, the bed shear stress in the case of the applied 3D numerical model is dependent on the turbulence as well as on the tangential velocity, whereas only the tangential velocity is taken into account in the case of the 2D numerical model.

2.3. Numerical Setup

2.3.1. 1.D Numerical Model

As a representative of 1D step-backwater models, the software HEC-RAS Version 4.1 [77] was used within this study. The spatial discretization in this software is provided by cross-sections including information about bed elevations and distances between cross-sections. Manning's value is used for the definition of roughness. The computational procedure for steady flow is based on the solution of the one-dimensional energy equation coupled with the momentum equation in situations where the water surface profile varies rapidly. For unsteady flow, the fully dynamic 1D Saint Vernant equation is solved using an implicit finite difference method.

2.3.2. 2D Numerical Model

The 2D numerical model Hydro_AS-2d [78], employing the depth-averaged Reynolds-equations, was applied in this study. The roughness is defined by Strickler's or Manning's value and the hydraulic conditions are calculated on a linear grid by a finite volume approach. The Surface-Water Modelling System (SMS) is used as a pre- and post-processing tool of the used 2D numerical model. The convective flow of the two-dimensional model is based on the Upwind-scheme by Pironneau [79] and the discretization of time is done by a second order explicit Runge Kutta method. The simulations were performed with the parabolic eddy viscosity model and a constant turbulent viscosity coefficient of 0.6.

2.3.3. 3D Numerical Model

The 3D river flow simulation model RSim-3D [73] was used for this comparative study. The spatial discretization in this software is provided by applying an unstructured computation grid consisting of polyhedral cells. Employing a generalized finite volume method, the Reynolds equations are approximated numerically. The SIMPLE method provides coupling of pressure and velocity fields, the standard k-ε model is implemented for turbulence closure, and the determination of the water surface elevation is achieved by evaluating the computed pressure at the water surface.

2.3.4. Spatial Discretization

An overview of the spatial discretization is given for all numerical models and all case studies in Table 2. The area of C.1 is covered by 109 cross-sections in the 1D model as well as 7980/47,880 computation nodes in 2D/3D, which are based on a horizontal, rectangular discretization of 0.1×0.03 m. For the numerical calculations of C.2, 43 cross-sections (1D), 18,918/59,220 (2D/3D) computation nodes were employed. The 2D computation mesh consists of rectangular cells with a node distance of 1.0×1.0 m, while in the case of 3D a polyhedral mesh of hexagonal cells (node distance: 1.0×1.0 m) was employed. The area of C.3 is covered by 22 cross-sections in the 1D model, 28,396/170,376 computation nodes in 2D/3D. The 2D computation mesh is based on triangular cells with a node distance of 1.0×1.0 m, while the 3D polyhedral mesh consists of 1.0×1.0 m hexagonal cells. Independent of the case study the water column was subdivided into six vertical layers. In general, several spatial discretizations of varying node distances (double and half the selected distance) were tested to ensure grid-independent solutions.

Table 2. Overview of the spatial discretization of all the numerical models (h. = hexagonal, r. = rectangular, and t. = triangular mesh discretization).

Case Study	1D	2D		3D		
	Cross-Sections	**Discretization (m × m)**	**Computation Nodes**	**Discretization (m × m)**	**Vertical Layers**	**Computation Nodes**
C.1	109	r. 0.1 × 0.03	7980	r. 0.1 × 0.03	6	47,880
C.2	43	r. 1.0 × 1.0	18,918	h. 1.0 × 1.0	6	59,220
C.3	22	t. 1.0 × 1.0	28,396	h. 1.0 × 1.0	6	170,376

2.3.5. Boundary and Initial Conditions

The inflow and outflow boundary conditions are listed in Table 3 for all simulation scenarios. At the inflow the normal velocities are derived from the given discharge and prescribed as boundary conditions. The normal velocities are calculated from a division of the total discharge and the boundary area, resulting in the specific discharge for 1D and 2D and the cell-based velocities in 3D, which are assigned orthogonal to the inflow cell faces. At the outflows energy slopes (1D and 2D models) or water surface levels (3D models) are used.

Table 3. Overview of boundary conditions for all simulation scenarios.

Acronym	Simulation Scenario	Inflow Boundary Discharge (m^3/s)	Outflow Boundary	
			Slope (-)	Water Level (m a.s.l.)
C.1	Calibration	0.09	0.005	0.394
C.2—Q_L	Calibration	1.64	0.005	99.140
C.2—Q_M	Validation	4.50	0.005	99.370
C.2—HQ_1	High flow scenario	64.00	0.005	101.073
C.3—Q_L	Calibration	2.66	0.0015	270.460
C.3—Q_M	Validation	8.85	0.0015	270.730

The 1D and 2D numerical simulations were performed without particular initial conditions (dry channel), while in the case of 3D numerical simulations the initial free surface elevations were set based on the results of the 1D numerical simulations.

2.4. Comparison of Modelling Results

In order to ensure comparability between the different model dimensions as well as case studies, result ratios $M_{r,d,c,q}$ were considered,

$$M_{r,d,c,q} = \left\{ \frac{r_{d,c,q,s,n}}{R_{c,q,s}} \right\} \tag{10}$$

where $r_{d,c,q,s,n}$ denotes the 2D or 3D simulated result (water depth, depth-averaged flow velocity or bed shear stress), while the 1D hydrodynamic result $R_{c,q,s}$ serves as reference in the calculation. The resulting ratio $M_{r,d,c,q}$ is the basis of the boxplot. The indices used point out the origin of individual sets distinguished by results r (water depth, depth-averaged flow velocity or bed shear stress), dimensions d (2D or 3D), case studies c (C.1 to C.3) and discharges q (Q_L to HQ_1). Additionally used indices representing the cross-section s and computational node n sliced by the cross-section are necessary for the ratio calculations. The average value is calculated separately for each individual set $M_{r,d,c,q}$.

3. Results

3.1. Calibration of Numerical Models

In the course of the calibration, the flow resistances were adjusted in the numerical models until a match with measured water surface elevations was achieved. Good agreements with average differences below 1 cm between measurements and simulations were obtained for C.1 (Figure 2a—solid lines) for all numerical models excluding the inlet area between distance 10.5 m and 9.5 m, where maximum differences of 3 cm occurred, which are related to issues of the physical model setup. The calibrated roughness values are k_{St} = 52 $m^{1/3}$ s^{-1} for the 1D and 2D model and k_S = 0.005 m for the 3D model.

In C.2 (Figure 2b—solid lines), comparisons of measurements and numerical calculations show average differences of 2 cm excluding maximum outliers of 12 cm (1D) and 9 cm (2D and 3D) at a distance of 90 m. The single occurrence of the outlier indicates a local measurement issue. Best agreement was achieved by employing roughness values in section 215–150 m of k_{St} = 9.1/9.0 $m^{1/3}$ s^{-1} (1D/2D-model) and k_S = 1.5 m (3D model), while the areas up- and downstream of this section (300–215 m and 150–0 m) received roughness values of k_{St} = 18.2/18.0 $m^{1/3}$ s^{-1} (1D/2D) and k_S = 0.25 m (3D model). The higher flow resistance in section 215–150 m reflects the stronger turbulent flow conditions in this area, which is characterized by a steeper bed slope and coarser bed sediments.

The average differences in C.3 (Figure 2c) are below 1 cm except the comparison at distance 220 m, where outliers of 9 cm (2D) and 8 cm (3D) occur. The single occurrence of the outlier again indicates a local measurement issue. The calibrated roughness values in C.3 are k_{St} = 25.0 m$^{1/3}$ s^{-1} for the 1D and 2D model and k_S = 0.20 m for the 3D model.

In the natural river sites (C.2 and C.3), additional discharge scenarios, defined by the mean discharge of each area (Table 3), have been simulated with the same roughness values and are depicted in Figure 2b,c with dotted lines. Due to lack of data, an evaluation with measurements was not possible, but the results were compared among each other. In C.2 (Figure 2b—dotted lines) the maximum average difference of 3 cm occurred between 2D and 3D simulations, while the maximum average difference in C.3 (Figure 2c—dotted lines) was 1 cm, determined between 1D and 3D simulations.

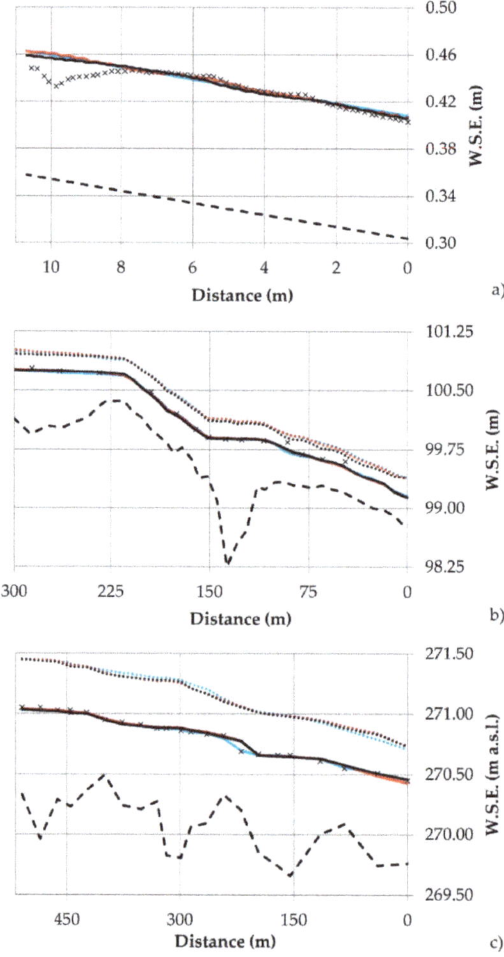

Figure 2. Calibration—Longitudinal profile plots of (**a**) C.1, (**b**) C.2 and (**c**) C.3 including bed elevations (dashed black line) and comparisons of measured (black crosses) and modelled water surface elevations (W.S.E.) at low flow (Q_L) conditions (results of 1D simulations = blue solid line, 2D simulations = red solid line and 3D simulations = black solid line); Comparison of modelled W.S.E. at an additional discharge scenario (mean flow Q_M) in C.2 and C.3 (results of 1D simulations = blue dotted line, 2D simulations = red dotted line and 3D simulations = black dotted line).

3.2. Sensitivity Analysis of Roughness Values in the Case of High Flow Conditions

The difficulties of using numerical simulations in cases of high flow conditions (HQ_1) without monitoring data is assessed in the form of a sensitivity analysis in C.2, demonstrating the differences between various models and the influences of changing roughness values. The water surface elevations simulated by 1D (blue solid line), 2D (red solid line) and 3D (black solid line) models is shown in Figure 3 including a bandwidth (dotted lines with corresponding colors) calculated with roughness values, which were changed by +/−20% referred to calibration (Section 3.1). Good agreements were found between 1D and 2D simulations, using roughness values of calibration resulting in average difference of around 5 cm, while the comparison to 3D simulations yields increasing differences starting with values of 1 cm at the outflow boundary (river section 0 m) up to 60 cm at the inflow boundary (river section 300 m). Following a roughness reduction of 20% in the case of 1D and 2D models, the highest differences compared to 3D simulations at reference conditions reduce to 40 cm, while a roughness increase of 20% in the case of 3D modelling leads to the smallest differences of 15 cm compared to 1D using roughness values of the calibration.

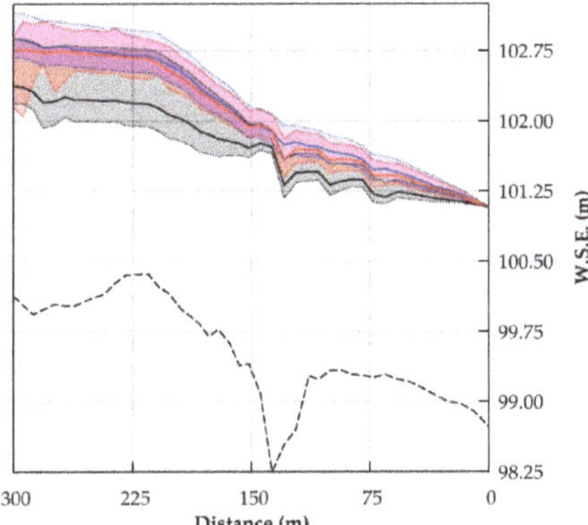

Figure 3. Sensitivity analysis—Longitudinal profile plot of C.2 at high flow conditions (HQ_1) including bed elevations (dashed black line) and modelled water surface elevations (W.S.E.) (results of 1D simulations = blue solid line, 2D simulations = red solid line and 3D simulations = black solid line) with a bandwidth considering +/−20% of roughness changes (dotted lines with corresponding colors).

3.3. Analysis of Bed Shear Stresses

Bed shear stresses simulated in C.1 are depicted in Figure 4a for all numerical models. In the case of the 1D simulation, the bed shear stresses are 4.9 Nm^{-2} without any differences in the lateral or longitudinal direction. Along the thalweg, similar values are obtained in the higher dimensional models, but due to the two consecutive bends higher bed shear stresses up to 6.5 Nm^{-2} and 10.0 Nm^{-2} occur along the inner banks of the 2D and 3D simulations, respectively, while the outer banks are characterized by lower values of around 2.0/0.4 Nm^{-2} (2D/3D).

In C.2 at mean flow conditions (Q_M; Figure 4b), the highest bed shear stresses arise in the area between cross-sections 21 and 30, which is a riffle section [72] with low water depths, high flow velocities as well as high roughness values, resulting in values of 79/108/93 Nm^{-2} depending on the model dimension (1D/2D/3D). In the other sections, the bed shear stresses are substantially lower with values between 2 and 16 Nm^{-2} upstream the riffle section and between 2 and 41 Nm^{-2} downstream of this section in all models. Independent of the section, it is obvious that the 1D model is not able to predict any lateral variability, leading to higher values at the river banks and lower bed shear stresses in the main stream compared to higher dimensional models. In addition, it is shown that the values in the main stream of the 2D model tend to be larger in comparison to the 3D model, while at the banks a slight reverse trend occurs.

The modelling results of C.3 at mean flow conditions (Q_M; Figure 4c) also yield the highest bed shear stresses in the riffle sections, which are located between cross-sections 7 and 11 (21/33/44 Nm^{-2} for 1D/2D/3D) as well as cross-section 16 and 19 (13/16/19 Nm^{-2} for 1D/2D/3D), and are independent of the model dimension. While in the 2D simulations, the peaks occur in the main stream of the river, the highest values in the 3D simulation arise at the river banks. The reduced spatial discretization in the case of 1D simulations is again shown by constant values in the lateral direction.

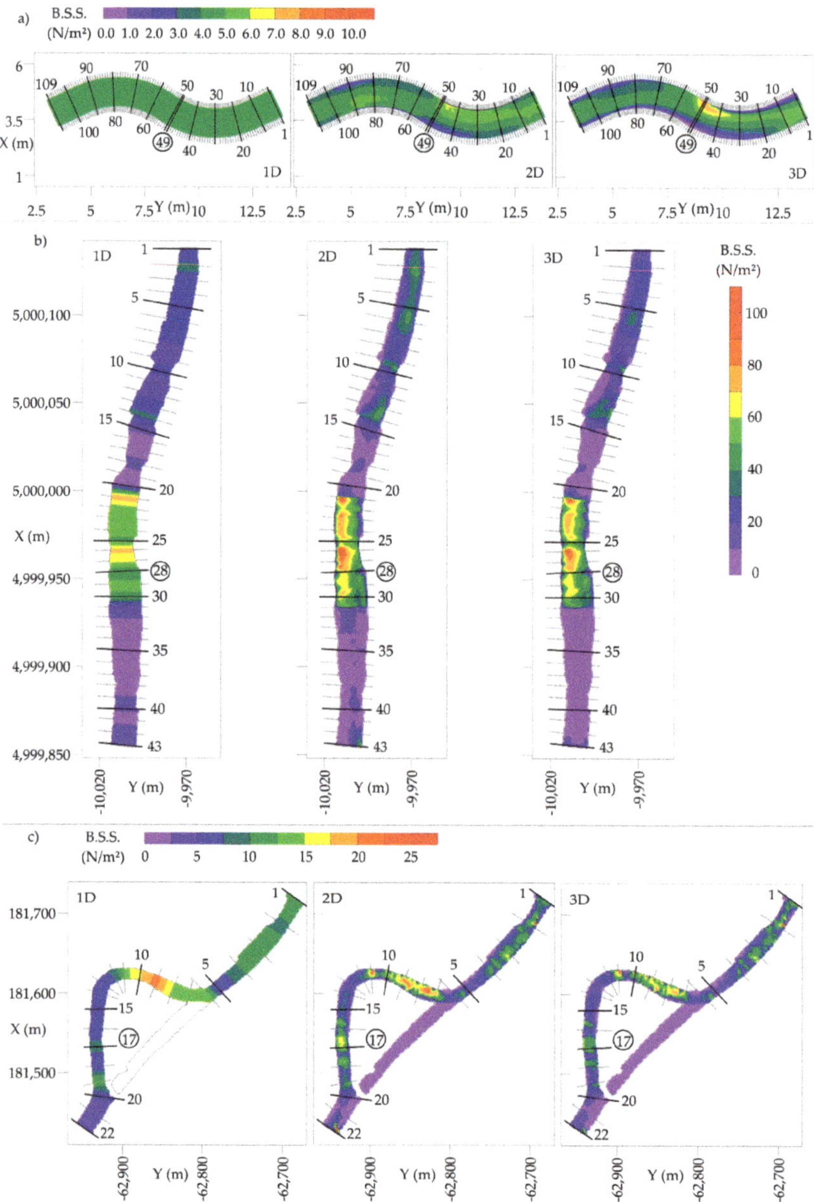

Figure 4. Comparison of bed shear stresses calculated by 1D/2D/3D models depicted for (**a**) C.1, (**b**) C.2 at Q_M and (**c**) C.3 at Q_M.

3.4. Generalized Comparison of Numerical Models

A generalized representation of modelling results including water depths, depth-averaged flow velocities and bed shear stresses is depicted in Figure 5 for all case studies and discharge scenarios. The water depth ratio between 2D/1D and 3D/1D (Figure 5a) at low flow and mean flow conditions is almost unity, stating that no substantial differences in the calculated water depths occurred by numerical models of different dimensions. An exception is the ratio 3D/1D at high flow conditions with quartiles of 0.80 and 0.90. Figure 5b shows the depth-averaged flow velocity ratio between 2D/1D and 3D/1D for all scenarios. The quartiles in the case of 2D/1D are between 0.64 and 1.23 and 0.61 and 1.43 in the case of 3D/1D. The bed shear stress ratios between 2D/1D and 3D/1D (Figure 5c) are characterized by quartiles of 0.45 and 1.28 and 0.43 and 1.15, respectively. Additionally, a summarized analysis (Figure 5d) including the averaged ratio between 2D and 3D values referred to 1D values shows almost no differences in the case of water depths (blue) at low flow and mean flow conditions resulting in values close to unity. An exception is the ratio 3D/1D at high flow conditions with a value of 0.85. In the case of flow velocities (grey), averaged ratios of 0.88 to 0.99 (2D/1D) and 0.88 to 1.13 (3D/1D) were obtained, whereby the ratios increase with the discharges in each case study. The largest deviations were found again for bed shear stresses (brown) with values of 0.90 to 1.00 (2D/1D) and 0.62 to 0.86 (3D/1D). Independent of the scenario, the calculated ratios are below the line of perfect agreement, indicating on average lower bed shear stresses in 3D models compared to 2D models.

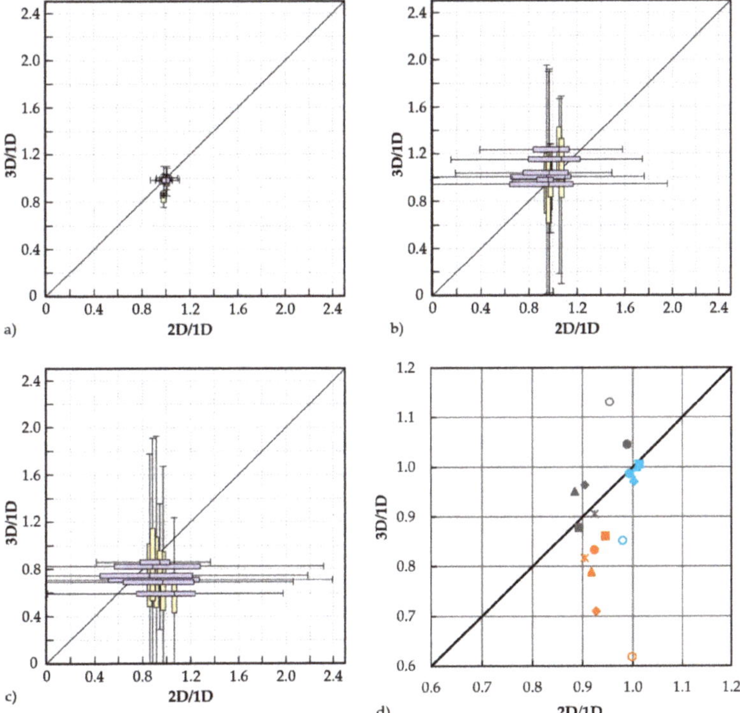

Figure 5. Relative comparison of the modeling results of (**a**) water depths, (**b**) depth-averaged flow velocities and (**c**) bed shear stresses consisting of ratios between 3D and 2D values, referring to 1D values depicted in box-plots for each case study and discharge scenario; (**d**) Average ratio between 3D and 2D values, referring to 1D values for water depths (blue), flow velocities (grey) and bed shear stresses (brown) depicted for each case study and discharge scenario (C.1 (*), C.2-Q_L (▲), C.2-Q_M (♦), C.2-HQ_1 (○), C.3-Q_L (■), C.3-Q_M (•)).

Although models were calibrated to water surface elevations, the results exhibit clear differences in bed shear stress ratios. While in the case of 2D/1D, minor differences are noticed between the different river sites and discharges, the ratio decreased in the case of 3D/1D from lower to higher discharges ($Q_L \rightarrow Q_M$ and $Q_L \rightarrow HQ_1$) as well as from small-sloped to steeper river sites (C.3→C.2).

3.5. Detailed Anaylsis of Simulated Hydrodynamics in Cross-Sections

Figure 6 depicts the simulated hydrodynamics of all model dimensions including bed shear stresses, flow velocities, turbulent kinetic energy (3D) and water surface elevations in three representative cross-sections of all case studies (CS 49 of C.1, CS 28 of C.2, CS 17 of C.3—encircled in Figure 1). The different model dimensions are based on various definitions of flow velocities. The area-averaged flow velocity (light grey solid line) in the case of 1D, the depth-averaged flow velocity (2D—dark grey solid line) and the near-wall tangential flow velocity (3D—black solid line), which equals the near-wall flow velocity in the case of small bed cell slopes.

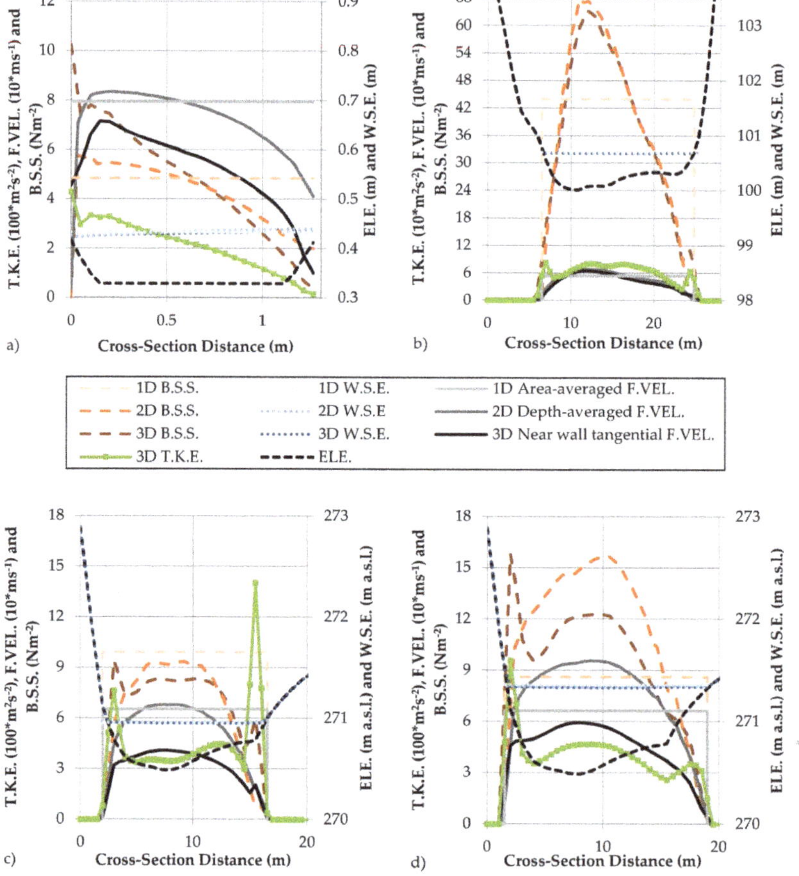

Figure 6. Detailed analysis of the hydrodynamics simulated by the 1D, 2D and 3D models in (**a**) CS 49 of C.1, (**b**) CS 28 of C.2 at Q_M, (**c**) CS 17 of C.3 at Q_L and (**d**) CS 17 of C.3 at Q_M including water surface elevations (blue dotted lines), bed shear stresses (brown large-spaced dashed lines), various flow velocities (area-averaged flow velocities (grey lines), depth-averaged flow velocities (dark grey lines), near wall tangential flow velocities (black lines)) and turbulent kinetic energies (green lines with rectangles) as well as bed elevations (black small-spaced dashed lines).

The area-averaged flow velocity in CS 49 of C.1 (Figure 6a) is constant with a value of 0.79 m s^{-1}. The profile of the 2D depth-averaged flow velocities is characterized by a maximum of 0.84 m s^{-1} close to the left bank and a minimum of 0.41 m s^{-1} on the right bank. A similar profile was calculated for the near-wall tangential flow velocity in the 3D model, whereby the values are substantially lower, characterized by a peak of 0.72 m s^{-1} and a minimum of 0.10 m s^{-1}. The near-wall turbulent kinetic energy follows a falling trend starting with the highest value (0.04 m^2 s^{-2}) at the left bank and ending with the lowest value at the right bank (0.01 m^2 s^{-2}). The water depths are almost equal in all models with an average value of 0.43 m, whereby slight increasing trends from the left to the right bank occur in the 2D and 3D models. Due to the trapezoidal channel geometry, the lowest water depths arise at the banks. The comparison of bed shear stresses displays the dependency of this variable on flow velocities, turbulent kinetic energies and water depths. In the case of 2D, the profile of the bed shear stresses is similar to the profile of the flow velocities resulting in a maximum of 5.8 Nm^{-2} in the area of the left bank and a minimum of 2.0 Nm^{-2} close to the right bank, but due to the influence of the water depths an offset between the peaks of bed shear stress and flow velocity occurs. Comparable results were found in the case of 3D models, where similarities occur between the profiles of the near-wall tangential flow velocities and bed shear stresses, with a maximum (10.2 Nm^{-2}) on the left bank and a minimum (0.4 Nm^{-2}) on the right bank. However, the strong influence of the turbulent kinetic energy is obvious particularly at the left bank resulting in the highest bed shear stresses of the entire cross-section. In contrast, the one-dimensional approach results in a constant value of 4.8 Nm^{-2} in the whole cross-section.

In CS 28 of C.2 (Figure 6b), the area-averaged flow velocity is constant with a value of 0.54 m s^{-1}. In the case of 2D, the profile of the depth-averaged flow velocities is characterized by a maximum of 0.69 m s^{-1} in the main stream of the river located at distance 11 m of the cross-section. Similarities are present for the profile of the near-wall tangential flow velocity in the 3D model with a peak of 0.64 m s^{-1}. The near-wall turbulent kinetic energies are characterized by a relative constant plateau in the cross-section center (11–19 m) with values between 0.70 and 0.80 m^2 s^{-1}, a decreasing tendency to the banks and peaks up to 0.83 m^2 s^{-1} on the banks. The water surface elevations are almost equal in all models with an average value of 100.68 m. Due to the straight river course in this cross-section, the water surface elevations do not change between the banks, hence the water depths only depend on the channel geometry. The comparison of bed shear stresses again indicates the dependency of this variable of flow velocities, turbulent kinetic energies and water depths. In the case of 2D as well as 3D, the profile of the bed shear stresses is similar to the profile of the depth-averaged and tangential flow velocities, resulting in maxima of 65.9 and 63.3 Nm^{-2} in the main stream of the river. However, the influence of the turbulent kinetic energy is visible at the right bank, resulting in a peak of 11.0 Nm^{-2}. In contrast, the one-dimensional approach leads to a constant value of 44.0 Nm^{-2} in the whole cross-section.

The hydrodynamic variables for low flow Q_L (Figure 6c) and mean flow conditions Q_M (Figure 6d) are depicted for CS 17 of C.3. The area-averaged flow velocity is constant over the whole cross-section with values of 0.65 m s^{-1} (Q_L) and 0.66 m s^{-1} (Q_M). The profile of the 2D depth-averaged flow velocities is characterized by a maximum of 0.68 m s^{-1} (Q_L) and 0.96 m s^{-1} (Q_M), located in the main stream of the river. A similar profile was calculated for the near-wall tangential flow velocity in the 3D model, whereby the values are substantially lower with a peak of 0.41 m s^{-1} (Q_L) and 0.59 m s^{-1} (Q_M).

The near-wall turbulent kinetic energies are characterized by values of 0.03–0.04 m^2 s^{-2} (Q_L), and 0.03–0.045 m^2 s^{-2} (Q_M) in the main stream and peaks up to 0.14 m^2 s^{-2} (Q_L), and 0.10 m^2 s^{-2} (Q_M) at the banks. The water surface elevations are almost equal in all models with an average value of 270.95 m a.s.l. at low flow conditions and 271.36/271.34/271.33 m a.s.l. (1D/2D/3D) at mean flow conditions. Due to the straight river course in this cross-section, the water surface elevations do not vary between the banks, thus the water depths only depend on the present geometry. The comparison of bed shear stresses again displays the dependency of this variable on flow velocities, turbulent kinetic energies and water depths. Independent of the discharge, the profiles of the bed shear stresses calculated in the 2D model follow the profiles of the depth-averaged flow velocities, resulting in peak values of 9.4 Nm^{-2} (Q_L) and 15.7 Nm^{-2} (Q_M) in the main stream of the river. In the 3D model, comparable results were found in the main stream, where similarities between the profiles of the near-wall tangential flow velocities and bed shear stresses occur, with a maximum of 8.4 Nm^{-2} (Q_L) and 12.2 Nm^{-2} (Q_M). However, the strong influence of the turbulent kinetic energies close to the banks is obvious leading to the highest bed shear stresses of 9.4/15.8 Nm^{-2} (Q_L/Q_M) in the entire cross-section. In contrast, the one-dimensional approach results in a constant value of 9.9/8.6 Nm^{-2} (Q_L/Q_M) in the whole cross-section.

4. Discussion

The comparison of simulation results calculated in different model dimensions in general involves the risk of potential result dependencies on the applied spatial discretization. In order to overcome these issues, a sensitivity analysis including a variation of horizontal node distances was performed to ensure grid-independent solutions. However, in the case of 3D simulations the vertical discretization must be considered as well. In engineering practice, this discretization is usually a justified compromise between computational time and numerical accuracy. Hence, in sediment transport simulations, for which a correct prediction of the shear stress parameter is paramount, which depends on the vertical discretization, literature reports the usage of typically five to eight layers; good agreement with measurements has shown to justify this choice [53,58,59,69]. In line with this practice, the simulations in this study were performed with six layers.

Based on grid independency, the numerical models were successfully calibrated at low flow conditions. Although all site-specific heterogeneities (e.g., bed sediments, channel units and vegetation) have to be aggregated in roughness values, good agreement with measured water surface elevations was achieved in all areas investigated. The study shows that the roughness values are constant along the river course in all case studies except C.2, where a substantial coarsening of sediments in a riffle section led to a lower Strickler value and a higher equivalent sand roughness, respectively. The determined values are in line with guidelines [41–43].

However, the comparison of simulation results calculated by different model dimensions shows that the largest differences were found for bed shear stresses, due to the following reasons: (i) The averaged comparison ratio is lower than unity because of the area-averaged approach in the case of 1D models resulting in overestimations in bank areas. (ii) Moreover, the values are lower in general in the case of 3D compared to 2D simulations, which has also been reported by Lane et al. [80] and is based on the different calculation approaches. The bed shear stress calculations in 2D models are based on the depth-averaged flow velocities, which are in general higher than the near-wall flow velocities used in 3D models. (iii) Additionally, it was shown that the averaged comparison ratio between 3D/2D and 3D/1D is decreasing with increasing discharges, independent of the case study. This fact is again related to the different calculation approaches based on the corresponding roughness definitions.

A closer look into the different roughness definitions reveals the known fact that the Strickler value is a variable depending on the water depth and thus actually has to be adjusted according to the discharge (Ferguson [1]), while the equivalent sand roughness can be considered as constant as long as the river bed is stable. Jäggi [40] suggested an improved calculation of the Strickler value depending on the water depth, which leads to increasing k_{St} values—exemplarily calculated for case study C.2 to be 16/19/27 m$^{1/3}$ s^{-1} ($Q_L/Q_M/HQ_1$)—while in this study, in line with predominant hydraulic engineering practice, the roughness is kept constant at 18 m$^{1/3}$ s^{-1}. In particular, the large differences in the case of high flow conditions have to be highlighted, which result in substantially higher bed shear stresses as well as water surface elevations calculated by the 1D and 2D models compared to the 3D model.

According to Jäggi [40], Lamb et al. [81] and Parker et al. [82], the Strickler value is additionally increasing with decreasing bed slopes. Considering this and the calculation of bed shear stresses in

1D and 2D models, where high k_{St} values result in low bed shear stresses, the decreasing 3D/1D ratios from small-sloped to steeper river sites (C.3→C.2) become obvious. Due to the fact that several parameters (e.g., water depth, equivalent sand roughness, etc.) vary between the different case studies, only this overall tendency can be identified.

A comparison to measure bed shear stresses could be a remedy to overcome the issue of considerable differences between the models. However, monitoring this parameter itself [83] is an ongoing challenge. Besides issues of handling the devices (installation, operation and service), concerns about the validity and explanatory power of time-averaged bed shear stresses were raised. Gmeiner et al. [84] and Liedermann et al. [85] showed that bed load transport takes place below expected discharges considering the concepts of Shields [86] and Zanke [87] for critical shear stresses. It is thus recommended to consider time series of bed shear stresses including fluctuations for which probabilistic approaches [88–91] are suitable. A part of these fluctuations is expected to be covered by considering the turbulent kinetic energies for calculating bed shear stresses in the case of 3D models. In this study, the dependency of bed shear stresses on turbulent kinetic energies are highlighted by peaks at the banks, which correlate with peaks of the turbulent kinetic energy. A proportional relationship between turbulent kinetic energy and bed shear stress for high strains has already been described by Menter [92]. Additionally, Rodi [93] stated an increasing influence of turbulence on shear stresses upon larger roughness values.

The aforementioned uncertainties of using deterministic bed shear stress calculations, as well as the additional limitations when employing these concepts in numerical models should always be considered by addressing the issues related to dependent processes including bed load transport or morphological changes. Particularly in the case of man-made interventions [94], including engineering measures [95] in rivers, reservoirs or oceans as well as management procedures [96] such as flushing, dredging or depositing sediments, it is recommended to focus on the evaluation of differences rather than on absolute values. In general, improving the results of 1D and 2D models by applying discharge-dependent Strickler values and preferring higher dimensional models over lower ones has the potential to enhance the validity of numerical simulations.

5. Conclusions

The application of numerical models in hydraulic engineering has become routine over recent decades, resulting in numerous published studies. Standard practice is the calibration of a single numerical model by comparing simulation results to monitoring data (e.g., water surface elevations, flow velocities, etc.) followed by an application of the models for any kind of work. Despite known uncertainties (e.g., model simplifications, mesh generation, roughness definition, etc.), it is generally assumed that the simulated results are valid for all hydrodynamic parameters, including properties for which no measurements are available. The present study addressed the differences between model dimensions, focusing on bed shear stresses considering different discharges as well as study sites.

The results show that, on average, the simulated bed shear stresses are 10% and 14–38% lower in the applied 2D and 3D model, respectively, as compared to the 1D model. At the same time, almost no differences were present in water surface elevations, which had been used for calibration. When comparing different river sites and discharges, minor differences in the ratio of bed shear stresses were found in 2D/1D models, while in the case of 3D/1D the differences became larger from lower to higher discharges as well as from small-sloped to steeper river sites. The major influence of different roughness definitions, as well as the various calculation approaches used in the numerical models, were identified as a cause of these differences. Moreover, reasons why, in general, the highest bed shear stresses in the main stream of the river occur in 2D models (e.g., depth-averaged versus near-bed flow velocities), and for bed shear stress peaks at the banks in 3D models were elaborated (e.g., consideration of turbulent kinetic energy).

The considerable differences between the numerical models present an issue in hydraulic engineering, due to the fact that bed shear stresses form the basis of many approaches including the calculation of sediment transport or evaluation of habitats. One of the key tasks of future works will be the comparison with measurement data considering the associated monitoring challenges. The question of whether the deterministic approach of calculating bed shear stresses, as well as the definition of bed shear stress, is adequately representing the occurring near-bed forces in an applied river context should also be addressed in future studies.

Author Contributions: K.G., M.T. and C.H. designed, set up and performed the numerical simulations and analyzed the results including the comparison with measurements as well as between the models. The draft text was prepared by K.G. and edited by M.T., C.H. and H.H.

Funding: This paper was written as a contribution to the Christian Doppler Laboratory for Sediment Research and Management. The financial support by the Austrian Federal Ministry for Digital and Economic Affairs and the National Foundation for Research, Technology and Development is gratefully acknowledged.

Acknowledgments: The authors thank the Hydraulic Engineering Institute of the University of Innsbruck for providing measurement data of the laboratory experiment.

Conflicts of Interest: The authors declare no conflict of interest.

References

1. Ferguson, R. Time to abandon the Manning equation? *Earth Surf. Process. Landf.* **2010**, *35*, 1873–1876. [CrossRef]
2. Powell, D.M. Flow resistance in gravel-bed rivers: Progress in research. *Earth-Sci. Rev.* **2014**, *136*, 301–338. [CrossRef]
3. Järvelä, J. *Flow Resistance in Environmental Channels: Focus on Vegetation*; Helsinki University of Technology Water Resources Publications: Helsinki, Finland, 2004.
4. Green, J.C. Modelling flow resistance in vegetated streams: Review and development of new theory. *Hydrol. Process.* **2004**, *19*, 1245–1259. [CrossRef]
5. Manga, M.; Kirchner, J.W. Stress partitioning in streams by large woody debris. *Water Resour. Res.* **2000**, *36*, 2373–2379. [CrossRef]
6. Wilcox, A.C.; Wohl, E.E. Flow resistance dynamics in step-pool stream channels: 1. Large woody debris and controls on total resistance. *Water Resour. Res.* **2006**, 42. [CrossRef]
7. Wilcox, A.C.; Nelson, J.M.; Wohl, E.E. Flow resistance dynamics in step-pool channels: 2. Partitioning between grain, spill, and woody debris resistance. *Water Resour. Res.* **2006**, *42*. [CrossRef]
8. Parker, G.; Peterson, A.W. Bar resistance of gravel-bed streams. *J. Hydraul. Div.* **1980**, *106*, 1559–1575.
9. Prestegaard, K.L. Bar resistance in gravel bed streams at bankfull stage. *Water Resour. Res.* **1983**, *19*, 472–476. [CrossRef]
10. Hey, R.D. Bar Form Resistance in Gravel-Bed Rivers. *J. Hydraul. Eng.* **1988**, *114*, 1498–1508. [CrossRef]
11. Millar, R.G. Grain and form resistance in gravel-bed rivers Résistances de grain et de forme dans les rivières à graviers. *J. Hydraul. Res.* **1999**, *37*, 303–312. [CrossRef]
12. Francalanci, S.; Solari, L.; Toffolon, M.; Parker, G. Do alternate bars affect sediment transport and flow resistance in gravel-bed rivers? *Earth Surf. Process. Landf.* **2012**, *37*, 866–875. [CrossRef]
13. Wohl, E.E.; Thompson, D.M. Velocity characteristics along a small step–pool channel. *Earth Surf. Process. Landf.* **2000**, *25*, 353–367. [CrossRef]
14. Curran, J.H.; Wohl, E.E. Large woody debris and flow resistance in step-pool channels, Cascade Range, Washington. *Geomorphology* **2003**, *51*, 141–157. [CrossRef]
15. Church, M.; Zimmermann, A. Form and stability of step-pool channels: Research progress. *Water Resour. Res.* **2007**, 43. [CrossRef]
16. Wilcox, A.C.; Wohl, E.E. Field measurements of three-dimensional hydraulics in a step-pool channel. *Geomorphology* **2007**, *83*, 215–231. [CrossRef]
17. Comiti, F.; Mao, L. Recent Advances in the Dynamics of Steep Channels. *Gravel Bed Rivers* **2012**. [CrossRef]
18. Comiti, F.; Cadol, D.; Wohl, E. Flow regimes, bed morphology, and flow resistance in self-formed step-pool channels. *Water Resour. Res.* **2009**, 45. [CrossRef]
19. Song, T.; Chiew, Y.M.; Chin, C.O. Effect of Bed-Load Movement on Flow Friction Factor. *J. Hydraul. Eng.* **1998**, *124*, 165–175. [CrossRef]

20. Bergeron, N.E.; Carbonneau, P. The effect of sediment concentration on bedload roughness. *Hydrol. Process.* **1999**, *13*, 2583–2589. [CrossRef]
21. Calomino, F.; Gaudio, R.; Miglio, A. Effect of bed-load concentration on friction factor in narrow channels. In Proceedings of the 2nd International Conference on Fluvial Hydraulics–River Flow, Naples, Italy, 23–25 June 2004; Taylor and Francis Group: London, UK, 2004; pp. 279–285.
22. Gao, P.; Abrahams, A.D. Bedload transport resistance in rough open-channel flows. *Earth Surf. Process. Landf.* **2004**, *29*, 423–435. [CrossRef]
23. Campbell, L.; McEwan, I.; Nikora, V.; Pokrajac, D.; Gallagher, M.; Manes, C. Bed-Load Effects on Hydrodynamics of Rough-Bed Open-Channel Flows. *J. Hydraul. Eng.* **2005**, *131*, 576–585. [CrossRef]
24. Recking, A.; Frey, P.; Paquier, A.; Belleudy, P.; Champagne, J.Y. Feedback between bed load transport and flow resistance in gravel and cobble bed rivers. *Water Resour. Res.* **2008**, *44*. [CrossRef]
25. Recking, A.; Frey, P.; Paquier, A.; Belleudy, P.; Champagne, J.Y. Bed-Load Transport Flume Experiments on Steep Slopes. *J. Hydraul. Eng.* **2008**, *134*, 1302–1310. [CrossRef]
26. Fischenich, C. *Robert Manning (A Historical Perspective)*; EMRRP Technical Notes Collection (ERDCTN-EMRRP-SR-10), U.S. Army Engineer Research and Development Center: Vicksburg, MS, USA, 2000.
27. Brown, G.O. The History of the Darcy-Weisbach Equation for Pipe Flow Resistance. In *Environmental and Water Resources History*; ASCE Civil Engineering Conference and Exposition: Washington, DC, USA, 2002; pp. 34–43.
28. Simmons, C.T. Henry Darcy (1803–1858): Immortalised by his scientific legacy. *Hydrogeol. J.* **2008**, *16*, 1023–1038. [CrossRef]
29. Zimmermann, A. Flow resistance in steep streams: An experimental study. *Water Resour. Res.* **2010**, 46. [CrossRef]
30. Bathurst, J.C. Flow resistance of large-scale roughness. *J. Hydraul. Div.* **1978**, *104*, 1587–1603.
31. Bathurst, J.C. At-a-site variation and minimum flow resistance for mountain rivers. *J. Hydrol.* **2002**, *269*, 11–26. [CrossRef]
32. Katul, G.; Wiberg, P.; Albertson, J.; Hornberger, G. A mixing layer theory for flow resistance in shallow streams. *Water Resour. Res.* **2002**, *38*, 32-1–32-8. [CrossRef]
33. Lee, A.J.; Ferguson, R.I. Velocity and flow resistance in step-pool streams. *Geomorphology* **2002**, *46*, 59–71. [CrossRef]
34. Smart, G.M.; Duncan, M.J.; Walsh, J.M. Relatively Rough Flow Resistance Equations. *J. Hydraul. Eng.* **2002**, *128*, 568–578. [CrossRef]
35. Aberle, J.; Smart, G.M. The influence of roughness structure on flow resistance on steep slopes. *J. Hydraul. Res.* **2003**, *41*, 259–269. [CrossRef]
36. Canovaro, F.; Solari, L. Dissipative analogies between a schematic macro-roughness arrangement and step–pool morphology. *Earth Surf. Process. Landf.* **2007**, *32*, 1628–1640. [CrossRef]
37. Ferguson, R. Flow resistance equations for gravel- and boulder-bed streams. *Water Resour. Res.* **2007**, *43*. [CrossRef]
38. Strickler, A. *Beiträge zur Frage der Geschwindigkeitsformel und der Rauhigkeitszahlen für Ströme, Kanäle und geschlossene Leitungen*; Eidgenossisches Amt für Wasserwirtschaft: Bern, Switzerland, 1923.
39. Meyer-Peter, E.; Müller, R. Formulas for bed-load transport. In Proceedings of the 2nd Meeting of IAHR, Stockholm, Sweden, 7–9 June 1948; IAHR: Stockholm, Sweden, 1948; pp. 39–64.
40. Jäggi, M. *Alternierende Kiesbänke: Untersuchungen über ihr Auftreten, den Zusammenhang mit der Bildung von Sohlenformen im allgemeinen, sowie ihre Auswirkungen auf Ufererosion und Fliesswiderstand*; Versuchsanstalt für Wasserbau, Hydrologie, und Glaziologie: ETH Zurich, Switzerland, 1983.
41. Chow, V.T. *Open Channel Hydraulics*; McGraw-Hill: New York, NY, USA, 1959.
42. Naudascher, E. *Hydraulik der Gerinne und Gerinnebauwerke*, 2nd ed.; Springer: Vienna, Austria, 1992.
43. US Army Corps of Engineers. *Engineering and Design—Hydraulic Design of Flood Control Channels*; American Society of Civil Engineers: New York, NY, USA, 1994; Volume 10.
44. Horton, R.E. Separate Roughness Coefficients for Channel Bottom and Sides. *Eng. News-Rec.* **1933**, *3*, 652–653.
45. Einstein, H.A. Der hydraulische oder Profil-Radius. *Schweiz. Bauztg.* **1934**, *103*, 89–91.
46. Krishnamurthy, M.; Christensen, B.A. Equivalent roughness for shallow channel. *J. Hydraul. Div.* **1972**, *98*, 2257–2263.
47. Yen Ben, C. Open Channel Flow Resistance. *J. Hydraul. Eng.* **2002**, *128*, 20–39. [CrossRef]

48. Järvelä, J. Effect of submerged flexible vegetation on flow structure and resistance. *J. Hydrol.* **2005**, *307*, 233–241. [CrossRef]
49. Nepf, H.M. Hydrodynamics of vegetated channels. *J. Hydraul. Res.* **2012**, *50*, 262–279. [CrossRef]
50. Aberle, J.; Järvelä, J. Flow resistance of emergent rigid and flexible floodplain vegetation. *J. Hydraul. Res.* **2013**, *51*, 33–45. [CrossRef]
51. Nikuradse, J. *Strömungsgesetze in Rauhen Rohren*; VDI-Verlag: Berlin, Germany, 1933; Volume 361.
52. Booker, D.J.; Sear, D.A.; Payne, A.J. Modelling three-dimensional flow structures and patterns of boundary shear stress in a natural pool–riffle sequence. *Earth Surf. Process. Landf.* **2001**, *26*, 553–576. [CrossRef]
53. Fischer-Antze, T.; Olsen, N.R.B.; Gutknecht, D. Three-dimensional CFD modeling of morphological bed changes in the Danube River. *Water Resour. Res.* **2008**, 44. [CrossRef]
54. Rüther, N.; Jacobsen, J.; Olsen, N.R.B.; Vatne, G. Prediction of the three-dimensional flow field and bed shear stresses in a regulated river in mid-Norway. *Hydrol. Res.* **2010**, *41*, 145–152. [CrossRef]
55. Guerrero, M.; Lamberti, A. Bed-roughness investigation for a 2-D model calibration: The San Martín case study at Lower Paranà. *Int. J. Sediment Res.* **2013**, *28*, 458–469. [CrossRef]
56. Guerrero, M.; Latosinski, F.; Nones, M.; Szupiany, R.N.; Re, M.; Gaeta, M.G. A sediment fluxes investigation for the 2-D modelling of large river morphodynamics. *Adv. Water Resour.* **2015**, *81*, 186–198. [CrossRef]
57. Paarlberg, J.A.; Guerrero, M.; Huthoff, F.; Re, M. Optimizing Dredge-and-Dump Activities for River Navigability Using a Hydro-Morphodynamic Model. *Water* **2015**, *7*, 3943–3962. [CrossRef]
58. Glas, M.; Glock, K.; Tritthart, M.; Liedermann, M.; Habersack, H. Hydrodynamic and morphodynamic sensitivity of a river's main channel to groyne geometry. *J. Hydraul. Res.* **2018**, *56*, 714–726. [CrossRef]
59. Glock, K.; Tritthart, M.; Gmeiner, P.; Pessenlehner, S.; Habersack, H. Evaluation of engineering measures on the Danube based on numerical analysis. *J. Appl. Water Eng. Res.* **2017**, 1–19. [CrossRef]
60. Nicholas, A.P.; Sandbach, S.D.; Ashworth, P.J.; Amsler, M.L.; Best, J.L.; Hardy, R.J.; Lane, S.N.; Orfeo, O.; Parsons, D.R.; Reesink, A.J.H.; et al. Modelling hydrodynamics in the Rio Paraná, Argentina: An evaluation and inter-comparison of reduced-complexity and physics based models applied to a large sand-bed river. *Geomorphology* **2012**, *169–170*, 192–211. [CrossRef]
61. Xie, Q.; Yang, J.; Lundström, S.; Dai, W. Understanding Morphodynamic Changes of a Tidal River Confluence through Field Measurements and Numerical Modeling. *Water* **2018**, *10*, 1424. [CrossRef]
62. Einstein, H.A. *The Bed-Load Function for Sediment Transportation in Open Channel Flows*; US Department of Agriculture: Washington, DC, USA, 1950; Volume 1026.
63. Van Rijn Leo, C. Sediment Transport, Part I: Bed Load Transport. *J. Hydraul. Eng.* **1984**, *110*, 1431–1456. [CrossRef]
64. Van Rijn, L.C. *Principles of Sediment Transport in Rivers, Estuaries and Coastal Seas*; Aqua Publications: Amsterdam, The Netherlands, 1993; Volume 1006.
65. Wu, W.; Wang, S.S.Y.; Jia, Y. Nonuniform sediment transport in alluvial rivers. *J. Hydraul. Res.* **2000**, *38*, 427–434. [CrossRef]
66. Bravo-Espinosa, M.; Osterkamp, W.R.; Lopes Vicente, L. Bedload Transport in Alluvial Channels. *J. Hydraul. Eng.* **2003**, *129*, 783–795. [CrossRef]
67. Habersack, H.; Seitz, H.; Laronne, J.B. Spatio-temporal variability of bedload transport rate: Analysis and 2D modelling approach. *Geodin. Acta* **2008**, *21*, 67–79. [CrossRef]
68. Riesterer, J.; Wenka, T.; Brudy-Zippelius, T.; Nestmann, F. Bed load transport modeling of a secondary flow influenced curved channel with 2D and 3D numerical models. *J. Appl. Water Eng. Res.* **2016**, *4*, 54–66. [CrossRef]
69. Tritthart, M.; Schober, B.; Habersack, H. Non-uniformity and layering in sediment transport modelling 1: Flume simulations. *J. Hydraul. Res.* **2011**, *49*, 325–334. [CrossRef]
70. Tritthart, M.; Liedermann, M.; Schober, B.; Habersack, H. Non-uniformity and layering in sediment transport modelling 2: River application. *J. Hydraul. Res.* **2011**, *49*, 335–344. [CrossRef]
71. Wu, W. Depth-Averaged Two-Dimensional Numerical Modeling of Unsteady Flow and Nonuniform Sediment Transport in Open Channels. *J. Hydraul. Eng.* **2004**, *130*, 1013–1024. [CrossRef]
72. Hauer, C.; Unfer, G.; Tritthart, M.; Formann, E.; Habersack, H. Variability of mesohabitat characteristics in riffle-pool reaches: Testing an integrative evaluation concept (FGC) for MEM-application. *River Res. Appl.* **2011**, *27*, 403–430. [CrossRef]
73. Tritthart, M. Three-dimensional numerical modelling of turbulent river flow using polyhedral finite volumes. *Wien. Mitt. Wasser-Abwasser-Gewässer* **2005**, *193*, 1–179.

74. Muhar, S.; Kainz, M.; Schwarz, M.; Jungwirth, M. *Ausweisung Flusstypspezifisch Erhaltener Fliessgewässerabschnitte in Österreich: Fliessgewässer mit Einem Einzugsgebiet Grösser als 500 km^2 Ohne Bundesflüsse*; Abt. I 3; Bundesministerium für Nachhaltigkeit und Tourismus (BMNT): Vienna, Austria, 1998.
75. Versteeg, H.K.; Malalasekera, W. *An Introduction to Computational Fluid Dynamics: The Finite Volume Method*; Pearson Education: Harlow, UK, 2007.
76. Schlichting, H.; Gersten, K. *Grenzschicht-Theorie*; Springer: Berlin, Germany, 2006.
77. US Army Corps of Engineers. *HEC-RAS River Analysis System, Hydraulic Reference Manual*; US Army Corps of Engieers, Hydrologic Engineering Center: Davis, CA, USA, 2010.
78. Nujic, M. *Praktischer Einsatz Eines Hochgenauen Verfahrens für die Berechnung von Tiefengemittelten Strömungen*; Mitteilungen des Institutes der Bundeswehr München: Munich, Germany, 1999.
79. Pironneau, O. *Finite Element Methods for Fluids*; Wiley: New York, NY, USA, 1989.
80. Lane, S.; Bradbrook, K.; Richards, K.; Biron, P.; Roy, A. The application of computational fluid dynamics to natural river channels: Three-dimensional versus two-dimensional approaches. *Geomorphology* **1999**, *29*, 1–20. [CrossRef]
81. Lamb, M.P.; Dietrich, W.E.; Venditti, J.G. Is the critical Shields stress for incipient sediment motion dependent on channel-bed slope? *J. Geophys. Res. Earth Surf.* **2008**, 113. [CrossRef]
82. Parker, C.; Clifford, N.J.; Thorne, C.R. Understanding the influence of slope on the threshold of coarse grain motion: Revisiting critical stream power. *Geomorphology* **2011**, *126*, 51–65. [CrossRef]
83. Gmeiner, P.; Liedermann, M.; Tritthart, M.; Habersack, H. Development and testing of a device for direct bed shear stress measurement. In Proceedings of the 2nd IAHR, Europe Conference, Munich, Germany, 27–29 June 2012.
84. Gmeiner, P.; Liedermann, M.; Haimann, M.; Tritthart, M.; Habersack, H. Grundlegende Prozesse betreffend Hydraulik, Sedimenttransport und Flussmorphologie an der Donau. *Österreichische Wasser Und Abfallwirtsch.* **2016**, *68*, 208–216. [CrossRef]
85. Liedermann, M.; Gmeiner, P.; Kreisler, A.; Tritthart, M.; Habersack, H. Insights into bedload transport processes of a large regulated gravel-bed river. *Earth Surf. Process. Landf.* **2017**, *43*, 514–523. [CrossRef]
86. Shields, A. *Application of Similarity Principles and Turbulence Research to Bed-Load Movement*; United States Department of Agriculture, Soil Conservation Service: Pasadena, CA, USA, 1936.
87. Zanke, U. Der Beginn der Geschiebebewegung als Wahrscheinlichkeitsproblem. *Wasser Und Boden* **1990**, *1*, 40–43.
88. Einstein, H.A.; El-Samni, E.-S.A. Hydrodynamic Forces on a Rough Wall. *Rev. Mod. Phys.* **1949**, *21*, 520–524. [CrossRef]
89. Grass, A.J. Initial instability of fine bed sand. *J. Hydraul. Div.* **1970**, *96*, 619–632.
90. McEwan, I.; Heald, J. Discrete Particle Modeling of Entrainment from Flat Uniformly Sized Sediment Beds. *J. Hydraul. Eng.* **2001**, *127*, 588–597. [CrossRef]
91. Ancey, C.; Heyman, J. A microstructural approach to bed load transport: Mean behaviour and fluctuations of particle transport rates. *J. Fluid Mech.* **2014**, *744*, 129–168. [CrossRef]
92. Menter, F.R. Two-equation eddy-viscosity turbulence models for engineering applications. *Aiaa J.* **1994**, *32*, 1598–1605. [CrossRef]
93. Rodi, W. Turbulence Modeling and Simulation in Hydraulics: A Historical Review. *J. Hydraul. Eng.* **2017**, *143*, 03117001. [CrossRef]
94. Giardino, A.; Schrijvershof, R.; Nederhoff, C.M.; de Vroeg, H.; Brière, C.; Tonnon, P.K.; Caires, S.; Walstra, D.J.; Sosa, J.; van Verseveld, W.; et al. A quantitative assessment of human interventions and climate change on the West African sediment budget. *Ocean Coast. Manag.* **2018**, *156*, 249–265. [CrossRef]
95. Habersack, H.; Piégay, H. 27 River restoration in the Alps and their surroundings: Past experience and future challenges. *Dev. Earth Surf. Process.* **2007**, *11*, 703–735.
96. Hauer, C.; Wagner, B.; Aigner, J.; Holzapfel, P.; Flödl, P.; Liedermann, M.; Tritthart, M.; Sindelar, C.; Pulg, U.; Klösch, M.; et al. State of the art, shortcomings and future challenges for a sustainable sediment management in hydropower: A review. *Renew. Sustain. Energy Rev.* **2018**, *98*, 40–55. [CrossRef]

Article

Modeling of River Channel Shading as a Factor for Changes in Hydromorphological Conditions of Small Lowland Rivers

Tomasz Kałuża [1], Mariusz Sojka [2], Rafał Wróżyński [2], Joanna Jaskuła [2], Stanisław Zaborowski [1] and Mateusz Hämmerling [1,*]

[1] Department of Hydraulic and Sanitary Engineering, Poznań University of Life Sciences, Piątkowska 94, 60-649 Poznań, Poland; tomasz.kaluza@up.poznan.pl (T.K.); zaborowski.stanislaw@gmail.com (S.Z.)

[2] Institute of Land Improvement, Environment Development and Geodesy, Poznań University of Life Sciences, Piatkowska 94, 60-649 Poznań, Poland; mariusz.sojka@up.poznan.pl (M.S.); rafal.wrozynski@up.poznan.pl (R.W.); joanna.jaskula@up.poznan.pl (J.J.)

* Correspondence: mateusz.hammerling@up.poznan.pl; Tel.: +48-61-846-6589

Received: 6 December 2019; Accepted: 12 February 2020; Published: 13 February 2020

Abstract: The ecological water quality in rivers and streams is influenced both by the morphological factors (within the watercourse channel and by the dynamic factors associated with flow), as well as biological factors (connected with the flora and fauna characteristic of its specific area). This paper presents an analysis of the effect of river channel shading by trees and shrubs on hydromorphological changes in a selected reach of the Wełna River, Poland. The analysis was conducted on two adjacent cross-sections (one in a reach lined with trees, the other in an open area with no tree or shrub vegetation). Data were collected during field surveys in the years 2014 and 2019. According to the Water Framework Directive, the Wełna River represents a watercourse with small and average-sized watershed areas, with sand being the dominant substrate of the river bottom. Flow volume, distributions of velocity in the sections, as well as substrate grain-size characteristics and river bottom morphology, were determined based on field measurements. In the study, the leaf area index (LAI) of vegetation was measured in the reach lined with trees, while the number and species composition of macrophytes were determined in the investigated river reaches. Moreover, a digital surface model (DSM) and Geoinformation Information System GIS tools were used to illustrate variability in shading within the tree-lined reach. The DSM model was based on Light Detection and Ranging (LIDAR) data. The results of this study enable us to establish the relationship between river shading by vegetation covering the bank zone, and changes in hydromorphological parameters of the river channel.

Keywords: small lowland rivers; flow conditions; riverbed shading; shading model; hydromorphological changes; LAI; LIDAR

---

1. Introduction

Present-day requirements imposed on river and canal maintenance are closely related to the assurance of good ecological water quality and adequate condition of the semiaquatic ecosystem comprising the river and its valley [1,2]. The ecological water quality of rivers and streams is affected both by the hydromorphological conditions of the watercourse channel and by biological factors, related to the flora and fauna characteristic of its specific area [3–5].

In turn, the condition of a semiaquatic ecosystem is connected with the spatial system of the watercourse and with the biological equilibrium within the valley [6]. This should be reflected in the approach to river maintenance, particularly the adopted flood control system [7]. Presently binding guidelines for the maintenance of rivers classified as natural (or semi-natural) not only accept vegetation within their channel cross-section, but also assign them numerous protective and environmental functions [8–10]. This relates to the comprehensive, modern understanding of water management, which for planned engineering measures requires an environmental impact analysis following the principles of sustainable development [1].

The legal framework for works in rivers is outlined, among others, in the Water Framework Directive. Recently, the European Union has enacted the Water Framework Directive with the aim of securing and, where possible, improving the ecological status of watercourses throughout all Member States. The fundamental legal issues are set out in the Acts of the European Parliament, which define the rights and responsibilities of all parties with respect to rivers and streams [1–3]. In addition, in many countries separate measures are being implemented to improve the natural condition of rivers (e.g. in the USA, Canada, Norway) [6,9].

The above-mentioned conditions need to be reconciled with economic concerns imposed by the analysis of efficiency for water management investments. These include primarily flood control measures, limiting the highest water stages and flows, as well as actions ensuring good ecological status of surface waters at the lowest water stages and flows [1]. Thus, these requirements jointly result in the need to precisely estimate hydraulic parameters of flow in rivers and canals [11]. It is the accuracy of assessment that determines the decision to adopt a specific solution, as well as the technology required to implement it. Referring this aspect to river hydraulics, it may be treated as an evaluation of river channel capacity under specific geometric parameters, river regulation infrastructure, and vegetation growth [12]. This pertains particularly to overgrown river channels and conditions found in rivers with floodplains.

Growth of aquatic vegetation is the main factor in determining flow conditions in small rivers and canals [9]. Aquatic vegetation in channel sections causes an increase in flow resistance. It is a potential source of flood risk and resulting losses [13], when heavy rains in the vegetation period may lead to flash floods in small watercourses, whose flow capacity is reduced as a result of vegetation overgrowth [12]. This in turn leads to a decrease in channel flow capacity. Growth of aquatic vegetation is the primary biological process affecting flow conditions. This growth is modified by hydraulic and geometric parameters of the channel, as well as other physical factors (light, temperature, changes in water levels) along with chemical, edaphic, and biotic factors [14,15]. However, the main factor initiating the whole cascade of resulting phenomena is connected with sunlight, which, in association with biogenic compounds availability, limits productivity of aquatic ecosystems [16–18].

Growth of aquatic plants (macrophytes) depends on the action of various ecological factors. The most important of these include light and heat, water quality [19], pH, water hardness, water current intensity [20], and bed substrate [21]. The response of aquatic plants to ecological factors is very strong, which has contributed to their utilization in the evaluation of the environmental condition. In most European countries, systems based on macrophytes are extensively applied to assess water status [22]. The value of macrophytes as bioindicators is related to their capacity to characterize more permanent changes in habitat characteristics [19].

Growth of aquatic plants is strongly dependent on light availability [23]; thus, all changes in water transparency lead to changes in the structure of plant communities, vegetation density, and depth of plant growth [24,25]. The maximum increase in biomass of aquatic vegetation is observed under particularly advantageous growth conditions, including insolation [26,27]. The light factor determines not only biomass of aquatic plants, but also their structure, as an increase in insolation results in the development of plant organization. Additionally, the light factor in the water affects growing plants through modification of water temperatures [27,28].

Numerous studies indicate that an important factor differentiating lighting conditions in watercourses is the vegetation-growing process on their banks, particularly forest and shrub communities [29]. In the case of small and medium-sized rivers flowing under the forest tree canopy, the amount of light reaching the water surface is several to several dozen percent lower than in rivers flowing through open areas. Such a situation may result in almost complete elimination of aquatic plants [6,30,31]. In other countries, the close dependence of aquatic plants on light conditions has already been used in practice to reduce excessive river channel vegetation overgrowth [18]. It has been suggested to increase shading of rivers in order to limit growth of common species of submerged plants [25]. This method has also been proposed to control excessive spread of invasive species in watercourses [24].

Trees and shrubs are natural landscape components in river valleys [32]. Factors affecting the richness of vegetation surrounding the channels and floodplains of lowland rivers include much slower water flow velocity, a greater width of the floodplain valley as well as river meandering and oxbow lake cutoff. The character of vegetation in the immediate vicinity of a watercourse channel and floodplains depends on the type of land use, groundwater levels, and frequency of flooding [12]. Typically, the area not utilized agriculturally is overgrown with floodplain forests. The areas with non-forest vegetation are usually various types of meadows and pastures, while in more moist locations, they are sedge rushes or fens [33]. Vegetation growing on riverbanks and floodplains has a considerable modifying effect on flow conditions. Apart from the obvious effect of direct factors on flow resistance during high flows (e.g. debris, bottom deposits, local obstacles) an essential role is also played by indirect factors, e.g. shading of the river channel and inhibition of aquatic vegetation growth. These changes are also reflected in river channel hydromorphology and the distribution of flow velocities [13].

When modeling flows in natural river channels, it is necessary to consider the effect of vegetation on flow resistance [34]. A problem typically found in vegetation-covered areas is related to the different flow dynamics compared to the dynamics of a stream free from vegetation. Velocities and depth gradients are then much smaller. This is naturally reflected in the dynamics of debris transport and hydromorphological changes in the watercourse channel [5,34,35].

In order to determine the shading area, the location of trees within the riverbank zone needs to be described. Energy reaching the ground in a given spot depends on the position of the sun, insolation time, and cloud cover. Due to the continuously changing cloud cover, the input energy fluctuates considerably, and the changes may be detected only through precise measurements. The weather service reports mean monthly diurnal total solar radiation for various weather stations. This facilitates the development of an average annual model for the area covered by a given station. Based on the available data, Rickert [36] proposed a procedure to calculate the effect of shading on macrophyte growth. Data include photosynthetically active radiation (PAR) reaching the water surface and the rest of the radiation, which may be expected in the river channel at various turbidity levels based on total radiation in % or expressed in W/m^2. This method is relatively labor-intensive and includes the need for analyses of individual hourly sequences. At a tree height of, e.g., 10 m, this is equivalent to shade range of 8 m in length. Solar radiation reaching the water surface is partly reflected and partly intercepted. The percentage of solar light reflection depends on factors such as angle of incidence (which then depends on latitude end land relief), wavelength, and land structure. The smaller the angle of incidence is, the greater is the amount of light which is reflected.

Variation in solar radiation is a fundamental to control most processes on the Earth. The solar radiation reaching the water surface is controlled by location (latitude, longitude elevation), season, atmospheric conditions, topography, and vegetation. Shading is a key parameter due to the control on the amount of direct radiation reaching the water surface. The factors influencing the fraction of solar radiation reaching the water surface can be taken into account by using tools and models working in a GIS environment. Nowadays calculations and analyses are carried out in different scales on the basis of data with different spatial resolution and type. Increasingly often high-resolution LIDAR data and GIS tools are used in shading analyses or assessments of incoming radiation reaching surface water bodies [37–40].

LIDAR data provide information on riparian attributes related to elevation, biomass overhanging the river, and vertical tree structure. In addition, high-resolution products generated by LIDAR data processing, such as the digital surface model (DSM) and digital elevation model (DEM), are used [41]. Bachiller-Jareno et al. [42] presented a methodology to estimate tree height and canopy extent, and a model that simulates the position of the sun across the sky for hourly or sub-hourly intervals to calculate the daily shade over the river surface. In addition, Loicq et al. [40] described the use of a spatially explicit method applying LIDAR-derived data to compute riparian shading based on direct and diffuse solar radiation. Karrasch and Hunger [41] demonstrated that during shading modeling based on DSM and DEM the illumination of the water body is underestimated, while shading is overestimated. In the United Kingdom, the Environment Agency (EA) has developed catchment 'shade maps' for every catchment managed under the Water Framework Directive in England and Wales [43].

The aim of this study was to determine the effect of vegetation growing on the Wełna riverbanks on changes in flow conditions within a relatively short river reach. Analyses were conducted from July to August.

2. Materials and Methods

Field studies were conducted in the years 2014 and 2019 on the Wełna River, located in the belt of central European lowlands (Figure 1a). The Wełna River is a right bank tributary of the Warta River (ranking third in Poland in terms of river size). The Wełna River is 118 km in length, and its catchment area is 2621 km^2. It flows from a lake located 10 km north-east of the town of Gniezno. Analyses were conducted on a selected river reach of 250 m in length, situated near the town of Cotoń. The study object is between a tree-covered area and an area with no tree or shrub cover, no water bodies, and no hydraulic structures, and there is no inflow from another watercourse. The investigated cross-sections no. 1 and 2 were shown on Figure 1a. In this way, any potential effect on disturbance of flow between the investigated cross-sections was eliminated (Figure 1b, Figure 1c).

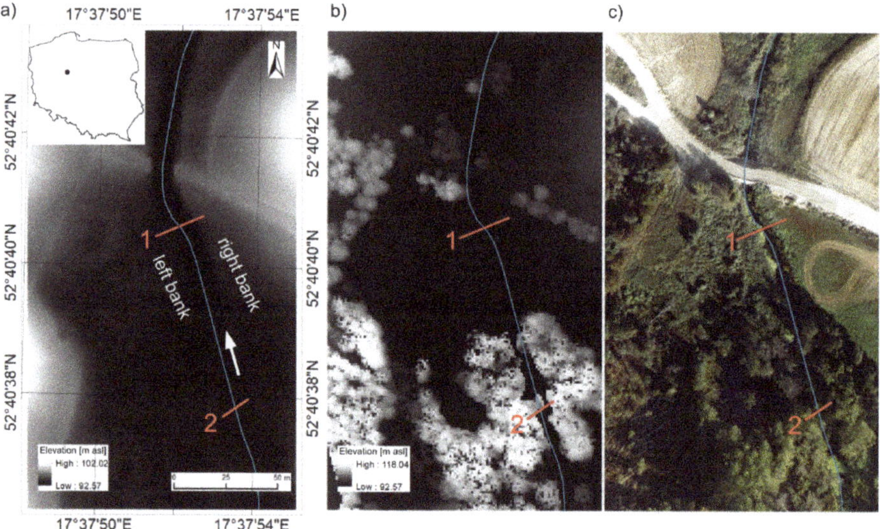

Figure 1. Location of the studied site in: (**a**) the digital elevation model (DEM), (**b**) the digital surface model (DSM), and (**c**) an aerial photograph.

Field measurements consisted of the assessment of watercourse hydrometric parameters. Within this study, the leaf area index (LAI) was also determined for vegetation in a tree-lined river reach, while the number and species composition of macrophytes found in the investigated river cross-sections were recorded [44]. Based on field measurements, the volume of flow and velocity distributions in cross-sections were determined along with the substrate grain-size characteristics and river bottom morphology. Moreover, a model was constructed to illustrate variability in the shade range within a river reach lined with trees. The shading analysis was based on DEM and a DSM developed on airborne LIDAR data. The results of this study enable us to establish the relationship between river shading by vegetation covering the bank zone and changes in the hydromorphological parameters of the river channel.

Based on LIDAR data, a DEM of the investigated river reach was developed (Figure 1a), while the location of trees was given in the DSM (Figure 1b). The current vegetation status in the analyzed Wełna River reach is presented in an aerial photograph (Figure 1c). Based on the forest stand description and a forest map, on the right bank of the river reach there is a forest subcompartment of 0.17 ha. Black alder (*Alnus glutinosa*) is the dominant species. The stand is approximately 70 years old. Common sallow (*Salix cinerea*) is the shrub species found in that area. In turn, a forest subcompartment of 7.7 ha is located on the left bank of the river. That subcompartment is a belt of swamps adjacent to the river at a length of approximately 50 m.

Another subcompartment of 12.35 ha is located farther from the riverbank. It is a mixed forest with the predominance of oak (*Quercus* L.) and pine (*Pinus sylvestris* L.). Two measurement cross-sections were established in the selected river reach. Cross-section 1 was situated in a location with no trees or shrubs, while cross-section 2 was located in a site with banks lined with trees and shrubs. The distance between cross-section 1 and cross-section 2 is 93 m. The locations of the cross-sections are marked in the presented maps (Figure 1). Figure 2 gives available aerial photographs of the investigated Wełna River in the years 2009–2019. They show an unchanged status of tree cover in this river reach.

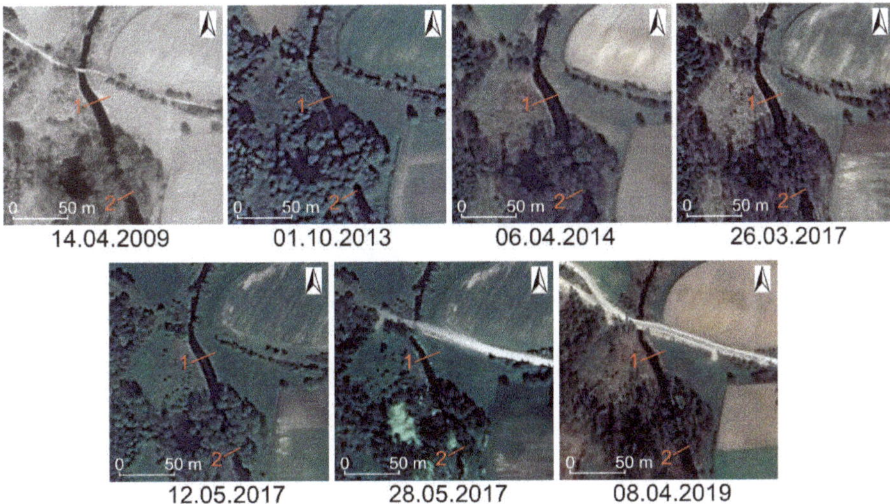

Figure 2. Changes of land cover and shading of the watercourse in different periods and years (images obtained from GoogleEarthPro).

Flow measurements were taken in the individual cross-sections. For this purpose, a miniature ADCP StreamPro stream flow meter was used. The ADCP StreamPro instrument measures distributions of velocity, discharge intensity, and river bottom geometry. It operates based on the emission of acoustic signals.

Waves reach suspensions floating in water and are reflected by them, as a result returning to the emitter. The instrument processes thus obtained data and next determined the hydraulic parameters of discharge in the river bed. Analyses of river channel geometry consisted in measurements of the location and ordinates of points on the riverbanks and the river bottom using a GPS real-time kinematic (RTK) device accurate to ± 3 mm for vertical measurements.

In order to determine the sediment composition analyses were conducted on the material collected from the river bottom. The grain-size distribution was analyzed using the sieve method according to the PN-B-00481:1988 (Polish standard) guidelines. To determine the grain-size composition, dried and ground samples were passed through a set of sieves (0.063, 0.125, 0.25, 0.5, 1.0, 2.0, 3.15, 4.0, 6.0, 8.0, 10.0, 16.0 mm). Samples were dried in 105–110 °C. In order to determine organic matter content, the dried samples were precisely weighed accurate to 0.01 g. Next, they were again subjected to heat—they were incinerated (600–800 °C) in a furnace. In this way the weight of the prepared samples made it possible to determine the content of organic matter. Samples were sieved again. Thus, the contents of minerals could be determined more precisely, excluding organic matter. Based on the recorded results, curves for fraction contents were plotted, and grain-size index (C_u) and grain-shape index (C_c) were calculated.

Thus, collected data were implemented in a graphic computer program, in which the shape of the cross-sections was plotted. Flow velocity was determined using an ADCP StreamPro device [45], applying the phenomenon of sound wave propagation in the water environment. Measurement results were used to determine flow values and to plot river channel cross-sections; next, wetted perimeter O, hydraulic radius R_h, and water table gradient I were calculated. At cross-section 2, with dense tree and shrub cover, regular measurements were taken for LAI on both riverbanks. These measurements were recorded during the vegetation season using a LAI 2000 device. The results provided the basis for an assessment of area cover by the tree canopy (including leaves). In this manner information was obtained on the availability of light-reaching plants, including macrophytes, at the ground level.

In the next stage, model calculations for solar radiation were made in the ArcGIS environment. The input data for the analysis were the Digital Surface Model (DSM) (Figure 1a). The DSM model was developed on the basis of the LIDAR data. The DSM is characterized with spatial resolution of 1 m and was provided by the Head Office of Geodesy and Cartography, Poland. The DSM contains information concerning objects on the surface, such as vegetation and buildings, which are crucial for shading analysis. The area for analysis was extended to make sure all of the shadows cast on the riverbed were included in the calculations. Solar radiation was calculated using the Area Solar Radiation tool. The output of the tool was a global radiation raster calculated for the whole year, vegetation period, and separate months from April to September with a one-hour interval in WH·m^{-2}. The results were assigned to the points of the cross-sections in the river.

3. Results

Results of flow measurements recorded using a miniature ADCP StreamPro flow meter were analyzed using the WinRiverII program. On this basis, values of flow, river bottom structure in the cross-sections, and the distribution of velocity were determined. Using the ADCP results and geodesy measures, graphs were prepared for the river cross-sections (Figure 3, Figure 4).

For the purpose of further calculations based on the above figures for each cross-section, their wetted perimeter, O, and cross-section, area A, were calculated. In the course of the measurements of water table ordinates, it was found that the water table gradient is so small that it is impossible to determine the slope at the river reach between the cross-sections. In order to determine slope, precise levelling was performed and the distance between the recorded measurements of water table height was increased to 380 m. At the river reach of that length, the mean hydraulic gradient was $I = 0.32‰$.

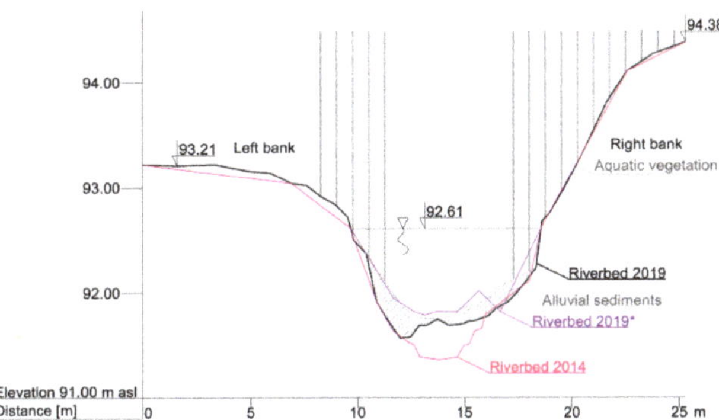

Figure 3. Bed changes in cross-section no. 1 (in an area with no tree cover—with submerged and emergent vegetation).

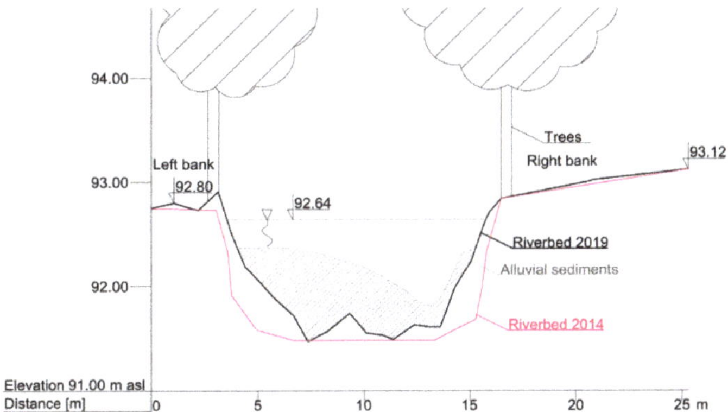

Figure 4. Bed changes in cross-section no. 2 (in a forested area).

In Figures 3 and 4 the cross-section was plotted for the status in 2014 and 2019. In the course of the study the cross-sections were also measured using a range pole. Cross-section 2 is wider, but slightly shallower than cross-section 1, in which, due to the bank overgrowth, flow is concentrated in the stream centerline of the watercourse. As a result of the large deposition of organic sediments the cross-sections measured using a range pole are slightly deeper (as they were probed to reach the so-called hard bottom). In this manner, the thickness of organic sediments deposited in the river channel could be assessed. Sediment thickness in cross-section 2 was much greater than in cross-section 1, with a maximum of 0.85 m. In contrast, in cross-section 1, the sediment thickness did not exceed 0.4 m. The analysis of the cross-sections also shows that within the previous five years channel shallowing was observed over the entire river reach. This trend is particularly evident in cross-section 2, where the bottom in the riverbank zone was elevated by as much as 0.6 m. Also, in cross-section 1 channel shallowing by 0.4 m was recorded. River bottom elevation and shallowing of the cross-sections may be connected with small hydraulic gradients in the investigated river reach, which is reflected in small flow velocities, at mean values for both cross-sections ranging from 0.06 m/s (cross-section 1) to 0.07 m/s (cross-section 2).

This is related, among other things, to vegetation overgrowth in the river reach in the vicinity of cross-section 2, being the downstream measuring site for the investigated reach. Thus, in the reach the predominant processes are the accumulation of sediments, to a considerable extent composed of organic matter.

A large proportion of the riverbank vegetation (trees and shrubs), i.e. those that are casting shadow over the entire river channel, causes shading, eliminating any aquatic vegetation within the river channel and banks. On the other hand, we may observe considerable accumulation of organic matter in the river reach. This also results from the small flow velocities. In cross-section 1, with the greatest share of riverbank and bottom vegetation, respective changes were observed in flow conditions. The flow velocity concentrates in the stream centerline zone, which also results in intensified erosion processes. This cross-section is the most compact.

Values of LAI were measured on the riverbanks at the tree-lined reach immediately above the ground. LAI is a measure for the total area of leaves per unit ground area, and it is directly related to the amount of light that can be intercepted by plants [44]. It is an important variable used to predict photosynthetic primary production and evapotranspiration, while it is also as a reference tool for crop growth. As such, LAI plays an essential role in theoretical production ecology. An inverse exponential relation has been established between LAI and light interception, which is linearly proportional to the primary production rate. Following the adopted methodology, the first measurement was taken in an open area with no vegetation (the reference measurement), while the next 12 were taken at the riverbanks, where trees and shrubs were found. For the right bank, the averaged LAI value was 3.03 [m^2/m^2], whereas for the left bank it was 3.06 [m^2/m^2]. The values indicate dense overgrowth reducing access to light in the investigated cross-section.

In each of the cross-sections the river bottom was raked using garden rakes in order to identify species of the bottom vegetation. The belt of the raked river bottom was approximately 1 m wide. Plant samples were weighed using scales. For cross-section 1 the obtained plant sample was approximately 1.5 kg. In cross-section 2 no bottom vegetation was found due to considerable shading. Samples of floating vegetation and plants growing on the banks were also collected. Identified plant species in the individual cross-sections are listed in Table 1.

Table 1. Plant species in individual sampling cross-sections.

No.	2014 Bottom Vegetation	Floating Vegetation	Rush Vegetation	2019 Bottom Vegetation	Floating Vegetation	Rush Vegetation
P1	*Potamogeton pectinatus*	*Hydrocharis morsus-ranae L., Spirodela polyrhiza*	*Sparganium erectum, Phragmites australis*	*Potamogeton pectinatus, Ceratophyllum demersum L.*	*Hydrocharis morsus-ranae L., Lemna minor L., Spirodela polyrhiza*	*Sparganium erectum, Phragmites australis, Typha latifolia L. Typha angustifolia, L. Humulus lupulus*
P2	no vegetation	no vegetation	*Sparganium erectum*	no vegetation	no vegetation	*Sparganium erectum*

Debris samples were collected from the river bottom. Substrate samples were subjected to sieve analysis in order to determine the grain-size composition. The sieve analysis was performed for samples from cross-sections 1 and 2. On this basis the uniformity index C_u was determined:

$$C_u = \frac{d_{60}}{d_{10}} \tag{1}$$

along with the curvature index C_c:

$$C_c = \frac{{d_{30}}^2}{d_{10} \cdot d_{60}} \tag{2}$$

where: d_{10}, d_{30} and d_{60} are diameters of particles, which together with smaller ones, account for 10, 30, and 60% of soil mass, respectively (Table 2).

Table 2. Debris parameters.

No.	2014		2019	
	C_u [-]	C_c [-]	C_u [-]	C_c [-]
P1	3.02	0.91	2.29	0.86
P2	2.51	1.83	2.39	0.84

For cross-section P1 the uniformity index C_u value classified the substrate as a grained substrate ($1 \leq C_u \leq 5$). The curvature index C_c indicates a poorly grained substrate. For cross-section P2 the recorded C_u value corresponds to a uniformly grained substrate. The C_c value corresponds to the values from the range for well-grained soils (C_c = 1–3). Results obtained after five years are similar and indicate a certain balance of river channel forming processes. Individual samples were incinerated in a furnace to determine organic matter content (MO). The following results were recorded: cross-section P1–2.4% MO (2014) and 4.1% (2019); cross-section P2–26.5% MO (2014) and 31.9% (2019).

The percentage share of organic matter in the sample from cross-section P2 is considerably greater compared to the sample from cross-section P1. This results from the fact that sample 2 was collected from the cross-section, in which trees and shrubs were growing on both banks, while dying plant parts accumulated on the watercourse channel bottom. The increase of organic matter content after five years indicates intensive accumulation of organic substance in the investigated cross-sections of that river.

The mean roughness coefficient was obtained using the results of hydrometric and geodesy analyses, applying the Manning formula for individual cross-sections. The results for the calculations of measurements both from 2014 and 2019 are presented in Table 3. When field measurements were taken, the water table was found in the main river channel. In cross-section 1, aquatic vegetation was growing on the channel banks and the river bottom. Paradoxically, the values of the roughness coefficient for that cross-section are lower than those obtained for cross-section 2, in which no aquatic vegetation was found either on the bottom or in the bank zone. This results from the parameters of that cross-section (greater width) and the damming effect due to the cross-section with a high proportion of vegetation, found in the immediate vicinity. This results in a reduced flow velocity, which under the assumption of the constant averaged value of the hydraulic gradient provides a higher value of the roughness coefficient. The resulting roughness coefficients are typical of overgrown river channels, in which the presence of vegetation drastically reduces the flow capacity of the river channel (Table 3).

Table 3. Results of calculations of roughness coefficient.

No.	2014 n [$m^{-1/3}\cdot s$]	2019 n [$m^{-1/3}\cdot s$]
P1	0.090	0.139
P2	0.156	0.188

Solar radiation was calculated using the Area Solar Radiation tool in ArcGIS 10.5. The results were assigned to the points of the cross-sections of the river in the monthly configuration (Figure 5) as well as the vegetation period (Figure 6a) and annual configurations (Figure 6b).

Calculations showed that solar radiation in the annual period at cross-sections no. 1 and no. 2 amounted to 847×10^3 and 135×10^3 $WH \cdot m^{-2}$, respectively (Table 4). In the analyzed river reach, total solar radiation varied, and ranged from 50×10^3 to 857×10^3 $WH \cdot m^{-2}$. From point no. 11, solar radiation was greater and ranged from 514 to 857×10^3 $WH \cdot m^{-2}$ with the mean value of 743×10^3 $WH \cdot m^{-2}$, which was connected with the exposure of the river channel. In the tree-lined river reach, total radiation ranged from 50 to 512×10^3 $WH \cdot m^{-2}$ with the mean of 245×10^3 $WH \cdot m^{-2}$. Results of calculations for individual months and the vegetation period are presented in Table 4. The value of radiation in the period from April to September ranged from 38×10^3 $WH \cdot m^{-2}$ at point no. 6 to 687 × 10^3 WH·m^{-2} at point no. 13. In cross-section no. 2, monthly values of solar radiation in the period from April to September ranged from 14 to 21 × 10^3 WH·m^{-2} at the total value of 110 × 10^3 WH·m^{-2}. In turn, in cross-section no. 1 solar radiation in individual months was almost five- to over eightfold higher, with a mean of over sixfold higher. Greater insolation in cross-section no. 1 contributed to more intensive growth of river channel vegetation.

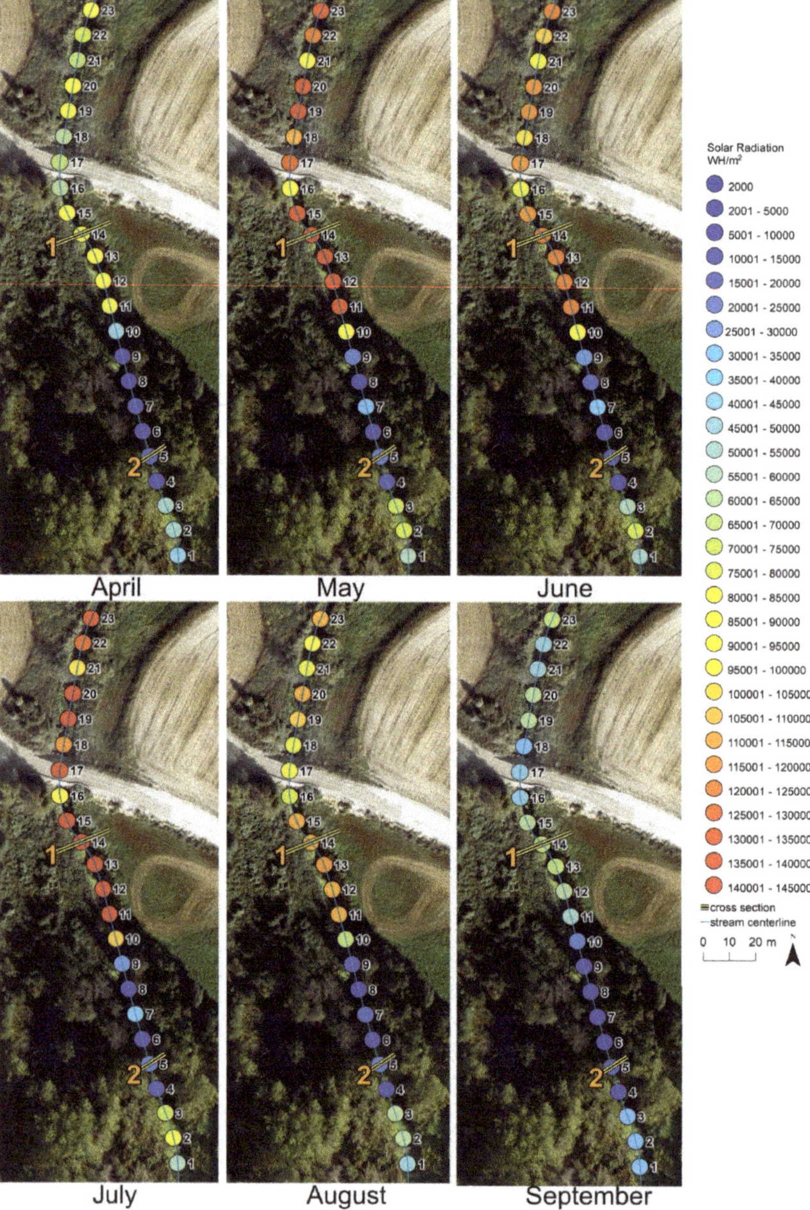

Figure 5. Distribution of total solar radiation (WH·m^{-2}) reaching the substrate in the analyzed Wełna River reach in individual months.

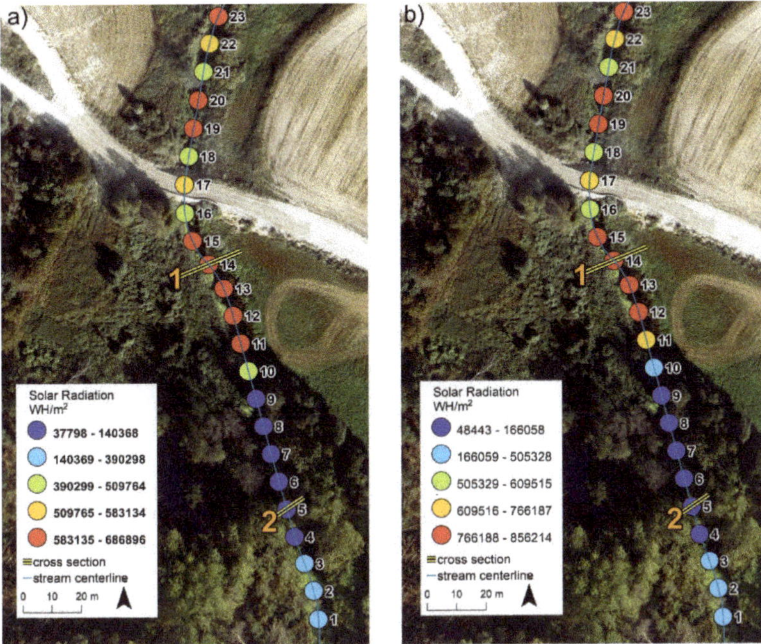

Figure 6. Distribution of total solar radiation (WH·m^{-2}) reaching the substrate in the investigated reach of the Wełna River for: (**a**) the period of April–September, (**b**) the annual period.

Table 4. Solar radiation (10^{3}·WH·m^{-2}) at points located along the investigated reach of the Wełna River.

No.	April	May	June	July	August	September	April-September
1	42	58	55	59	49	31	293
2	52	78	80	85	63	32	390
3	52	70	59	71	60	32	344
4	7	10	8	10	9	5	50
5 (P2)	19	21	15	20	21	14	110
6	5	7	8	8	6	4	38
7	11	33	36	38	19	4	140
8	3	4	18	7	3	2	38
9	10	20	26	26	12	7	101
10	47	99	93	106	69	21	434
11	91	137	120	142	112	51	652
12	95	138	122	143	113	64	675
13	97	140	123	145	115	67	687
14 (P1)	95	138	121	143	113	65	677
15	95	136	119	140	112	65	666
16	63	92	81	96	75	43	451
17	68	127	117	137	88	45	583
18	62	110	103	119	81	35	510
19	90	135	120	141	108	61	655
20	94	136	119	141	112	63	665
21	68	99	87	103	81	46	483
22	79	121	109	126	97	46	579
23	96	139	122	144	114	65	680

4. Discussion

The obtained results indicate the importance of shading for macrophyte growth. Similarly, as in a study by Jusik and Staniszewski [17], it was shown that increased shading perhaps limited macrophyte biodiversity and total cover in the river channels. The development of aquatic plants in the river channels is limited by shading [23]. Ali et al. [23] suggested that the management of tree vegetation might control incoming solar radiation, affecting submerged macrophytes. Similarly to the present case study, other studies shown that the impact of trees on the growth of aquatic plants depends on the group of macrophytes and the range of shadow impact [18,46]. Also, Jusik and Szoszkiewicz [47] observed significant correlations between the level of morphological modifications and shadowing of lowland river channels. Together with the increase in modifications of watercourses, the rate of shadowing decreased due to the changes of land use and simplification of the riparian species structure, commonly related to river regulation works. Kurtz et al. [48] conducted a shading experiment over a vegetation period to measure the effects of light reduction on *Vallisneria americana* in Perdido Bay (Florida–Alabama). The results showed that a 92% light reduction led to a decrease in chlorophyll *a* concentration, biomass, and leaf dimensions of macrophytes. Schneider et al. [49] showed that submerged macroscopic algae (*Chara intermedia* and *C. contraria*), which grow in an upright position, are taller at higher light intensities. In turn, Tan et al. [50] showed that reduced light reduced the growth of *M. aquaticum*; all of the analyzed growth indicators were significantly higher in sunlight treatment compared to the shading treatments. Kankanamge et al. [51] observed differences in shade tolerance between native and non-native species. In high shade ($\geq$ 90%), the reduction in lateral spread and an increase in main stem length for non-native macrophytes was observed, while native species showed no response for these traits.

The river's shading is spatially and temporally heterogeneous. Spatial variations are related to landscape (channel width, orientation, and hillshade) and canopy characteristics (canopy extent, structure, and height) [41,52]. Temporal variations are related to temporal variability of canopy structure and sun position in terms of days, months, and years [42]. As it was shown in this study, only the use of a suitable model allows reliable assessment of the amount of energy reaching the ground.

During the planning of river conservation or regulation works, it is crucial to maintain proper light conditions for macrophytes in view of the potential problems with both absence and limitation of aquatic plant development [4]. Excess light, which can even cause full overgrowth of the river channel, respectively changes water flow conditions [34] and causes a decrease in channel permeability [14]. The phenomenon of river channel overgrowth is very intensive in lowland rivers flowing through agricultural areas [16], with the simultaneous presence of strong sunlight conditions (lack of trees, removed during regulation works) and significant input of nutrients due to the surface flow [14].

Flow resistance for a specific type of aquatic vegetation is a function of many variables, including flow velocity, plant shape (habit), and roughness of the river bottom and channel walls. In recent decades hydraulic flow conditions in open river channels overgrown with vegetation have been investigated by many research centers [7,53,54]. Similarly, as in the present study, most research on the subject has been experimental [55,56]. For example, the authors of numerous papers analyzed flow conditions in overgrown river channels at an increase in the roughness coefficient [57] or in terms of changes in debris transport processes [55]. Many papers also indicate the need to combine the stream transport capacity with the degree of shading of the cross-section by aquatic vegetation [58]. In literature on the subject [58], an association was noted between the degree of cross-section shading by vegetation and values of the roughness coefficient, and biomass of aquatic plants [57]. Similarly, as in the case of these results, Łoboda et al. [59] showed that the presence of plants within the river channel cross-section slightly reduces the cross-section surface area, but markedly changes the distribution of flow velocity. This observation was also confirmed by studies on the effect of river channel dredging on values of the roughness coefficient and distributions of velocity [57,60].

In a similar way as by Przyborowski et al. [61], results obtained in this study may be referred to hydromorphological changes, in the analyzed case connected with transport and sedimentation of fine organic particles. The presence of plants has a considerable effect on the above-mentioned processes.

Calculations conducted according to the method proposed by Rickert [36] for the location at $\varphi = 52.5°$ northern latitude (comparable to the location of the reach in Cotoń with ordinates 52.678°N 17.631°E) for the cross-section shaded by trees showed that the actual solar radiation reaching the water surface decreases within 24-hour periods to values comparable to those obtained in the model for the shaded river reach. The values of total photosynthetically active radiation for the shaded river reach confirmed a lack of potential conditions for hydrophyte growth in that part of the river reach.

5. Conclusions

This paper presents an example of studies on the effect of factors causing watercourse channel shading on water flow conditions and the dynamics of hydromorphological processes. The presented analysis indicates the importance of local conditions determining availability of light energy for plants on the hydromorphological conditions found in a watercourse. The selected reach of the Wełna River is characterized by high variability in river channel shading, resulting from the presence of a dense forest complex growing in the upstream river stretch (LAI = 3). Immediately outside the forest edge, the river flows among meadows and arable fields, where there are no trees. These conditions affect river channel morphology and flow conditions. In the forested river reach, bottom and riverbank vegetation is completely absent. The river channel is wide; however, because of the vicinity of trees and shrubs, considerable accumulation of organic matter is observed in the river channel (26.5% MO compared to 2.4% MO in the treeless cross-section in 2014, and 31.9% MO to 4.1% in 2019). This is also promoted by the very low flow velocity (approximately 0.1 m/s). This velocity results from the damming effect as a result of overgrowth in the river reach located immediately outside the forest. The downstream reach is characterized by a large amount of vegetation overgrowing the banks and the river bottom. This cross-section is more compact, and flow velocities in the streamline are slightly higher (although with reference to the entire cross-section together with the vegetation (dead) zones the mean velocity is also approximately 0.1 m/s). The developed model of light availability makes it possible to determine river channel overgrowth conditions, and thus to predict potential dynamics of hydrodynamic processes. Analysis of total radiation reaching the water surface through trees, ranging from 50 to 512×10^3 WH·m^{-2} at the mean value of 245×10^3 WH·m^{-2}, confirmed a significant role of shading, as limiting macrophyte growth, which in turn causes changes in the character of hydromorphological processes.

Author Contributions: Conceptualization, T.K.; methodology, T.K. and R.W.; validation, T.K., S.Z. and M.H.; investigation, T.K., S.Z. and M.H.; writing—original draft preparation, T.K., M.S., R.W., J.J., S.Z. and M.H.; writing—review and editing, T.K., M.S. and J.J.; visualization, S.Z. and R.W.; supervision, T.K. and M.S.; project administration, T.K. All authors have read and agreed to the published version of the manuscript.

Funding: The publication was co-financed within the framework of the Ministry of Science and Higher Education program "Regional Initiative Excellence" in the years 2019–2022, Project No. 005/RID/2018/19.

Conflicts of Interest: The authors declare no conflict of interest.

References

1. Angelopoulos, N.V.; Cowx, I.G.; Buijse, A.D. Integrated planning framework for successful river restoration projects: Upscaling lessons learnt from European case studies. *Environ. Sci. Policy* **2017**, *76*, 12–22. [CrossRef]
2. Kałuża, T.; Pietruczuk, K.; Szoszkiewicz, K.; Tymiński, T. Assessment and classification of the ecological status of rivers in Poland according to the requirements of the water framework directive. *Wasserwirtsch. Wassertech.* **2014**, *104*, 24–29. [CrossRef]
3. Birk, S.; Willby, N. Towards harmonization of ecological quality classification: Establishing common grounds in European macrophyte assessment for rivers. *Hydrobiologia* **2010**, *652*, 149–163. [CrossRef]

4. Hachoł, J.; Bondar-Nowakowska, E. An assessment of the ecological status of diverse watercourses of Lower Silesia, Poland. *Pol. J. Environ Stud.* **2012**, *21*, 75–81.
5. Hajdukiewicz, H.; Wyżga, B.; Zawiejska, J.; Amirowicz, A.; Oglęcki, P.; Radecki-Pawlik, A. Assessment of river hydromorphological quality for restoration purposes: An example of the application of RHQ method to a Polish Carpathian river. *Acta Geophys.* **2017**, *65*, 423–440. [CrossRef]
6. Wiegleb, G.; Herr, B.; Zander, B.; Bröring, U.; Brux, H.; Van De Weyer, K. Natural variation of macrophyte vegetation of lowland streams at the regional level. *Limnologica* **2015**, *51*, 5362. [CrossRef]
7. Laks, I.; Sojka, M.; Walczak, Z.; Wróżyński, R. Possibilities of using low quality digital elevation models of floodplains in hydraulic numerical models. *Water* **2017**, *9*, 283. [CrossRef]
8. Mazur, R.; Kałuża, T.; Chmist, J.; Walczak, N.; Laks, I.; Strzeliński, P. Influence of deposition of fine plant debris in river floodplain shrubs on flood flow conditions- The Warta River case study. *Phys. Chem. Earth.* **2016**, *94*, 106–113. [CrossRef]
9. Vermaat, J.E.; Van Viersen, W. Growth potential for weeds in rivers. *H2O* **1990**, *23*, 534–536.
10. Walczak, N.; Walczak, Z.; Kałuża, T.; Hämmerling, M.; Stachowski, P. The impact of shrubby floodplain vegetation growth on the discharge capacity of river valleys. *Water* **2018**, *10*, 556. [CrossRef]
11. Szałkiewicz, E.; Dysarz, T.; Kałuża, T.; Malinger, A.; Radecki-Pawlik, A. Analysis of in-stream restoration structures impact on hydraulic condition and sedimentation in the Flinta River, Poland. *Carpathian J. Earth Env.* **2019**, *14*, 275–286. [CrossRef]
12. Laks, I.; Kałuża, T.; Sojka, M.; Walczak, Z.; Wróżyński, R. Problems with modelling water distribution in open channels with hydraulic engineering structures. *Annu. Set Environ. Prot.* **2013**, *15*, 245–257.
13. Mikuś, P.; Wyżga, B.; Radecki-Pawlik, A.; Zawiejska, J.; Amirowicz, A.; Oglęcki, P. Environment-friendly reduction of flood risk and infrastructure damage in a mountain river: Case study of the Czarny Dunajec. *Geomorphology* **2016**, *272*, 43–54. [CrossRef]
14. Dawson, F.H. The reduction of light as a technique for the control of aquatic plants an assessment. In Proceedings of the Association of Applied Biologists Symposium Aquatic Weeds and their Control, Oxford, UK, 7–8 April 1981.
15. Ptak, M.; Sojka, M.; Kałuża, T.; Choiński, A.; Nowak, B. Long-term water temperature trends of the Warta River in the years 1960–2009. *Ecohydrol. Hydrobiol.* **2019**, *19*, 441–451. [CrossRef]
16. Burrell, T.K.; O'Brien, M.; Graham, S.E.; Simon, K.S.; Harding, J.S.; Mcintosh, A.R. Riparian shading mitigates stream eutrophication in agricultural catchments. *Freshw. Sci.* **2014**, *33*, 73–84. [CrossRef]
17. Jusik, S.; Staniszewski, R. Shading of river channels as an important factor reducing macrophyte biodiversity. *Pol. J. Environ. Stud.* **2019**, *28*, 1215–1222. [CrossRef]
18. Sender, J. The effect of riparian forest shade on the structural characteristics of macrophytes in a mid-forest lake. *Appl. Ecol. Env. Res.* **2016**, *14*, 249–261. [CrossRef]
19. Szoszkiewicz, K.; Ferreira, T.; Korte, T.; Baattrup-Pedersen, A.; Davy-Bowker, J.; O'Hare, M. European river plant communities: The importance of organic pollution and the usefulness of existing macrophyte metrics. *Hydrobiologia* **2006**, *566*, 211–234. [CrossRef]
20. Radecki-Pawlik, A.; Kuboń, P.; Radecki-Pawlik, B.; Plesiński, K. Bed-load transport in two different-sized mountain catchments: Mlynne and Lososina streams, Polish Carpathians. *Water* **2019**, *11*, 272. [CrossRef]
21. Pedersen, M.L.; Baatrup-Pedersen, A.; Roth, F.R.; Madsen, T.V.; Larsen, S.E. Short-term impacts of weed cutting on physical habitats in lowland rivers—The importance of initial environmental conditions. *Pol. J. Environ. Stud.* **2011**, *20*, 1271–1280.
22. Szoszkiewicz, K.; Zbierska, J.; Staniszewski, R.; Jusik, S. The variability of macrophyte metrics used in river monitoring. *Oceanol. Hydrobiol. Stud.* **2009**, *38*, 117–126. [CrossRef]
23. Ali, M.M.; Hassan, S.A.; Shaheen, A.S.M. Impact of riparian trees shade on aquatic plant abundance in conservation islands. *Acta Bot. Croat.* **2011**, *70*, 245–258. [CrossRef]
24. Schooler, S. Shade as a management tool for the invasive submerged macrophyte, Cabomba caroliniana. *J. Aquat. Plant Manag.* **2008**, *46*, 168–171.
25. Abernethy, V. Response of *Elodea canadensis Michx and Myriophyllum spicatum* L. to shade, cutting and competition in experimental culture. *Hydrobiologia* **1996**, *340*, 219–224. [CrossRef]
26. Dawson, F.H. Water flow and vegetation of running waters. In *Vegetation of Inland Waters*; Symoens, J.J., Ed.; Kluwer Dordrecht/Junk: The Hague, The Netherlands, 1988. [CrossRef]

27. Kohler, J.; Hachoł, J.; Hilt, S. Regulation of submersed macrophyte biomass in a temperate lowland river: Interactions between shading by bank vegetation, epiphyton and water turbidity. *Aquat. Bot.* **2010**, *92*, 129–136. [CrossRef]
28. Wersal, R.M.; Madsen, J.D. Influences of light intensity variations on growth characteristics of *Myriophyllum aquaticum*. *J. Freshw. Ecol.* **2013**, *28*, 147–164. [CrossRef]
29. Lin, Y.; Herold, M. Tree species classification based on explicit tree structure feature parameters derived from static terrestrial laser scanning data. *Agric. For. Meteorol.* **2016**, *216*, 105–114. [CrossRef]
30. Wilcock, R.J.; Scarsbrook, M.R.; Cooke, J.G.; Costley, K.J.; Nagels, J.W. Shade and flow effects on ammonia retention in macrophyte-rich streams: Implications for water quality. *Environ. Pollut.* **2004**, *132*, 95–100. [CrossRef]
31. Zefferman, E. Increasing canopy shading reduces growth but not establishment of *Elodea nuttallii* and *Myriophyllum spicatum* in stream channels. *Hydrobiologia* **2014**, *734*, 159–170. [CrossRef]
32. Kałuża, T.; Sojka, M.; Strzeliński, P.; Wróżyński, R. Application of terrestrial laser scanning to tree trunk bark structure characteristics evaluation and analysis of their effect on the flow resistance coefficient. *Water* **2018**, *10*, 753. [CrossRef]
33. Lu, J.; Wang, Z.; Xing, W.; Liu, G. Effects of substrate and shading on the growth of two submerged macrophytes. *Hydrobiologia* **2013**, *700*, 157–167. [CrossRef]
34. Kałuża, T. The influence of the trees and bushes shadow on the changes of flow conditions in the lowland watercourse. *Acta Sci. Pol. Form. Circumiectus* **2015**, *14*, 29. [CrossRef]
35. Dysarz, T.; Szałkiewicz, E.; Wicher-Dysarz, J. Long-Term impact of sediment deposition and erosion on water surface profiles in the Ner River. *Water* **2017**, *9*, 168. [CrossRef]
36. Rickert, K. Der Einfluss von Gehölz auf die Lichtverhältnisse und das Abflussverhalten in Fließgewässern. Mitteilungen des Instituts für Wasser-wirtschaft, Hydrologie und landwirtsch Wasserbau der Univesität Hannover: Hannover, Germany, 1986.
37. Bode, C.A.; Limm, M.P.; Power, M.E.; Finlay, J.C. Subcanopy solar radiation model: Predicting solar radiation across a heavily vegetated landscape using LiDAR and GIS solar radiation models. *Remote Sens. Environ.* **2014**, *154*, 387–397. [CrossRef]
38. Johnson, M.F.; Wilby, R.L. Seeing the landscape for the trees: Metrics to guide riparian shade management in river catchments. *Water Resour. Res.* **2015**, *51*, 3754–3769. [CrossRef]
39. Wawrzyniak, V.; Allemand, P.; Bailly, S.; Lejot, J.; Piégay, H. Coupling LiDAR and thermal imagery to model the effects of riparian vegetation shade and groundwater inputs on summer river temperature. *Sci. Total Environ.* **2017**, *592*, 616–626. [CrossRef]
40. Loicq, P.; Moatar, F.; Jullian, Y.; Dugdale, S.J.; Hannah, D.M. Improving representation of riparian vegetation shading in a regional stream temperature model using LiDAR data. *Sci. Total Environ.* **2018**, *624*, 480–490. [CrossRef]
41. Karrasch, P.; Hunger, S. Simulation of vegetation and relief induced shadows on rivers with remote sensing data. Proceedings of Earth Resources and Environmental Remote Sensing/GIS Applications VIII, Warsaw, Poland, 5 October 2017.
42. Bachiller-Jareno, N.; Hutchins, M.G.; Bowes, M.J.; Charlton, M.B.; Orr, H.G. A novel application of remote sensing for modelling impacts of tree shading on water quality. *J. Environ. Manag.* **2019**, *230*, 33–42. [CrossRef]
43. Environment Agency. *Keeping Rivers Cool: Getting Ready for Climate Change by Creating Riparian Shade*; Environment Agency Horizon House: Bristol, UK, 2012.
44. Houborg, R.; McCabe, M.F. Daily Retrieval of NDVI and LAI at 3 m resolution via the fusion of CubeSat, Landsat, and MODIS Data. *Remote Sens.* **2018**, *10*, 890. [CrossRef]
45. Lee, K.; Ho, H.C.; Marian, M.; Wu, C.H. Uncertainty in open channel discharge measurements acquired with StreamPro ADCP. *J. Hydrol.* **2014**, *509*, 101–114. [CrossRef]
46. Zhu, Z.; Song, S.; Yan, Y.; Li, P.; Jeelani, N.; Wang, P.; An, S.; Leng, X. Combined effects of light reduction and ammonia nitrogen enrichment on the submerged macrophyte *Vallisneria natans*. *Mar. Freshw. Res.* **2018**, *69*, 764–770. [CrossRef]
47. Jusik, S.; Szoszkiewicz, K. Biodiversity of water plants in diverse conditions of morphological modifications in lowland rivers of western Poland. *Nauka Przyroda Technologie* **2009**, *3*, 84.

48. Kurtz, J.C.; Yates, D.F.; Macauley, J.M.; Quarles, R.L.; Genthner, F.J.; Chancy, C.A.; Devereux, R. Effects of light reduction on growth of the submerged macrophyte *Vallisneria americana* and the community of root-associated heterotrophic bacteria. *J. Exp. Mar. Biol. Ecol.* **2003**, *291*, 199–218. [CrossRef]
49. Schneider, S.C.; Pichler, D.E.; Andersen, T.; Melzer, A. Light acclimation in submerged macrophytes: The roles of plant elongation, pigmentation and branch orientation differ among Chara species. *Aquat. Bot.* **2015**, *120*, 121–128. [CrossRef]
50. Tan, B.C.; He, H.; Gu, J.; Li, K.Y. Effects of nutrient levels and light intensity on aquatic macrophyte (*Myriophyllum aquaticum*) grown in floating-bed platform. *Ecol. Eng.* **2019**, *128*, 27–32. [CrossRef]
51. Kankanamge, C.E.; Matheson, F.E.; Riis, T. Shading constrains the growth of invasive submerged macrophytes in streams. *Aquat. Bot.* **2019**, *158*, 103125. [CrossRef]
52. Li, G.; Jackson, C.R.; Kraseski, K.A. Modeled riparian stream shading: Agreement with field measurements and sensitivity to riparian conditions. *J. Hydrol.* **2012**, *428*, 142–151. [CrossRef]
53. Antonarakis, A.S.; Richards, K.S.; Brasington, J.; Muller, E. Determining leaf area index and leafy tree roughness using terrestrial laser scanning. *Water Resour. Res.* **2010**, *46*. [CrossRef]
54. Laks, I.; Szoszkiewicz, K.; Kałuża, T. Analysis of in situ water velocity distributions in the lowland river floodplain covered by grassland and reed marsh habitats—A case study of the bypass channel of Warta River (Western Poland). *J. Hydrol. Hydromech.* **2017**, *65*, 325–332. [CrossRef]
55. Przyborowski, Ł.; Łoboda, A.M.; Bialik, R.J. Experimental Investigations of Interactions between Sand Wave Movements, Flow Structure, and Individual Aquatic Plants in Natural Rivers: A Case Study of *Potamogeton pectinatus* L. *Water* **2018**, *10*, 1166. [CrossRef]
56. Przyborowski, Ł.; Łoboda, A.M.; Karpiński, M.; Bialik, R.J. Characteristics of flow around aquatic plants in natural conditions: Experimental setup, challenges and difficulties. In *Free Surface Flows and Transport Processes*; Kalinowska, M.B., Mrokowska, M.M., Rowiński, P.M., Eds.; Springer: Cham, Switzerland, 2018; pp. 347–361. [CrossRef]
57. Kenel, B.; Uehlinger, U. Effects of plant cutting and dredging on habitat conditions in streams. *Arch. Hydrobiology* **1998**, *143*, 257–273. [CrossRef]
58. Querner, E.P. A model to estimate timing of aquatic weed control in drainage canals. *Irrig. Drain. Syst.* **1997**, *11*, 157–169. [CrossRef]
59. Łoboda, A.M.; Bialik, R.J.; Karpiński, M.; Przyborowski, Ł. Two simultaneously occurring *Potamogeton* species: Similarities and differences in seasonal changes of biomechanical properties. *Pol. J. Environ Stud.* **2019**, *28*, 237–253. [CrossRef]
60. Tymiński, T. Hydraulic Model Investigation of Flow Conditions for Floodplains with Coniferous and Deciduous Shrubs. *Pol. J. Environ Stud.* **2012**, *21*, 1047–1052.
61. Przyborowski, Ł.; Łoboda, A.M.; Bialik, R.J.; Västilä, K. Flow field downstream of individual aquatic plants —Experiments in a natural river with *Potamogeton crispus* L. and *Myriophyllum spicatum* L. *Hydrol. Processes* **2019**, *33*, 1324–1337. [CrossRef]

Article

A Non-Tuned Machine Learning Technique for Abutment Scour Depth in Clear Water Condition

Hossein Bonakdari [1,*], Fatemeh Moradi [2], Isa Ebtehaj [2], Bahram Gharabaghi [3], Ahmed A. Sattar [4,5], Amir Hossein Azimi [6] and Artur Radecki-Pawlik [7]

1 Department of Soils and Agri-Food Engineering, Laval University, Québec, QC G1V 0A6, Canada
2 Environmental Research Center, Razi University, Kermanshah 6714414971, Iran; f.moradi23@yahoo.com (F.M.); isa.ebtehaj@gmail.com (I.E.)
3 School of Engineering, University of Guelph, Guelph, ON NIG 2W1, Canada; bgharaba@uoguelph.ca
4 Department of Irrigation & Hydraulics, Faculty of Engineering, Cairo University, Giza 12316, Egypt; ahmoudy77@yahoo.com
5 Faculty of Civil Engineering, German University in Cairo, New Cairo 13611, Egypt
6 Department of Civil Engineering, Lakehead University, Thunder Bay, ON P7B 5E1, Canada; azimi@lakeheadu.ca
7 Division of Structural Mechanics and Material Mechanics, Faculty of Civil Engineering, Cracow University of Technology, 31-155 Krakow, Poland; rmradeck@cyf-kr.edu.pl
* Correspondence: hossein.bonakdari@fsaa.ulaval.ca; Tel.: +1-418-656-2131; Fax: +1-418-656-3723

Received: 8 December 2019; Accepted: 16 January 2020; Published: 20 January 2020

Abstract: Abutment scour is a complex three-dimensional phenomenon, which is one of the leading causes of marine structure damage. Structural integrity is potentially attainable through the precise estimation of local scour depth. Due to the high complexity of scouring hydrodynamics, existing regression-based relations cannot make accurate predictions. Therefore, this study presented a novel expansion of extreme learning machines (ELM) to predict abutment scour depth (d_s) in clear water conditions. The model was built using the relative flow depth (h/L), excess abutment Froude number (F_e), abutment shape factor (K_s), and relative sediment size (d_{50}/L). A wide range of experimental samples was collected from the literature, and data was utilized to develop the ELM model. The ELM model reliability was evaluated based on the estimation results and several statistical indices. According to the results, the sigmoid activation function (correlation coefficient, R = 0.97; root mean square error, $RMSE$ = 0.162; mean absolute percentage error, $MAPE$ = 7.69; and scatter index, SI = 0.088) performed the best compared with the hard limit, triangular bias, radial basis, and sine activation functions. Eleven input combinations were considered to investigate the impact of each dimensionless variable on the abutment scour depth. It was found that $d_s/L = f\ (F_e, h/L, d_{50}/L, K_s)$ was the best ELM model, indicating that the dimensional analysis of the original data properly reflected the underlying physics of the problem. Also, the absence of one variable from this input combination resulted in a significant accuracy reduction. The results also demonstrated that the proposed ELM model significantly outperformed the regression-based equations derived from the literature. The ELM model presented a fundamental equation for abutment scours depth prediction. Based on the simulation results, it appeared the ELM model could be used effectively in practical engineering applications of predicting abutment scour depth. The estimated uncertainty of the developed ELM model was calculated and compared with the conventional and artificial intelligence-based models. The lowest uncertainty with a value of ±0.026 was found in the proposed model in comparison with ±0.50 as the best uncertainty of the other models.

Keywords: abutment; clear water; equation; sensitivity analysis; scour

---

1. Introduction

The presence of hydraulic structures obstructs the flow and leads to subsequent scour around structure foundations. Scour is recently recognized as the most common cause of bridge foundation failure, leading to property losses, business failure, and the disruption of economic activities [1,2]. For accurate scour depth prediction, knowledge of scouring is necessary. Scour depth overestimation and underestimation lead to higher construction costs and abutment foundation damage, respectively [3–5]. Hence, an accurate method of predicting scour depth is necessary to reduce economic costs and achieve high-stability confidence coefficients for foundations.

When river flow comes in contact with bridge abutments, the pressure creates a downward flow around the structures. This downward flow forms bed cavities containing vortices. Shear stress-generated upstream of an abutment leads to primary vortices, which withdraw to the abutment and create secondary vortices that are less able than the initial vortices. Besides, unstable shear layers produced according to the separation of flow upstream and downstream of bridge abutments rotate in the form of vortex structures known as wake vortices. Wake vortices act similarly to small eddies and sediment rising from the bed [6–9].

Numerous experimental studies have been done to survey the abutment scour and deposition process in clear water conditions [6,10–21]. Dey and Barbhuiya [22] proposed different regression-based equations based on the abutments' type as follows:

$$\frac{d_s}{L} = 7.281\ ({F_e}^{0.314})\left(\frac{h}{L}\right)^{0.128}\left(\frac{L}{d_{50}}\right)^{-0.167} \qquad Vertical-Wall \tag{1}$$

$$\frac{d_s}{L} = 8.319\ ({F_e}^{0.312})\left(\frac{h}{L}\right)^{0.101}\left(\frac{L}{d_{50}}\right)^{-0.231} \qquad 45^o Wing-Wall \tag{2}$$

$$\frac{d_s}{L} = 8.689\ ({F_e}^{0.192})\left(\frac{h}{L}\right)^{0.103}\left(\frac{L}{d_{50}}\right)^{-0.296} \qquad Semicircular \tag{3}$$

Moreover, the conventional regression-based model of the Muzzammil [23] is defined as follows:

$$\frac{d_s}{L} = 9.694\, K_s\ ({F_e}^{0.648})\left(\frac{h}{L}\right)^{0.04}\left(\frac{d_{50}}{L}\right)^{-0.075} \tag{4}$$

where d_{50} denotes as the median diameter of sediment particles, d_s is the equilibrium scour depth, h is the approaching flow depth, L is the radius of scour hole at original bed level upstream, and F_e is the excess abutment Froude number, which is defined as:

$$F_e = \frac{U_e}{\sqrt{g(s-1)l}} \tag{5}$$

where l is the transverse length of an abutment, g is the acceleration due to gravity, U_e is the excess approaching flow velocity, and s is the relative density of sediment particles.

The majority of studies offer several empirical equations that are highly dependent on experimental conditions; hence, the results obtained with various methods differ significantly. There is much disagreement on the selection of a comprehensive approach to designing bridge abutments that are protected against scour. Moreover, the equations obtained cannot serve as general and reliable equations for estimating bridge abutment scour [24,25].

Various studies have addressed scour prediction using feedforward neural networks (FFNN) [26]; gene expression programming (GEP) [27–29]; adaptive neuro-fuzzy inference systems (ANFIS) [30]; support vector machines (SVM) [31]; group method of data handling (GMDH) [32–35], and model trees [36].

Bateni and Jeng [37] applied an ANFIS-based method to estimate the scour at pile groups. They found that the ANFIS model outperformed conventional techniques. Muzzammil [38] employed a neural network to evaluate the scour at abutments and concluded that neural networks could serve as alternative experimental and regression methods for predicting the complex flow structures around bridge abutments. Their artificial neural network (ANN) model produced better results than a conventional regression model. The raw parameters also exhibited superior performance to dimensionless parameters in scour prediction. Ghani and Azamathulla [39] employed an ANFIS model to investigate the scour process at culvert outlets. They found that ANFIS predicted scour depth more accurately than artificial neural networks and regression equations.

Begum [40] employed an ANN model to estimate the scour depth around semi-circular bridge abutments. They applied sensitivity analysis to the input parameters to investigate the impact of each independent variable and determined that the ANN model was more precise than other experimental formulas available. Azamathulla [28] used gene-expression programming (GEP) and artificial neural network (ANN) models to assess the scour process at bridge abutments. According to the findings, the GEP model produced more satisfactory results compared to the ANN model in predicting scour depth. Etemad-Shahidi et al. [36] employed model trees (MT) and evaluated datasets with a broad range of variables for predicting scour around circular piers. In terms of relative scour depth prediction, their model was presented as a function of Froude number, relative flow depth, and flow velocity. The model tree (MT) results were compared with an experimental formula, and it was found that the MT could estimate scouring depth more accurately.

One of the most popular techniques for scour depth prediction is the feedforward neural networks (FFNN). It has recently been employed as a classic method in data mining to validate the superiority of artificial intelligence methods, such as ANFIS, extreme learning machine (ELM), ANN, etc. [26,41–45]. The classic FFNN is trained with a backpropagation algorithm. The ANFIS, which is a combination of neural network and fuzzy logic, is employed by Moradi et al. [46] to calculate the abutment scour depth. The authors compared different techniques for generating the fuzzy inference systems (FIS) (i.e., sub-clustering (SC), fuzzy c-means clustering (FCM), and grid partitioning (GP)). The results indicated that the ANFIS-SC outperformed all other existing ANFIS techniques. Besides, Azimi et al. [47] proposed a multi-objective design of ANFIS optimized by genetic algorithm. The authors indicated that this technique not only provided more accurate results but also enhanced the generalizability of the ANFIS model in abutment scour depth prediction. The main problems of the mentioned ANFIS-based techniques are the failure to provide a specific relationship to apply in practical tasks.

The shortcomings of feedforward neural networks back propagation (FFNN-BP) are the low learning rate, the slow convergence rate, and the presence of local minima [48]. Besides, other artificial intelligence-based methods like ANFIS are black-box methods, which do not yield equations for use in practical work. Classification-based methods, such as model trees (MT), classify data, and, in particular, conditions yield equations. However, the main problems with such methods are the lack of flexibility and adequate performance in different hydraulic conditions. Although the GEP method provides an equation for use in practical engineering problems, the problems with this method are the high central processing unit (CPU) time consumption and that the equations presented are often achieved using several highly sophisticated functions.

For these reasons, Huang et al. [49] proposed a new algorithm, the extreme learning machine (ELM), for FFNN training based on single hidden layer feedforward neural networks (SLFNs). ELM requires adjusting only the activation function type and some hidden layers, while there are several user-defined parameters, such as hidden layer bias adjustment, during algorithm and input weight adjustment [50]. Moreover, using ELM leads to produce equations for practical problems [51].

Compared to other learning algorithms, such as back propagation (BP), ELM achieves swift learning and performs well in generation function processing [52–55]. Using ELM in various engineering science fields, such as feature selection [56], classification [57], and regression [51,58], has provided acceptable results.

Since ELM has not been used for scour depth modeling, this method was applied here to predict the abutment scour depth in clear water conditions. This approach overcame many issues in modeling scour depth by using the existing artificial intelligence methods. For this purpose, the parameters affecting the abutment scour depth were determined first. Then, different activation functions were considered in predicting the abutment scour depth. Upon selecting the best activation function and carrying out a dimensional analysis, the influence of each parameter affecting the scour depth was evaluated from eleven input combinations. Moreover, the current study results and two empirical formulae were compared with previous studies. Finally, the optimum combination was used to present a relationship for practical engineering problems.

2. Materials and Methods

2.1. Data Collection

To estimate the scour depth around bridge abutments in clear water with a uniform sediment bed, a set of 295 experimental data were collected from the study of Dey and Barbhuiya [22]. They performed experiments in a lab flume (20 × 0.7 × 0.9 m long, deep, and wide) using a vertical wall, 45° wing-wall, and semicircular abutments, which are schematically illustrated in Figure 1.

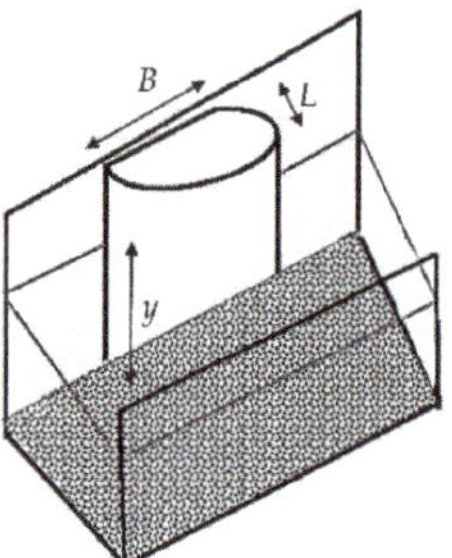

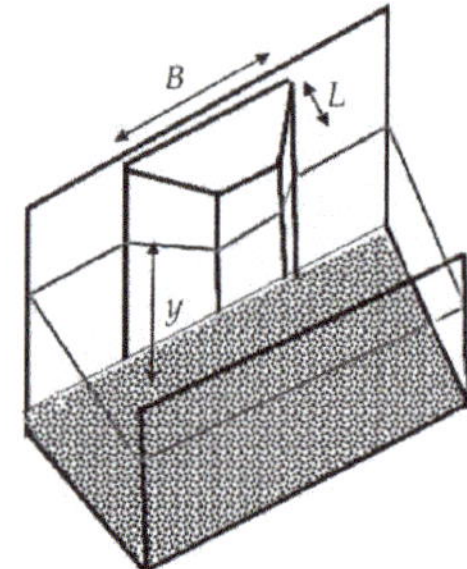

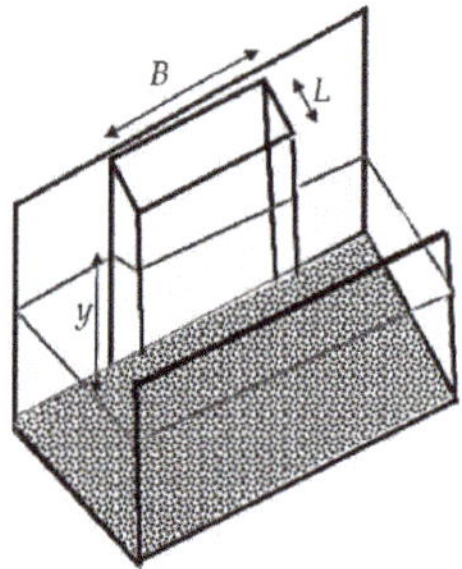

Figure 1. Schematic of the different abutment in the flume. L and B are the abutment length and width respectively.

During the tests, an abutment was set in the bed, sustained by a sediment intromission 0.3 m deep, 2.4 m long, and 0.9 m wide, and appended to the flume sidewall. Recess sediments were placed 9 m downstream of the flume, and the discharge was adjusted by the entrance inlet. The flow depth was measured using a Vernier point gauge with ±0.1 mm accuracy. All tests were done in clear water conditions and $U/U_c = 0.95$, where U_c is the critical velocity of bed particles. Table 1 presents the sediment particle sizes, abutment geometric shapes, and flow conditions. Of the 295 experimental samples, approximately 50% of the samples were randomly selected for model training (148 samples), and the rest of the samples were applied for the testing stage (147 samples).

Table 1. Statistical indices for the collected dataset.

Variable	Min	Max	Mean	Variance	Standard Deviation
d_{50} (mm)	0.26	3.1	0.92	0.51	0.71
h (m)	0.05	0.25	0.16	0.004	0.06
U (m/s)	0.219	0.67	0.35	0.013	0.11
d_s (m)	0.024	0.293	0.13	0.003	0.06
L (m)	0.04	0.13	0.08	0.001	0.03
B (m)	0.08	0.36	0.18	0.006	0.08

Note: d_{50} is the average diameter of particles, h is the stream depth of approach, U is the average stream velocity, U_c is the critical velocity of sediment, d_s is the scour depth, L and B are the abutment length and width respectively.

2.2. Extreme Learning Machine

The ELM was first introduced by Huang et al. [49] as a new robust learning algorithm for single-layer feedforward neural networks (SLFNN). The ELM is very simple to use since only the network architecture should be defined. Hence, the ELM does not have the complexity of gradient algorithms, such as local minima, learning epochs, and learning rate. Moreover, it is proven that the speed of ELM is much greater than SVM and gradient-based algorithms, such as back-propagation. Thus, ELM training often takes a few seconds or a few minutes in very complex cases with important data where modeling with the classical neural network is not easy. In this algorithm, the weight vector is associated with the input and hidden layers. Moreover, the primary neurons in the hidden layer are generated randomly, and a unique optimal solution is produced to determine the number of hidden layer neurons during training. If m is the number of nodes in the input layer, M is the number of nodes in the hidden layer, n is the number of nodes in the output layer, b_i is the bias, and $g(x)$ is the activation function of neurons in the hidden layer, for N arbitrary separate samples (x_i, t_i), $x_i = [x_{i1}, x_{i2}, \ldots, x_{im}]^T \in R^m$, $_i = [t_{i1}, t_{i2}, \ldots, t_{in}]^T \in R^n$, model network training by ELM is presented as follows (Figure 2):

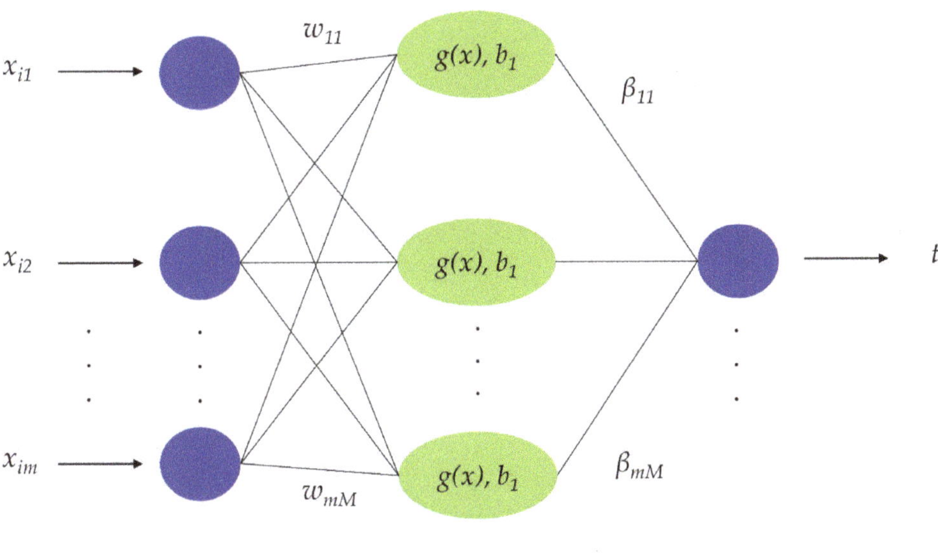

Figure 2. Extreme learning machine (ELM) structure.

The standard SLFNN forms with M neurons in the hidden layer, and activation function $g(x)$ is modeled as follows [59,60]:

$$\sum_{i=1}^{M} \beta_i g(W_i \cdot X_i + b_i) = O_j \quad j = 1, 2, \ldots, N \tag{6}$$

where $O_i = [O_{i1}, O_{i2}, \ldots, O_{in}]^T$ is the output vector from SLFNN, W_i. X_i is the inner product of X_i and W_i, b_i is the bias of the ith hidden layer neuron, $\beta_i = [\beta_{i1}, \beta_{i2}, \ldots, \beta_{in}]^T$ is a gravimetric vector associated with the ith hidden layer neuron and output nodes, and $W_i = [W_{1i}, W_{2i}, \ldots, W_{mi}]^T$ is the weight vector related to the hidden layer neuron and input nodes.

Activation functions are used to calculate the weight of neuron response output. The neurons consist of two parts, including the weighted sums of inputs and activation functions. When a set of weighted input signals is applied, activation functions are used to obtain the layer response. For neurons in the same layer, the same activity functions are used, which can be linear or nonlinear. In linear functions, a straight line is drawn, and in nonlinear functions, a curved line is drawn. The ELM nonlinear activity functions investigated in this study included hard limit sine, sigmoid, radial basis, and triangular basis: they are briefly defined as follows (see Figure 3):

- ➢ A sine function inserts the actual amount that has −1 to +1 efficiency.
- ➢ Sigmoid is a continuous function that gradually changes between asymptotic values of 0 and 1 or −1 and +1.
- ➢ When neurons use a hard limit transfer function, and if the input threshold net is gained, the output is 1, otherwise 0.
- ➢ The radial basis function output is based on the distance from the origin points.
- ➢ A triangular basis function can serve as a neuron transfer function. This function calculates the output layer from a given input layer.

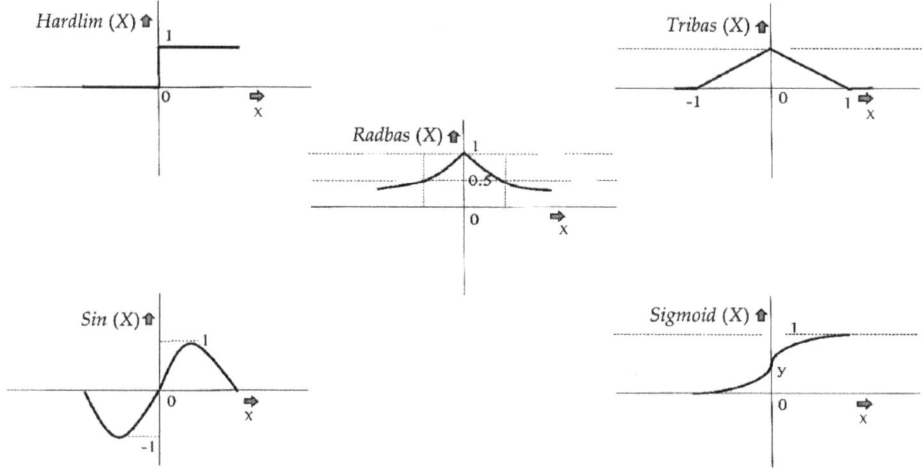

Figure 3. The different activation functions in the ELM model.

The standard SLFNN form with an activation function $g(x)$ and M hidden neurons can approximate all N samples with 0 error.

The relation between w_i, β_i, and $\mathrm{b_i}$ is defined as follows:

Equation (6) can be expressed in the form of the following matrix:

$$\boldsymbol{H}\beta = Y \tag{7}$$

where H is the output matrix related to the hidden layer. When the input weights are selected randomly, and the hidden layer neurons are determined, the network training equivalent to the search for the least-squares $(\hat{\beta})$ related to Equation (7) is minimized as follows:

$$\min_{\beta} \|\boldsymbol{H}\beta - Y\| \tag{8}$$

The smallest-norm least-squares solution is:

$$\hat{\beta} = \boldsymbol{H}^{+}Y \tag{9}$$

where H^{+} is the Moore–Penrose generalized inversion of hidden layer matrix H.

Therefore, the ELM method is summarized in three steps:

(i) Generate the hidden layer bias (b); randomly generate the weight vector (w) that links the input and hidden layers;
(ii) Compute the hidden layer output matrix (H);
(iii) Compute the output weight.

2.3. Methodology

Scour around abutments based on the flow's capacity to transfer bed materials is classified into two categories: moving bed scour and clear water scour. Moving bed scour when the flow moves the general bedload, while clear water scour happens when the flow does not move sediment downstream. The effective parameter on scour depth d_s at an abutment located on a bed with uniform sediment in clear water is a function of sediment characteristics, flow parameters, and hydraulic structure geometry [61–63].

$$d_s = f(U, U_c, L, g, K_s, d_{50}, h, \rho, \rho_s, \nu) \tag{10}$$

where U is the average stream velocity, U_c is the critical velocity of sediment, L is the abutment length, g is the gravitational acceleration, d_{50} is the average diameter of particles, h is the stream depth of approach, ρ and ρ_s represent the fluid and sediment density, respectively, and ν is the fluid kinematic velocity. The K_s is the abutment form factor that was determined by a trial-and-error procedure for the best collapsed of all data by Mellville [16]. It is 1 for vertical-wall abutments, 0.82 for 45° wing-wall abutments, and 0.75 for semicircular abutments. Dey and Barbhuiya [17] neglected the effect of viscosity and considered the relative density constant to express the following relationship:

$$d_s = f(U, U_c, L, g, K_s, d_{50}, h) \tag{11}$$

Abutment scour is occurred when the additional stream velocity U_e is greater than 0, where $U_e = U - \xi U_c$, and ξ is a constant coefficient for vertical wall, 45° wing-wall, and semicircular wall abutments of 0.5, 0.55, and 0.6, respectively.

$$d_s = f(U_e, L, g, K_s, d_{50}, h) \tag{12}$$

Most studies in the field of artificial intelligence consider the effective parameters in output parameter estimation to be dimensionless, which leads to good results [64–66]. According to the Buckingham Π theorem, the abutment scours depth is normalized as:

$$\frac{d_s}{L} = f\left(F_e, \frac{d_{50}}{L}, \frac{h}{L}, K_s, R_e\right) \tag{13}$$

where $F_e = U_e/\sqrt{\Delta g L}$ is the excess abutment Froude number and $\Delta = s - 1$, where s is the specific gravity of sediment, and R_e is Reynolds number. As the flow around an abutment is quite turbulent, the impact of R_e is neglected, and the following equation is ultimately recommended as:

$$\frac{d_s}{L} = f\left(F_e, \frac{d_{50}}{L}, \frac{h}{L}, K_s\right) \tag{14}$$

To predict the relative scour, the activation functions used in the ELM model were initially controlled to select the best one. Subsequently, the effect of each input variable on relative scours depth in clear water condition (i.e., Equation (9)) was investigated. Various input parameter combinations were also evaluated, and the results are listed in Table 2.

Table 2. Different input combinations, containing the effective variables on abutment scour depth.

Model. No	Fe	h/L	d_{50}/L	K_s
ELM1	green	green	green	green
ELM2	blue	blue	blue	
ELM3	blue	blue		blue
ELM4	blue		blue	blue
ELM5		blue	blue	blue
ELM6	orange	orange		
ELM7	orange		orange	
ELM8	orange			orange
ELM9		orange	orange	
ELM10		orange		orange
ELM11			orange	orange

Note: ELM, indicates extreme learning machine; relative flow depth (h/L); excess abutment Froude number (Fe); abutment shape factor (Ks); relative sediment size (d_{50}/L); green, blue and orange colors show the model with four, three and two inputs.

2.4. The Goodness of Fit Statistics

To evaluate the performance of each model presented in this study, scour depth was estimated using various statistical indices: correlation coefficient (R), root mean square error ($RMSE$), mean absolute percentage error ($MAPE$), and scatter index (SI), which are calculated as follows:

$$R = \frac{\sum_{i=1}^{n} (E_i - \overline{E})(M_i - \overline{M})}{\sqrt{\sum_{i=1}^{n} (E_i - \overline{E})^2 \sum_{i=1}^{n} (M_i - \overline{M})^2}} \tag{15}$$

$$RMSE = \sqrt{\left(\frac{1}{n}\right)\sum_{i=1}^{n} (E_i - M_i)^2} \tag{16}$$

$$MAPE = \left(\frac{100}{n}\right)\sum_{i=1}^{n}\left(\frac{|E_i - M_i|}{E_i}\right) \tag{17}$$

$$SI = \frac{RMSE}{\overline{E}} \tag{18}$$

where M_i and E_i are the modeled and experimented d_s/L values, respectively, and $\overline{M}$ and $\overline{E}$ are the mean modeled and experimented d_s/L values, respectively. Recently, a new goodness-of-fit index was defined by Eray et al. [67] as combined accuracy (CA). It is a combination of $RMSE$, MAE, and R^2 as follows:

$$CA = 0.33\left((1 - R^2) + MAE + RMSE\right) \tag{19}$$

where R^2 and MAE are determination coefficient and mean absolute error, respectively.

3. Results

The ELM activation functions' performance in predicting abutment scour depth in clear water was examined using Dey and Barbhuiya [22] dataset. According to Figure 4, it could be concluded that the sigmoid activation function predicted scour around an abutment well ($R = 0.97$), whereby most data points had an average error below 8% ($MAPE = 7.69\%$) and the lowest data scattering ($SI = 0.088$).

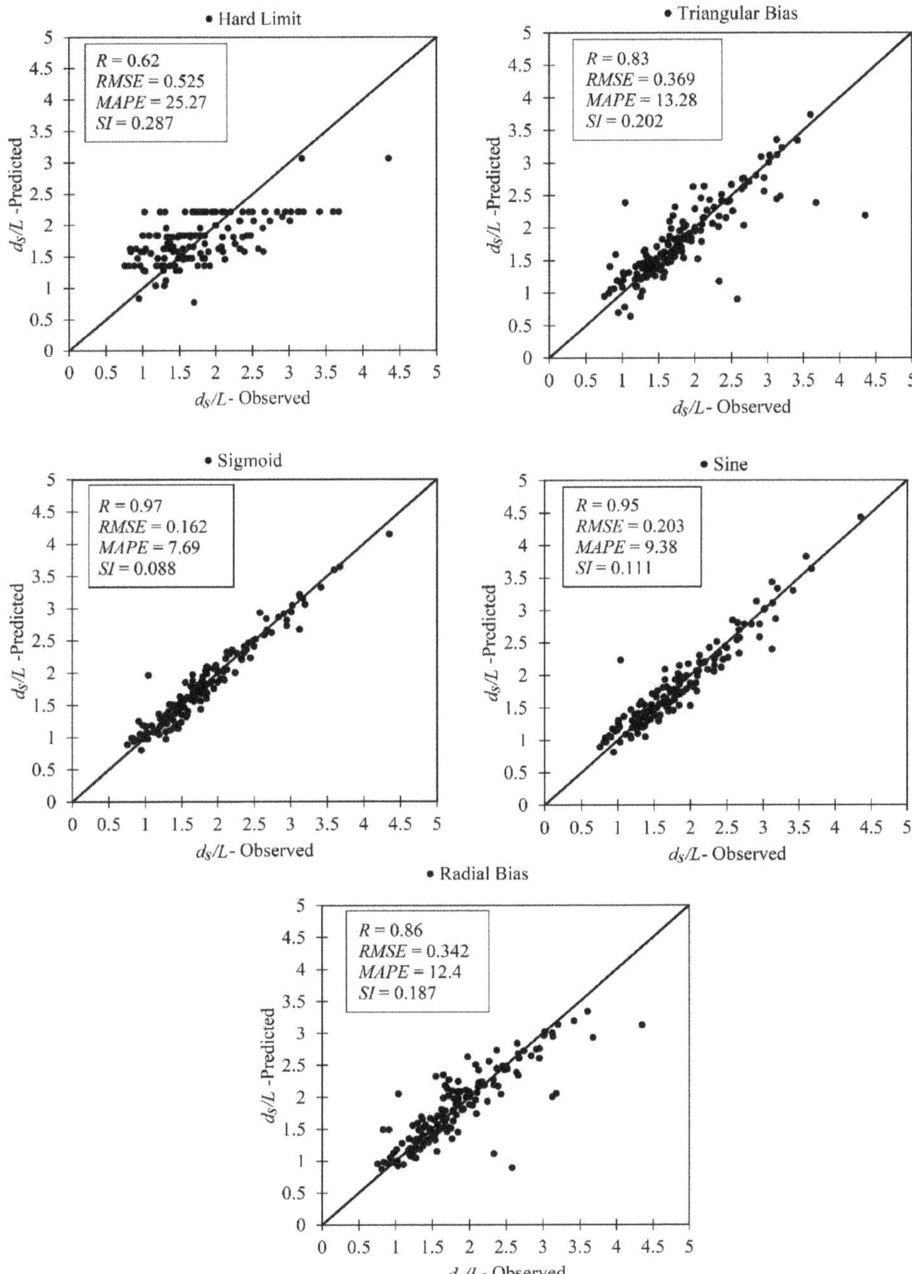

Figure 4. Scatter plot of different ELM activation functions.

The *MAPE* and *SI* of the sine activation function were over 2% and 4% higher than the sigmoid activation function, respectively, though the correlation coefficient for the sigmoid function had a small margin ($R = 0.95$). According to the statistical indices for the hard limit activation function, there was a significant difference between the estimated data points and the observed data ($R = 0.62$). Moreover, with an average error of 26% ($MAPE = 25.27$), the hard limit activation function had the highest error value among activation functions, signifying the poor performance of the hard limit activation function in scouring estimation. According to Figure 4, the mean error for the triangular basis and radial basis functions was 6% and 5% higher, respectively, than the sigmoid activation function. Therefore, the sigmoid activation function outperformed the other activation functions in scour depth estimation. The sigmoid function was, therefore, applied to abutment scour depth modeling with the input combinations presented in Table 2.

Figure 5 presents a comparison of 11 input combinations' performance (see Table 2) according to various statistical indices. The ELM1 input combination included all effective parameters on abutment scour depth in clear water conditions (Equation (14)). The ELM results exhibited greater ability in the prediction of the abutment scour depth by using the ELM1 input combination. This was because the difference between estimated and experimental data was negligible ($R = 0.97$), and the estimated values had less than 8% relative error ($MAPE = 7.69$). Besides, the relative error (RE) distributions in Figure 5 indicated that 46% and 74% of all samples were estimated with less than 5% and 10% (respectively) RE, and only 5% of samples were estimated with more than 20% RE, (see Tables S1 and S2 for more detail).

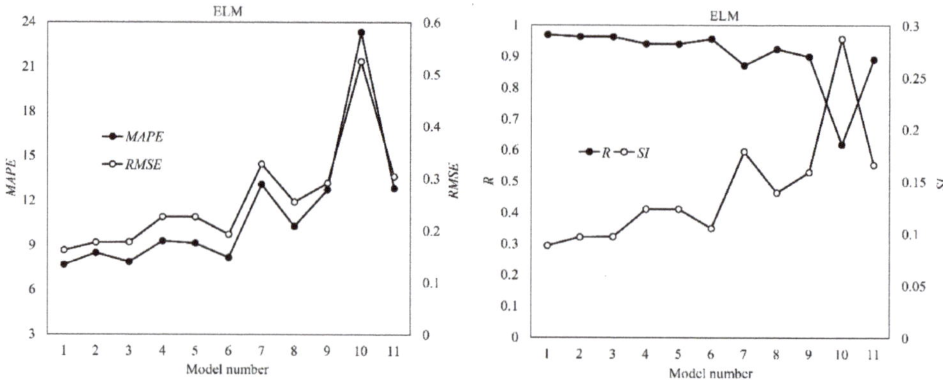

Figure 5. Comparison of different input combinations in abutment scours depth prediction.

Next, by neglecting various parameters at a time and assessing 10 new input combinations (ELM2 to ELM11), the effect of each parameter on d_s/L was examined, based on Equation (9). The input combinations ELM2 to ELM5 used three input parameters for predicting abutment scour depth. According to Figure 5, the *R* and *SI* values for ELM2 were 0.96 and 0.096, respectively. The difference between ELM2 and ELM1 was the absence of parameter K_s, which led to a higher mean relative error ($MAPE = 8.49$) compared to ELM1 ($MAPE = 7.69$). Besides, the RE distribution provided in Figure 5 demonstrated that ELM2 increased the RE distribution so that more than 74% estimated samples with less than 10% RE decreased to 67.46%. Consequently, the results indicated that neglecting the abutment geometry parameter in scour depth evaluation was significant, whereby ignoring it reduced d_s/L performance and accuracy.

Figure 5. *Cont.*

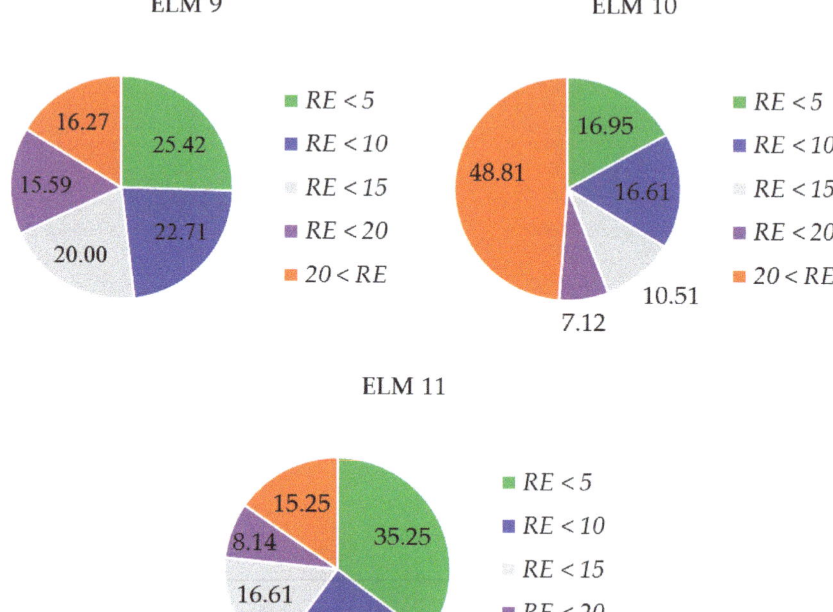

Figure 5. Relative error distribution for different input combinations.

The ELM3 input combination did not include the effect of input variable d_{50}/L on relative scour depth. By eliminating the effect of d_{50}/L on the abutment scour depth, the accuracy of scour depth estimation reduced, and this increased the *RMSE* compared to ELM1 ($R = 0.96$, $RMSE = 0.177$). ELM4 included parameters d_{50}/L, K_s, and F_e as the input combination. The statistical indices signified that eliminating the relative flow depth parameter led to an over 1% increase in mean relative error ($MAPE = 9.29$) over the model that includes all effective parameters in scour depth estimation ($MAPE = 7.69$).

Thus, it could be concluded that the incoming flow around abutments was an effective parameter in relative scour depth (d_s/L) estimation. By eliminating the parameter F_e from the relative scour depth estimation (ELM5), the mean relative error ($MAPE = 9.14$) and the root mean square error ($RMSE = 0.226$) increased more than ELM1. Among models with three input variables, model five exhibited the weakest performance. In fact, among the four dimensionless parameters in Equation (14), parameter F_e appeared to be the most effective parameter on d_s/L estimation. The results of RE distribution for ELM3 to ELM6 showed that the percentage of the estimated samples with less than 5% and 10% in these models was less than ELM1. It should be noted that the lower performance to estimate d_s/L was related to ELM5. According to the explanation given, the first rank of the most effective parameters in abutment scour depth predicting was related to *Fe*, and the h/L, K_s, and d_{50}/L were in the second to fourth rank, respectively.

In estimating d_s/L, ELM6 to ELM11 only used two dimensionless parameters (i.e., from a total of four parameters in Equation (14)). Parameter F_e in ELM6, ELM7, and ELM8 was one of the two constant parameters. Concerning all indices presented in Figure 5 for the three models, in addition to F_e, ELM6 also considered h/L, and most estimated values showed little difference from the observed values ($R = 0.96$). The mean relative error for ELM6 was less than 9% ($MAPE = 8.17\%$). Figure 5 indicates that ELM7 ($RMSE = 0.254$) and ELM8 ($RMSE = 0.327$) exhibited weaker performance than

ELM6 ($RMSE$ = 0.192); in addition to F_e, ELM7 considered the impact of d_{50}/L, and the majority of data points had a mean relative error of 14% ($MAPE$ = 13.1).

Finally, among models that only considered two parameters in estimating the relative scour depth, ELM6 ($d_s/L = f$ (F_e, h/L)) was selected as the best. Among other two-parameter models that did not include parameter F_e, ELM10 performed the worst. This model included the effect of h/L and K_s. ELM10 exhibited the highest scattering and lowest accuracy (SI = 0.287, R = 0.61), with the majority of data points having a relative error of 24% (Figure 5). The RE distribution indicated that the lowest performance of models with only three input parameters was related to the ELM5 so that only 16.94% of all samples were estimated by this model and had relative error less than 5%.

Based on the given explanation, it could be concluded that in the ELM method, the F_e and h/L parameters were effective inputs on relative scour depth in clear water conditions (when the bed sediment particles are not in motion from upstream toward downstream). On the other hand, not considering the excess abutment Froude number and relative sediment size in relative scour depth estimation led to a significant decrease in accuracy and a significant increase in relative error. According to the results, ELM1 with h/L, d_{50}/L, K_s, and F_e as effective parameters in the estimation of scour depth at abutments was deemed the best ELM model.

To evaluate the proposed ELM model's accuracy in predicting the scour depth around abutments in clear water condition, the performance of the developed ELM-based model was compared with conventional regression-based models [22,23] and recently artificial intelligence (AI) based techniques [46,47] by combined accuracy (CA) index for training and testing phases. The lowest CA at both phases is related to Muzammil [23], who provided his model using conventional nonlinear repression. Another conventional equation is Dey and Barbhuiya [22]. The CA index value of this model (CA = 0.56) was almost half the value obtained for the Muzammil [23] model (CA = 1.27).

However, there was a significant difference between the performance of regression methods and artificial intelligence methods. In addition to the developed ELM model, Moradi et al. [46] and Azimi et al. [47] models were also classified as AI-based models. Both of these methods were presented based on ANFIS so that the Moradi et al. [46] was provided as classical ANFIS, while the model of Azimi et al. [47] was introduced as a multi-objective evolutionary optimization of ANFIS.

The optimized model [47] performed better than the classical model [46], as shown in Figure 6. The model of Azimi et al. [47] provided excellent results in the training phase, but comparing the accuracy of this model with the ELM in the test phase, which indicated the generalizability of a model, demonstrated the superior performance of the developed ELM. Also, the developed method in the current study not only had a higher accuracy than existing ones but also eliminated the underlying problem of existing AI-based models [46,47] in providing a specific relationship for practical applications.

Figure 7 indicates the Taylor diagram [68] for developed ELM model and conventional regression-based and AI-based techniques by considering the results of the standard deviation. The correlation coefficient and $RMSE$ were provided simultaneously for a model in a single diagram using the Taylor diagram. The very low performance of the conventional regression-based techniques in comparison with the AI-based models and developed ELM was clear. Owing to this figure, the developed ELM model outperformed than existing models (conventional and AI-based models).

Table 3 presents the quantitative evaluation of the uncertainties in the abutment scours depth estimation using the developed ELM model versus the existing ones. It should be noted that the uncertainty analysis (UA) was done for the testing phase [69,70]. To calculate the UA, the individual estimation error (IEE) was $e_j = E_j - T_j$, where E and T are the estimated and target values, respectively. Using IEE, the mean of IEE (MIEE) and standard deviation of IEE (SDIEE) were computed as $\bar{e} = \sum_{j=1}^{n} e_j$ and $S_e = \sqrt{\sum_{j=1}^{n} \left(e_j - \bar{e}\right)^2 / n - 1}$, respectively (where n denotes to a number of the testing samples). The positive (or negative) value of the MIEE indicated that the desired model over-estimated (under-estimated) the target values of the abutment scour depth. In the following, the MIEE and SDIEE were employed to define confidence bound (CB) by the Wilson score method [71] without continuity correction so that the ±1.96 S_e yielded an almost 95% CB.

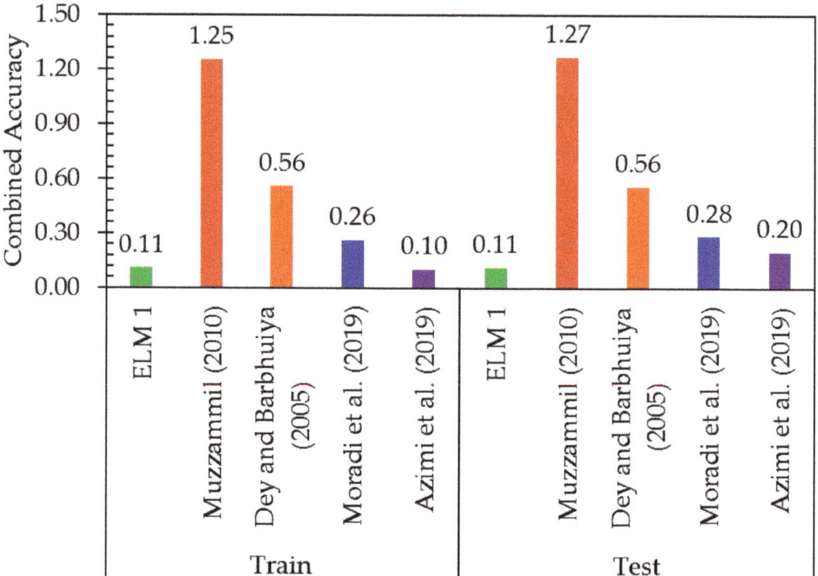

Figure 6. Comparison of the developed ELM model for abutment scour depth prediction with existing conventional and artificial intelligence (AI)-based techniques by combined accuracy (*CA*) index.

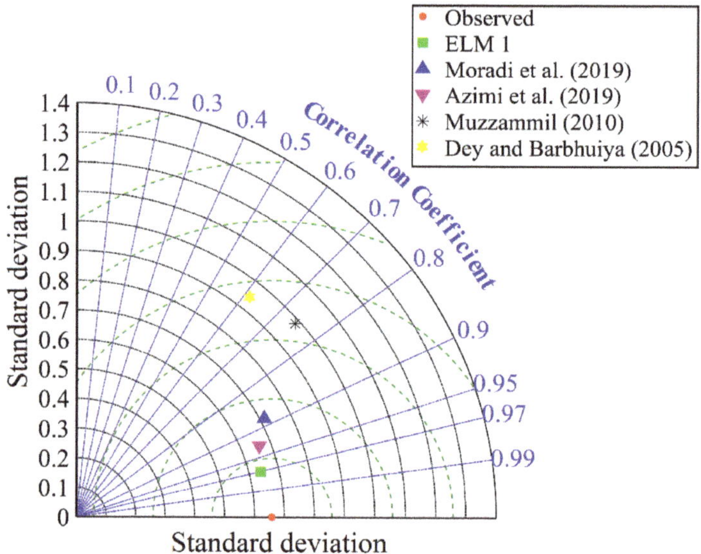

Figure 7. Taylor diagram for developed ELM model and conventional regression-based and AI-based techniques.

Owing to the results of Table 3, it was clear that the developed ELM outperformed than conventional [22,23] and AI-based models [46,47] with the lowest computed uncertainty. The MIEE for the developed ELM model was computed as +0.001 compared to 0.044, −0.044, −1.613, and −0.171 for Moradi et al. [46], Azimi et al. [47], Muzzammil [23], and Dey and Barbhuiya [22], respectively. It was observed that the developed ELM model and the results of Moradi et al. (2019) over-estimated the abutment scour depth, while the other prediction models underestimated it. Similar to MEE, the lowest value of the standard deviation of estimated error (SDEE), 95% estimation error interval (EEI), and width of uncertainty band (WUB) were related to the developed ELM model.

Table 3. Uncertainty analysis of the developed ELM model versus existing regression and artificial intelligence (AI)-based techniques.

Model	MIEE	SDEE	95% CB	WUB
ELM1	0.001	0.162	(−0.025 0.028)	±0.026
Moradi et al. [46]	0.044	0.344	(−0.012 0.100)	±0.056
Azimi et al. [47]	−0.044	0.304	(−0.093 0.006)	±0.050
Muzzammil [23]	−1.613	0.702	(−1.728 −1.499)	±0.114
Dey and Barbhuiya [22]	−0.171	0.757	(−0.294 −0.047)	±0.123

Note: MIEE, mean of individual estimation error; SDEE, standard deviation of individual estimation error; CB, confidence bound; WUB, width of uncertainty bound.

Due to the best performance of the ELM1 in comparison with other ELM models (ELM2 to ELM11) and existing model, the equation extracted from ELM1, which was selected as the best model in this study, is presented as follows:

$$\frac{d_s}{L} = \left| \left[\frac{1}{(1+\exp(InW \times InV + BHN))} \right]^T \times OutW \right| \tag{20}$$

where InW is the input weight, InV is the input variable vector, BHN is the bias of hidden neurons, and $OutW$ is the output weight. The matrices of each variable in Equation (20) are defined as follows:

$$InV = \begin{bmatrix} Fe \\ \frac{h}{L} \\ \frac{d_{50}}{L} \\ k_s \end{bmatrix} InW = \begin{bmatrix} +0.79 & -0.98 & +0.46 & -0.03 \\ +0.55 & -0.40 & -0.24 & -0.20 \\ +0.65 & -0.75 & -0.35 & -0.77 \\ -0.15 & +0.25 & -0.96 & -0.89 \\ +0.33 & +0.21 & +0.20 & -0.29 \\ +0.62 & +0.37 & -0.33 & +0.26 \\ +0.62 & -0.88 & -0.51 & +0.09 \\ -0.98 & +0.85 & +0.72 & -0.85 \\ -0.80 & +0.29 & +0.46 & +0.89 \\ +0.73 & +0.59 & -0.24 & -0.98 \\ -0.90 & +0.11 & -0.07 & +0.27 \\ -0.98 & -0.24 & -0.75 & +0.87 \\ -0.64 & +0.70 & +0.22 & +0.02 \\ -0.86 & +0.94 & +0.23 & +0.54 \\ +0.42 & -0.84 & +0.07 & +0.10 \\ -0.63 & +0.34 & +0.89 & -0.53 \\ +0.61 & -0.14 & -0.01 & -0.27 \\ +0.02 & +0.24 & +0.48 & -0.20 \\ +0.49 & +0.17 & +0.49 & +0.54 \\ +0.10 & -0.71 & +1.00 & -0.58 \end{bmatrix} OutW = \begin{bmatrix} 7.36 \\ 47.73 \\ 77.48 \\ -267.24 \\ 1998.12 \\ 921.93 \\ -185.62 \\ -138.59 \\ -36.49 \\ -206.82 \\ 57.95 \\ 305.21 \\ 384.68 \\ 46.36 \\ -102.69 \\ 727.28 \\ 192.17 \\ -3163.74 \\ -323.08 \\ 445.98 \end{bmatrix} BHN = \begin{bmatrix} 0.26 \\ 0.91 \\ 0.67 \\ 0.77 \\ 0.14 \\ 0.52 \\ 0.23 \\ 0.37 \\ 0.94 \\ 0.60 \\ 0.16 \\ 0.45 \\ 0.32 \\ 0.35 \\ 0.77 \\ 0.69 \\ 0.22 \\ 0.97 \\ 0.20 \\ 0.44 \end{bmatrix}$$

To evaluate the performance of the developed ELM-based model (Equation (20)) in scour depth prediction, the Ballio et al. [72] dataset was applied. The scatter plot of the predicted relative scours depth (d_s/l) versus observed ones that were collected from Ballio et al. [72] are provided in Figure 8. Owing to this figure, all samples were estimated with a relative error of less than 10%, so that the means relative error for these samples was less than 5% (*MAPE* = 4.407%). The other statistical indices related to estimated samples (*R* = 0.948; *RMSE* = 0.036; *SI* = 0.05; *CA* = 0.16) indicated the high prediction level of the developed model at a dataset, which was different from the data employed in model training.

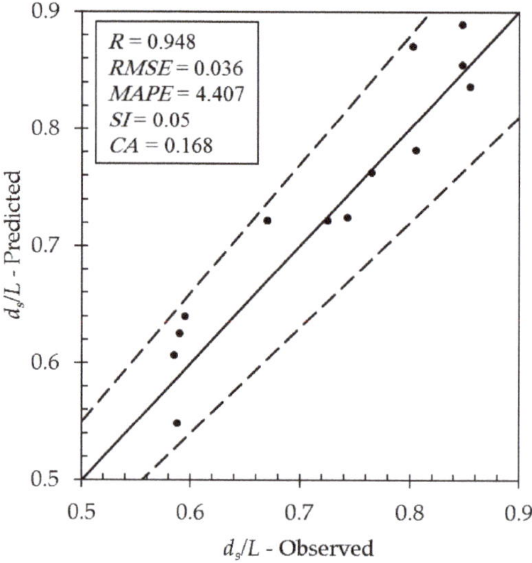

Figure 8. Scatter plot of developed ELM-based model for Ballio et al. [72] model.

4. Discussion

The performance of the ELM machine learning model developed in the current study was compared with conventional models [22,23] and recently developed AI-based techniques [46,47]. The model's performance was examined in terms of five statistical indices of correlation coefficient (*R*), root means square error (*RMSE*), scatter index (*SI*), mean absolute percentage error (*MAPE*), and recently introduced combined accuracy (*CA*). Besides, the comparison of the developed model was also compared with the existing one in terms of the Taylor diagram and uncertainty analysis. The results of all comparisons demonstrated that the developed model provided better performance than conventional and AI-based regression models. Over-fitting is a well-known problem in machine learning modeling. Over-fitting occurred when the developed model fitted exactly through the training stage, leading to low performance of the developed model at unseen samples (i.e., training stage). It is well-known that an organized model with lower adjustable parameters through the training stage could prevent over-fitting challenge. It should be noted that in the current study, data splitting was considered as 50%–50% so that 50% of all samples were randomly selected for training and testing, while the rest of the ones applied to check the performance of the model for unseen samples. The results indicated the high level of accuracy of the developed model at both the training and testing stage simultaneously. Consequently, the developed model in the current study did not experience the over-fitting problem and produced an acceptable prediction for unseen samples in the testing stages. Moreover, the generalizability of the developed model was examined by Ballio et al. [72] dataset, which differed from Dey and Barbhuiya [22] that was applied for training and testing stages.

To investigate the alteration trend of the scour depth around abutment to each input variable provided in the Equation (20) (F_e, h/L, d_{50}/L, k_s), the partial derivative sensitivity analysis (PDSA) was performed [73,74]. In the PDSA, the partial differential of the development model was calculated between output and each input variable, and the results of the partial difference were reported as the sensitivity of the developed model to x_i input (in the current study, I = 1, 2, 3, 4). The positive (or negative) value of the sensitivity demonstrated that an increase in the x_i resulted in an increase (or decrease) in the output variable. Besides, the high (or low) value of the sensitivity implied the higher (or lower) efficiency of the x_i on the output variable.

Figure 9 indicates the results of the PDSA of developed ELM-based model (Equation (20)) to input variables (F_e, h/L, d_{50}/L, K_s). All of the sensitivity values for F_e were positive. Indeed, the increasing (or decreasing) of the F_e led to enhance (or reduction) the d_s/L. The highest value of sensitivity of the model to F_e was related to the lower value of these variables. The sensitivity value for F_e indicated that the developed ELM model was the most sensitive to F_e in comparison to others. Similar to F_e, the highest sensitivity for h/L and d_{50}/L attained at the lowest value of each variable. The sensitivity sign in the h/L was not constant, so the trend of the developed ELM model to h/L alteration was not clear, while, for most samples, an increasing (or decreasing) in the d_{50}/L resulted to reduction (or enhancing) of the output variable. The results of PDSA for K_s were similar to h/L without any constant alteration trend.

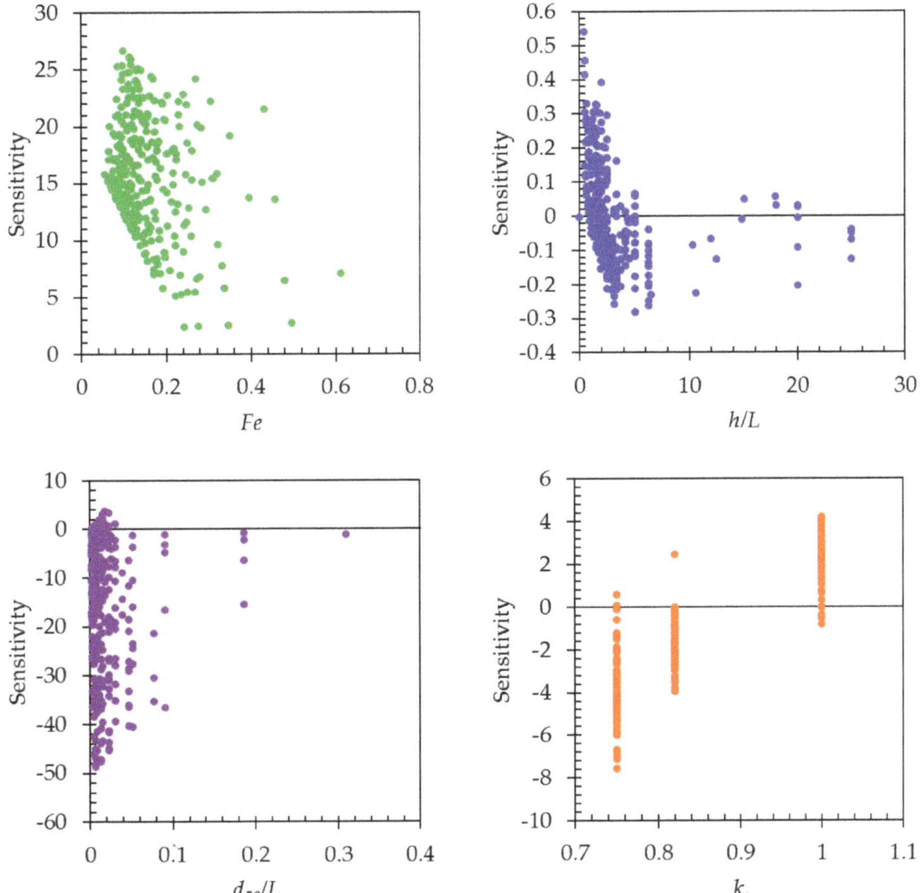

Figure 9. Partial derivative sensitivity analysis (PDSA) results for each input variable in the developed ELM-based model.

Figure 10 represents the effect of altering the input variables on the compressibility indices. This was accomplished by changing one of the input variable (i.e., F_e, h/L, d_{50}/L, k_s) at a constant rate (−10 to

10%), while the other ones were constant. The *MAPE* of the developed ELM model in abutment scours depth estimation was calculated for each input variable variations in the defined range. Figure 10 shows that the compressibility indices estimated by the developed ELM model were significantly affected by an error in measuring F_e and k_s so that the *MAPE* increased drastically. For instance, a 10% error in measuring F_e and K_s resulted in more than 14% *MAPE*, which was more than two times the *MAPE* for applying these variables without measuring error. In contrast, the error in measuring h/L and d_{50}/L was not significant, so the 10% error in measuring h/L and d_{50}/L led to increasing the *MAPE* less than 1% related to applying these variables without any measuring error.

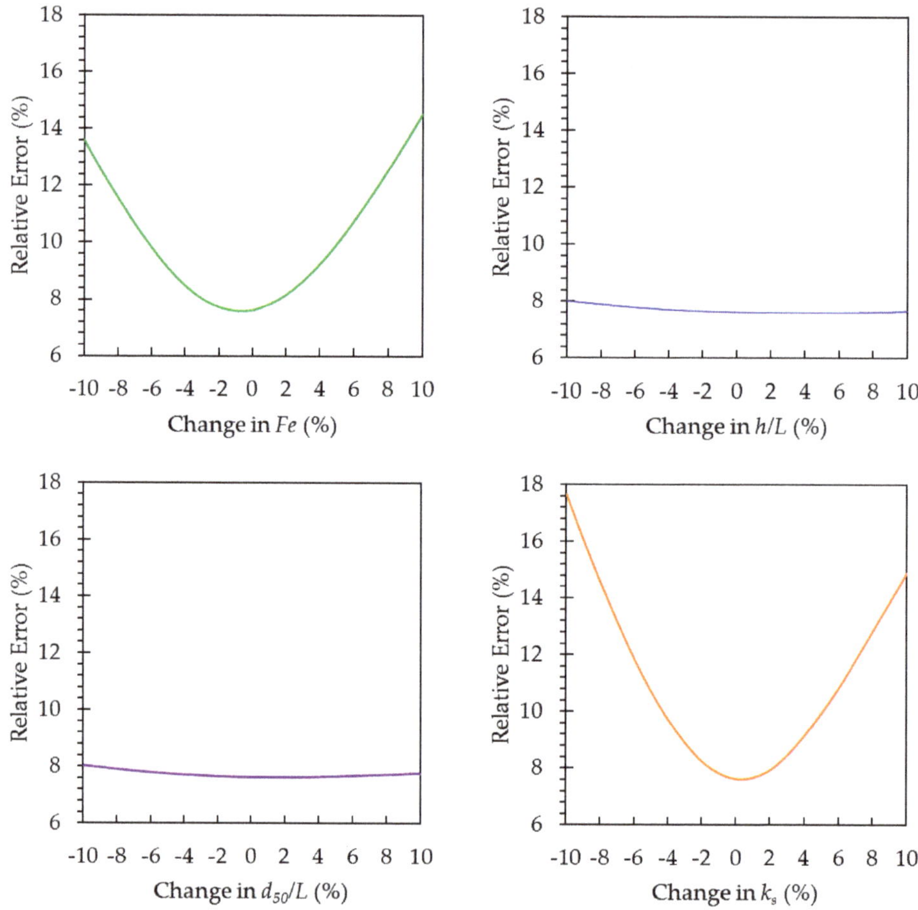

Figure 10. Alteration effect of the input variables on the output mean absolute percentage error (*MAPE*).

5. Conclusions

According to the significance of scour at bridge abutments and the prevention of damage caused by incorrect scour depth estimation, the local scour depth at abutments in clear water was studied in this work using extreme learning machines (ELM). The results of this study are presented below.

- ➢ Predicting the normalized scour depth over abutment length (d_s/L) at bridge abutments in clear water condition and on a uniform sediment bed was a function of relative depth (h/L), excess abutment Froude number (F_e), relative sediment size (d_{50}/L), and the structure's geometric coefficient (K_s).

- ➢ Among the nonlinear activation functions used in extreme learning machines, the sigmoid activation function exhibited good accuracy (R = 0.97) and mean error below 8% ($MAPE$ = 7.69%) in estimating the scour depth at bridge abutments.
- ➢ Among the four parameters affecting scour depth, elimination of F_e and d_{50}/L led to a 24% increase in relative error ($MAPE$ = 23.57). The combination of h/L and Fe performed the best compared to the other two-parameter models.
- ➢ According to the ELM method, Model 1 that included all parameters (F_e, d_{50}/L, K_s, h/L) for predicting the relative scour depth around abutments embedded on a bed with uniform materials in clear water condition was deemed the best model.
- ➢ The best ELM model (ELM1) predicted the relative scour depth around abutments with greater accuracy than the conventional regression and AI-based techniques.

Supplementary Materials: The following are available online at http://www.mdpi.com/2073-4441/12/1/301/s1. Table S1: *MAPE*, *RMSE*, *R* and *SI* values of different input combinations in abutment scour depth prediction, Table S2: Relative error (RE) values for different input combinations verified in this study.

Author Contributions: Conceptualization, H.B. and F.M.; methodology, H.B., F.M. and I.E.; formal analysis, F.M. and I.E.; visualization, F.M., A.A.S. and A.H.A.; writing—original draft preparation, H.B., F.M. and I.E.; writing—review and editing, H.B., A.A.S., A.H.A., B.G. and A.R.-P.; project administration, H.B. All authors have read and agreed to the published version of the manuscript.

Funding: This research was funded by The Natural Sciences and Engineering Research Council of Canada.

Conflicts of Interest: The authors declare no conflict of interest.

References

1. Melville, B.W.; Sutherland, A.J. Design method for local scour at bridge piers. *J. Hydraul. Eng.* **1988**, *114*, 1210–1226. [CrossRef]
2. Macky, G.H. *Survey of Roading Expenditure Due to Scour*; C.R. 90.09; Department of Scientific and Industrial Research (DSIR) Hydrology Centre: Christchurch, New Zealand, 1990.
3. Cardoso, A.H.; Bettess, R. Effects of time and channel geometry on scour at bridge abutments. *J. Hydraul. Eng.* **1999**, *125*, 388–399. [CrossRef]
4. Azamathulla, H.M.; Ghani, A.A.; Zakaria, N.A.; Guven, A. Genetic programming to predict bridge pier scour. *J. Hydraul. Eng.* **2009**, *136*, 165–169. [CrossRef]
5. Ghazvinei, P.T.; Mohamed, T.A.; Ghazali, A.H.; Huat, B.K. Scour hazard assessment and bridge abutment instability analysis. *Electron. J. Geotech. Eng.* **2012**, *17*, 2213–2224.
6. Lim, S.Y. Equilibrium clear-water scour around an abutment. *J. Hydraul. Eng.* **1997**, *123*, 237–243. [CrossRef]
7. Kwan, T.F. *A Study of Abutment Scour*; University of Auckland: Auckland, New Zealand, 1988.
8. May, R.W.; Ackers, J.C.; Kirby, A.M. *Manual on Scour at Bridges and Other Hydraulic Structures*; Pub. No. 551; Construction Industry Research and Information Association: London, UK, 2002.
9. Hosseini, K.; Karami, H.; Hosseinjanzadeh, H.; Ardeshir, A. Prediction of time-varying maximum scour depth around short abutments using soft computing methodologies-A comparative study. *KSCE J. Civ. Eng.* **2015**, *20*, 2070–2081. [CrossRef]
10. Manzouri, M.; Azimi, A.H. Laboratory experiments evaluating sedimentation and mound formation of obliquely discharged sand particles in stagnant water. *Int. J. Sediment Res.* **2019**, *34*, 564–576. [CrossRef]
11. Azimi, A.H.; Qian, Y.; Zhu, D.Z.; Rajaratnam, N. An experimental study on circular slurry wall jets. *Int. J. Multiph. Flow* **2015**, *74*, 34–44. [CrossRef]
12. Laursen, E.M. Analysis of relief bridge scour. *J. Hydraul. Div.* **1963**, *89*, 93–118.
13. Gill, M.A. Erosion of sand beds around spur dikes. *J. Hydraul. Div. ASCE* **1972**, *98*, 1587–1602.
14. Tey, C.B. *Local Scour at Bridge Abutments*; University of Auckland: Auckland, New Zealand, 1984.
15. Azimi, A.H.; Zhu, D.Z.; Rajaratnam, N. An experimental study of sand deposition from slurry wall jets. *J. Eng. Mech. ASCE* **2014**, *140*, 296–314. [CrossRef]
16. Melville, B.W. Local scour at bridge abutments. *J. Hydraul. Eng.* **1992**, *118*, 615–631. [CrossRef]
17. Dey, S.; Barbhuiya, A.K. Clear water scour at abutments. *Proc. Inst Civ Eng. Water Manag.* **2004**, *157*, 77–97. [CrossRef]

18. Fael, C.M.; Simarro-Grande, G.; Martín-Vide, J.P.; Cardoso, A.H. Local scour at vertical-wall abutments under clear-water flow conditions. *Water Resour. Res.* **2006**, *42*, W10408. [CrossRef]
19. Abou-Seida, M.M.; Elsaeed, G.H.; Mostafa, T.M.; Elzahry, E.F. Local scour at bridge abutments in cohesive soil. *J. Hydraul. Res.* **2012**, *50*, 171–180. [CrossRef]
20. Kumcu, S.Y.; Kokpinar, M.A.; Gogus, M. Scour protection around vertical-wall bridge abutments with collars. *KSCE J. Civ. Eng.* **2014**, *18*, 1884–1895. [CrossRef]
21. Barbhuiya, A.K.; Mazumder, M.H. Live-bed local scour around vertical-wall abutments. *ISH J. Hydraul. Eng.* **2014**, *20*, 339–351. [CrossRef]
22. Dey, S.; Barbhuiya, A.K. Time variation of scour at abutments. *J. Hydraul. Eng.* **2005**, *131*, 11–23. [CrossRef]
23. Muzzammil, M. ANFIS approach to the scour depth prediction at a bridge Abutment. *J. Hydroinformatics* **2010**, *12*, 474–485. [CrossRef]
24. Ettema, R.; Melville, B.W.; Barkdoll, B. Scale effect in pier-scour experiments. *J. Hydraul. Eng.* **1998**, *124*, 639–642. [CrossRef]
25. Bateni, S.M.; Borghei, S.M.; Jeng, D.S. Neural network and neuro-fuzzy assessments for scour depth around bridge piers. *Eng. Appl. Artif. Intell.* **2007**, *20*, 401–414. [CrossRef]
26. Zanganeh, M.; Yeganeh-Bakhtiary, A.; Yamashita, T. ANFIS and ANN models for the estimation of wind and wave-induced current velocities at Joeutsu-Ogata coast. *J. Hydroinformat.* **2016**, *18*, 371–391. [CrossRef]
27. Khan, M.; Azamathulla, H.M.; Tufail, M. Gene-expression programming to predict pier scour depth using laboratory data. *J. Hydroinformat.* **2012**, *14*, 628–645. [CrossRef]
28. Azamathulla, H.M. Gene-expression programming to predict scour at a bridge abutment. *J. Hydroinformat.* **2012**, *14*, 324–331. [CrossRef]
29. Roushangar, K.; Akhgar, S.; Erfan, A.; Shiri, J. Modeling scour depth downstream of grade-control structures using data driven and empirical approaches. *J. Hydroinformat.* **2016**, *18*, 946–960. [CrossRef]
30. Azimi, H.; Bonakdari, H.; Ebtehaj, I.; Ashraf Talesh, S.H.; Jamali, A. Evolutionary Pareto Optimization of an ANFIS Network for Modeling Scour at Pile Groups in Clear Water Condition. *Fuzzy Sets Syst.* **2016**, *319*, 50–69. [CrossRef]
31. Sharafi, H.; Ebtehaj, I.; Bonakdari, H.; Zaji, A.H. Design of a Support Vector Machine with Different Kernel Functions to Predict Scour Depth around Bridge Piers. *Nat. Hazards* **2016**, *84*, 2145–2162. [CrossRef]
32. Najafzadeh, M.; Barani, G.A.; Azamathulla, H.M. Prediction of pipeline scour depth in clear-water and live-bed conditions using group method of data handling. *Neural Comput. Appl.* **2014**, *24*, 629–635. [CrossRef]
33. Najafzadeh, M.; Barani, G.A.; Hessami-Kermani, M.R. Group method of data handling to predict scour at downstream of a ski-jump bucket spillway. *Earth Sci. Inf.* **2014**, *7*, 231–248. [CrossRef]
34. Najafzadeh, M.; Barani, G.A.; Kermani, M.R. GMDH based back propagation algorithm to predict abutment scour in cohesive soils. *Ocean Eng.* **2013**, *59*, 100–106. [CrossRef]
35. Najafzadeh, M.; Lim, S.Y. Application of improved neuro-fuzzy GMDH to predict scour depth at sluice gates. *Earth Sci. Inf.* **2015**, *8*, 187–196. [CrossRef]
36. Etemad-Shahidi, A.; Bonakdar, L.; Jeng, D.S. Estimation of scour depth around circular piers: Applications of model tree. *J. Hydroinformat.* **2015**, *17*, 226–238. [CrossRef]
37. Bateni, S.M.; Jeng, D.S. Estimation of pile group scour using adaptive neuro-fuzzy approach. *Ocean Eng.* **2007**, *34*, 1344–1354. [CrossRef]
38. Muzzammil, M. Application of neural networks to scour depth prediction at the bridge abutments. *Eng. Appl. Comput. Fluid Mech.* **2008**, *2*, 30–40. [CrossRef]
39. Ghani, A.A.; Azamathulla, H. Gene-expression programming for sediment transport in sewer pipe systems. *J. Pipeline Syst. Eng. Pract.* **2010**, *2*, 102–106.
40. Begum, S.A.; Fujail, A.M.; Barbhuiya, A.K. Artificial neural network to predict equilibrium local scour depth around semicircular bridge abutments. In Proceedings of the 6th SASTech, Kuala Lumpur, Malaysia, 24–25 March 2012; Organized by Khavaran Institute of Higher Education: Kuala Lumpur, Malaysia, 2012.
41. Zamani, A.; Sorbi, M.R.; Safavi, A.A. Application of neural network and ANFIS model for earthquake occurrence in Iran. *Earth Sci. Inf.* **2013**, *6*, 71–85. [CrossRef]
42. Mehr, A.D.; Kahya, E.; Bagheri, F.; Deliktas, E. Successive-station monthly streamflow prediction using neuro-wavelet technique. *Earth Sci. Inf.* **2014**, *7*, 217–229. [CrossRef]

43. Hassan, M.; Shamim, M.A.; Hashmi, H.N.; Ashiq, S.Z.; Ahmed, I.; Pasha, G.A.; Naeem, U.A.; Ghumman, A.R.; Han, D. Predicting streamflows to a multipurpose reservoir using artificial neural networks and regression techniques. *Earth Sci. Inf.* **2015**, *8*, 337–352. [CrossRef]
44. Ebtehaj, I.; Bonakdari, H.; Gharabaghi, B. Closure to "An integrated framework of Extreme Learning Machines for predicting scour at pile groups in clear water condition by Ebtehaj, I., Bonakdari, H., Moradi, F., Gharabaghi, B., Khozani, Z.S.". *Coast. Eng.* **2019**, *147*, 135–137. [CrossRef]
45. Ebtehaj, I.; Bonakdari, H.; Shamshirband, S.; Ismail, Z.; Hashim, R. New Approach to Estimate Velocity at Limit of Deposition in Storm Sewers Using Vector Machine Coupled with Firefly Algorithm. *J. Pipeline Syst. Eng. Pract.* **2016**, *20*, 04016018. [CrossRef]
46. Moradi, F.; Bonakdari, H.; Kisi, O.; Ebtehaj, I.; Shiri, J.; Gharabaghi, B. Abutment scour depth modeling using neuro-fuzzy-embedded techniques. *Mar. Georesour. Geotechnol.* **2019**, *37*, 190–200. [CrossRef]
47. Azimi, H.; Bonakdari, H.; Ebtehaj, I.; Shabanlou, S.; Talesh, S.H.A.; Jamali, A. A pareto design of evolutionary hybrid optimization of ANFIS model in prediction abutment scour depth. *Sādhanā* **2019**, *44*, 169–182. [CrossRef]
48. Huang, G.B.; Zhou, H.; Ding, X.; Zhang, R. Extreme learning machine for regression and multiclass classification. *IEEE Trans. Syst. Man Cyber Part B* **2012**, *42*, 513–529. [CrossRef] [PubMed]
49. Huang, G.B.; Zhu, Q.Y.; Siew, C.K. Extreme learning machine: Theory and applications. *Neurocomputing* **2006**, *70*, 489–501. [CrossRef]
50. Cheng, G.J.; Cai, L.; Pan, H.X. Comparison of Extreme Learning Machine with Support Vector Regression for Reservoir Permeability Prediction. *Comput. Intell. Secur.* **2009**, *2*, 173–176.
51. Ebtehaj, I.; Bonakdari, H.; Shamshirband, S. Extreme learning machine assessment for estimating sediment transport in open channels. *Eng. Comput.* **2016**, *32*, 691–704. [CrossRef]
52. Naganna, S.R.; Deka, P.C.; Ghorbani, M.A.; Biazar, S.M.; Al-Ansari, N.; Yaseen, Z.M. Dew point temperature estimation: Application of artificial intelligence model integrated with nature-inspired optimization algorithms. *Water* **2019**, *11*, 742. [CrossRef]
53. Niu, W.J.; Feng, Z.K.; Feng, B.F.; Min, Y.W.; Cheng, C.T.; Zhou, J.Z. Comparison of multiple linear regression.; artificial neural network.; extreme learning machine.; and support vector machine in deriving operation rule of hydropower reservoir. *Water* **2019**, *11*, 88. [CrossRef]
54. Ren, J.; Ren, B.; Zhang, Q.; Zheng, X. A Novel Hybrid Extreme Learning Machine Approach Improved by K Nearest Neighbor Method and Fireworks Algorithm for Flood Forecasting in Medium and Small Watershed of Loess Region. *Water* **2019**, *11*, 1848. [CrossRef]
55. Zhu, R.; Yang, L.; Liu, T.; Wen, X.; Zhang, L.; Chang, Y. Hydrological Responses to the Future Climate Change in a Data Scarce Region.; Northwest China: Application of Machine Learning Models. *Water* **2019**, *11*, 1588. [CrossRef]
56. Bhasin, V.; Bedi, P.; Singh, N.; Aggarwal, C. FS-EHS: Harmony Search Based Feature Selection Algorithm for Steganalysis Using ELM. In *Innovations in Bio-Inspired Computing and Applications. Advances in Intelligent Systems and Computing*; Snášel, V., Abraham, A., Krömer, P., Pant, M., Muda, A., Eds.; Springer: Cham, Switzerland, 2016; Volume 424.
57. Shen, Y.; Xu, J.; Li, H.; Xiao, L. ELM-based spectral-spatial classification of hyperspectral images using bilateral filtering information on spectral band-subsets. In Proceedings of the IEEE International Geoscience and Remote Sensing Symposium (IGARSS), Beijing, China, 10–15 July 2016; pp. 497–500.
58. Ebtehaj, I.; Bonakdari, H. A Comparative Study of Extreme Learning Machines and Support Vector Machines in Prediction of Sediment Transport in Open Channels. *Int. J. Eng. Ttrans. B Appl.* **2016**, *29*, 1523–1530.
59. Huang, G.B.; Chen, L.; Siew, C.K. Universal approximation using incremental constructive feedforward networks with random hidden nodes. *IEEE Trans. Neural Netw. Learn. Syst.* **2006**, *17*, 879–892. [CrossRef] [PubMed]
60. Liang, N.Y.; Huang, G.B.; Saratchandran, P.; Sundararajan, N. A fast and accurate online sequential learning algorithm for feedforward networks. *IEEE Trans. Neural Netw. Learn. Syst.* **2006**, *17*, 1411–1423. [CrossRef] [PubMed]
61. Raudkivi, A.J.; Ettema, R. Clear-water scour at cylindrical piers. *J. Hydraul. Eng.* **1983**, *109*, 338–350. [CrossRef]
62. Sheppard, D.M.; Odeh, M.; Glasser, T. Large scale clear-water local pier scour experiments. *J. Hydraul. Eng.* **2004**, *130*, 957–963. [CrossRef]

63. Firat, M.; Gungor, M. Generalized regression neural networks and feedforward neural networks for prediction of scour depth around bridge piers. *Adv. Eng. Softw.* **2009**, *40*, 731–737. [CrossRef]
64. Azamathulla, H.; Abghani, A.M.; Zakaria, N.A. An ANFIS based approach for predicting the scour below Flip-Bucket spillway. In Proceedings of the 2nd International Conference on Managing River in the 21st Century: Solution towards Sustainable River Basins, Riverside Kuching, Sarawak, 6–8 June 2007.
65. Yasa, R.; Etemad-Shahidi, A. Classification and regression trees approach for predicting current-induced scour depth under pipelines. *J. Offshore Mech. Arct. Eng.* **2014**, *136*, 011702. [CrossRef]
66. Najafzadeh, M.; Etemad-Shahidi, A.; Lim, S.Y. Scour prediction in long contractions using ANFIS and SVM. *Ocean Eng.* **2016**, *111*, 128–135. [CrossRef]
67. Eray, O.; Mert, C.; Kisi, O. Comparison of multi-gene genetic programming and dynamic evolving neural-fuzzy inference system in modeling pan evaporation. *Hydrol. Res.* **2018**, *49*, 1221–1233. [CrossRef]
68. Taylor, K.E. Summarizing multiple aspects of model performance in a single diagram. *J. Geophys. Res. Atmos.* **2001**, *106*, 7183–7192. [CrossRef]
69. Ebtehaj, I.; Sattar, A.M.; Bonakdari, H.; Zaji, A.H. Prediction of scour depth around bridge piers using self-adaptive extreme learning machine. *J. Hydroinformat.* **2016**, *19*, 207–224. [CrossRef]
70. Khozani, Z.S.; Bonakdari, H.; Ebtehaj, I. An analysis of shear stress distribution in circular channels with sediment deposition based on Gene Expression Programming. *Int. J. Sediment Res.* **2017**, *32*, 575–584. [CrossRef]
71. Newcombe, R.G. Interval estimation for the difference between independent proportions: Comparison of eleven methods. *Stat. Med.* **1998**, *17*, 873–890. [CrossRef]
72. Ballio, F.; Teruzzi, A.; Radice, A. Constriction effects in clear-water scour at abutments. *J. Hydraul. Eng.* **2009**, *135*, 140–145. [CrossRef]
73. Azimi, H.; Bonakdari, H.; Ebtehaj, I.; Gharabaghi, B.; Khoshbin, F. Evolutionary design of generalized group method of data handling-type neural network for estimating the hydraulic jump roller length. *Acta Mech.* **2018**, *229*, 1197–1214. [CrossRef]
74. Gholami, A.; Bonakdari, H.; Zeynoddin, M.; Ebtehaj, I.; Gharabaghi, B.; Khodashenas, S.R. Reliable method of determining stable threshold channel shape using experimental and gene expression programming techniques. *Neural Comput. Appl.* **2019**, *31*, 5799–5817. [CrossRef]

Article

The Evolution of Gravel-Bed Rivers during the Post-Regulation Period in the Polish Carpathians

Elżbieta Gorczyca [1,*], Kazimierz Krzemień [1] and Krzysztof Jarzyna [2]

1 Institute of Geography and Spatial Management, Faculty of Geography and Geology, Jagiellonian University in Kraków, Gronostajowa Str. 7, 30-387 Kraków, Poland; kazimierz.krzemien@uj.edu.pl

2 Independent Researcher, 30-638 Kraków, Poland; jarzynak2@gmail.com

* Correspondence: elzbieta.gorczyca@uj.edu.pl

Received: 13 December 2019; Accepted: 14 January 2020; Published: 16 January 2020

Abstract: This study provides a conceptual model of the functioning of gravel-bed rivers during the post-regulation period in Poland and forecasts their subsequent evolution. The main difference between fluvial processes during the pre-regulation and post-regulation period is that they are limited to a zone that is currently several times narrower and trapped in a deep-cut channel. During the river post-regulation period, the construction of additional river training works was significantly limited in river channels. Moreover, all forms of economic activity were significantly reduced in the channel free migration zone, particularly bed gravel extraction operations. As a result of these changes, a limited recovery of the functioning and hydromorphology of the river channel occurred via a return to conditions in effect prior to river regulation. In recovering sections of river, the channel gradually broadens, and its sinuosity and number of threads increase. The overall process can be called spontaneous renaturalization, which yields a characteristic post-regulation river channel. The conceptual model was developed on the basis of the evolution of the gravel-bed river, the Raba River, during the post-regulation period in the Polish Carpathian Mountains.

Keywords: river channel; spontaneous renaturalization; post-regulation period; river training; Polish Carpathians

---

1. Introduction

Since the late 19th century, gravel-bed river channels in mountain areas and their forelands have been subject to strong human impact in the temperate climate zone. This is due to construction of levees, river training works, channel debris extraction, and construction of reservoir dams [1–3]. River channels have in effect changed substantially in terms of cross-section and longitudinal profile.

Similar change processes have been observed in the Polish Carpathians and their foreland areas [4,5]. Direct human impact on river channels has been accompanied by changes in land use, which varied from catchment to catchment. Decades of intensive river regulation have led to a simplified morphology and pattern of Carpathian rivers which, consequently, became more narrow and deeper, in addition to following a single thread instead of many, with little ability for the river channel to migrate [4,6–8]. Increasing channel depth due to human impact [9–11] has led to large changes in connectivity between the river channel and the floodplain. The outcome in terms of ecology has been the loss of the most valuable habitats along the channel including initial gravel bar ecosystems and alluvial forests in mountain areas [12]. The regulation of the Carpathian river channels has also had other negative outcomes including a reduction in their ability to retain water that could be used for commercial or farming purposes, as well as an increase in the erosion risk to bridges, buildings, and other structures found near the river channel [13].

The period of the largest changes in channel morphology and overall functioning of Carpathian river channels caused by river training works and channel debris extraction lasted from 1961 to 1990.

Today the drive to change river channels is much less substantial in the studied area. In addition, it is illegal to collect the gravel from river channels and various government agencies have allowed the development of the free migration zone. The period after 1990 is called herein the post-regulation period. Changes in river channels observed during this period of time have been termed spontaneous renaturalization [14]. Such changes occur most often in the Carpathian river channels with a high width to depth ratio (W/D) [8]. Similar evolving patterns are present in the channels of other rivers, for example, the Sacramento River in California, USA [15] and the Magra and Piava rivers in northern Italy [16,17].

Researchers perceive river channel renaturalization either as a hydro-geomorphologic process [5], biological process [18], or ecological process [19,20]. The authors of this paper treat the examined subject matter from a hydro-geomorphologic perspective.

The main purpose of this paper is to discuss the pattern of spontaneous renaturalization in a mountain area gravel-bed river channel. This process is believed to represent an important mechanism that explains the functioning of a channel system during the post-regulation period in Poland. The course of renaturalization is described for the example of the Raba River flowing across southern Poland. The river serves as a good representation of Carpathian gravel-bed rivers. River renaturalization is also discussed in terms of future change directions and rates of change.

The Introduction of this paper focuses on the chronology of river channel change as well as types of human impact in the Raba river channel in the 19th and 20th centuries. This history of human impact led to the creation of a regulated river channel by the 1970s. The paper focuses on the analysis of channel morphology and functioning during the post-regulation period using the 1970s and 1980s as a base period for the purpose of comparison. The changes discussed in this paper occur as a result of fluvial processes that vary in intensity from site to site depending on the local fluvial environment, including type and degree of human impact in the channel. Special attention is given to the role of natural fluvial processes that, in particular, include large floods which are considered to be the most important part of successful spontaneous renaturalization processes in the Raba River.

Large floods occurred along the Raba River every two to four years in the period from 1951 to 1972. These major flooding events affected the morphology and the channel pattern of the Raba River including the years 1951, 1958, 1960, 1962, 1963, 1965, 1970, and 1972. The period from 1973 to 1996 was characterized by a much smaller frequency of large and very large flood events. The only major floods in this period to affect the morphology of the Raba River occurred in 1987 and 1996. The year 1997 initiated a new period of increasingly frequent floods, of which the most critical events in terms of channel morphology change were the floods of 1997, 2001, 2010, and 2014 [8].

The analysis of flood effects on changes of functioning in the Raba river channel in post-regulation period employed the concept of the flood pulse [21,22]. This concept does not focus on the river channel but its relationship with its overall environment including the valley floor [23]. The concept assumes that a river channel follows an oscillating pattern of widening and narrowing. The former occurs most intensively after large flood events, which revitalize lost connectivity between the river channel and its floodplain and allow for a regeneration of hydro-geomorphologic processes [24–26]. Looking for proof of the formation of channel widening zones following floods and their narrowing during floodless periods constitutes an important part of this study.

The analysis of morphologic diversity in the longitudinal profile of a post-regulation channel in this paper uses the concept of beads on a string. This concept was proposed by Stanford et al. [27] and Ward et al. [28] and introduces the idea of beads representing zones of the most intensive morphologic, hydrologic, and biological diversity based on the capacity for the formation of a broad floodplain. These zones are characterized by multidimensional interrelationship between the river channel and the floodplain, with one affecting the other. Bead zones alternate with string zones that have a simplified structure and limited ability to affect their surroundings. In this study we identify bead zones along the Raba river channel, analyze conditions for their occurrence, and assess their significance in spontaneous renaturalization. This is another important part of the proposed conceptual model of the evolution of a gravel-bed river in a mountain area during a post-regulation period.

2. Study Area

The Raba River is one of the larger mountain area rivers in the upper Vistula basin. It is 131.7 km long, with an average gradient of 0.25°. The area of the Raba catchment is 1537 km^2. It is an elongated catchment, with a larger proportion of right-bank catchment area (Figure 1). The Raba catchment is located in the following two different geomorphologic regions: (1) the Carpathian Mountains and (2) the Carpathian foreland basins (Figure 2). The study area is formed of flysch rocks from the Cretaceous and Paleogene, in its Carpathian part. These are mostly sandstone, claystone, siltstone, and conglomerates [29]. Sub-Carpathian basin, on the other hand, is formed of thick clastic formations from the Miocene that lay atop Mesozoic, Paleozoic, and Precambrian formations to yield an unconformity. The Miocene formation includes mostly siltstone, claystone, and shale with sandstone inserts [29]. The Raba river channel cuts into relatively thin alluvial deposits or flysch parent material [8]. The highest point in the catchment is Mt. Turbacz (1310 m). The lowest point is the river's point of confluence with the Vistula River at 180 m above sea level. The lowest annual precipitation totals are noted in the northern part of the catchment across the Carpathian foreland and range from about 700 to 800 mm/yr. The highest totals are noted in the middle mountain area of the Gorce Range and equal about 1200 mm/yr. The mean annual discharge of the Raba River at the Proszówki gauging site covering more than 90% of the catchment area equals 17.6 m^3/s, while the maximum recorded discharge equals 1470 m^3/s [30].

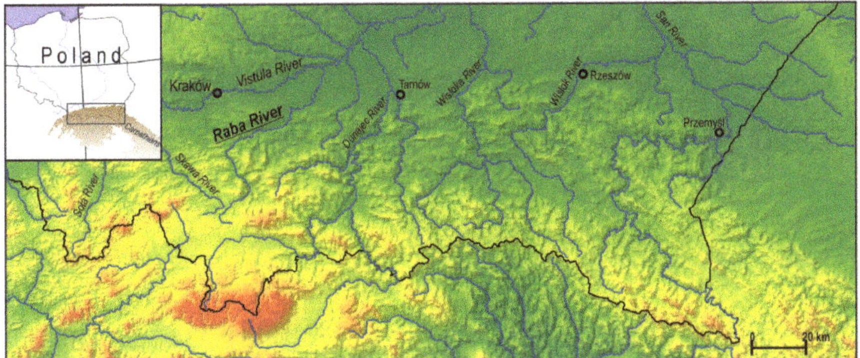

Figure 1. Study area.

The Raba catchment is a rural area characterized by the predominance of the agricultural type of land use [31]. Arable land covers 40% of the catchment area, forests cover 35.6%, meadows 12%, and fruit growing areas about 2%. Built-up areas cover the remaining 10.4% of the catchment area [32]. Many municipalities are located in the near vicinity of the Raba River and produce a significant impact on land use and management across the valley floor. This consequently affects the functioning of the channel. The high human impact on the valley floor and river channel is due to the presence of vital transportation links such as roads and railways. One of the most important roads in southern Poland runs through the Raba valley.

Figure 2. Location and intensities of downcutting in the Raba river channel.

The Raba river channel has been gradually regulated since the beginning of the 20th century. Today about 60% of the river channel is lined with longitudinal structures, which were built during every decade since the 1960s. About 600 m of regulated banks are noted, on average, for each one kilometer of the Raba river channel. About two-thirds of the structures were deemed to be well preserved, while the rest as poorly preserved. River training works are found most often along concave banks, and less often along convex banks. The structures are meant to protect against erosion. In areas characterized by a high density of buildings situated close to the riverbanks, both banks are usually regulated in order to provide flood protection.

Transverse training structures such as straight drop structures are much less impactful than longitudinal structures.

The most important hydrotechnical structure found along the Raba River is the dam at Dobczyce. The resulting Dobczyce Reservoir was built in 1986 and is located at the 60th kilometer of the Raba River. The purpose of the reservoir is to help local communities obtain drinking water, protect local communities from flooding, equilibrate river discharge, and generate electricity.

3. Research Methods

Research on channel structure of the Raba River was performed by employing a channel mapping procedure developed at the Institute of Geography and Spatial Management, Jagiellonian University [33,34]. The main assumption employed in this research method is an understanding of the river channel system as a structure consisting of homogenous morphodynamic reaches. The log used to collect data covers the following four groups of data: (1) initial information, (2) channel parameters, (3) hydrodynamic parameters, and (4) hydrometeorologic parameters of the studied period of time. Principal log entries included information on geographic location, geologic structure, channel parameters, cross sections, longitudinal profile, banks, floor type, sediments, buildings along the watercourse, and channel type. The data log included 57 types of quantitative information and 48 types of qualitative information. Quantitative information is used to calculate indices characterizing the studied channel [34]. The entire river channel of the Raba River was surveyed in line with the above guidelines, in 2015. In addition, some reaches were surveyed in 2009, 2010, and 2014. Grain-size distribution of bed material were assessed using a method proposed by Wolman [35].

Information on the degree of river regulation works and their current state was also collected during the field surveys and supplemented with corresponding information from the Regional Water Management Board in Kraków.

The Raba river channel was also examined using a digital elevation model produced from 2010 to 2013. Orthophotos were also examined for the years 2009 and 2012. These are available from the Cartography Collection at Institute of Geography and Spatial Management, Jagiellonian University and were received from the Central Geodesic and Cartographic Documentation Bureau. The analysis in the study was enhanced using orthophotos available from Google Earth for the period 2014 to 2019. Cartographic analysis and morphometric analysis were performed using QGIS software.

Data provided by the Institute of Meteorology and Water Management (Polish, IMGW) were used to examine the hydrodynamic characteristics of the studied watercourses. Six gauging sites were analyzed in total (Figure 1).

This study compares survey data for the Raba river channel from the years 1976 to 2015. A survey of the Raba river channel performed in 2015 utilized the boundaries of homogenous morphodynamic reaches identified during the 1976 survey. Therefore, the zones of spontaneous renaturalization ("beads on a string") identified in the Raba river channel in 2015 do not overlap with the boundaries of the analyzed channel reaches. The procedure for the survey performed in 2015 was described earlier in this paper. In 1976 the Raba river channel survey was performed as part of a project called "Dynamic river channel typology for the Carpathians and their foreland." The 1976 project was run by the Institute of Geography of Jagiellonian University, Department of Geomorphology [36]. The survey data on the Raba river channel morphology and pattern from 1976 make it possible to make viable comparisons with analogical survey data collected in 2015. The data logs used in 1976 were somewhat different with respect to selected parameters, i.e., more descriptive and less quantitative. This means that not all information contained in the data logs could be compared. In this study we also looked, in detail, at the morphology and pattern of the Raba river channel, and changes therein, for major flood events in 2010 and 2014. The observed changes make it possible to determine change directions in the evolution of gravel-bed rivers in mountain areas large floods.

4. Human Impact on Carpathian Rivers

Carpathian river channels that were regulated starting in the late 19th century had evolved in earlier periods to high bedload transport conditions resulting from intensive agriculture and the development of a dense network of unpaved roads. In the 19th century the Carpathians were a densely populated region, largely deforested, with arable land found as high as 1000 m above sea level. The region had a very high road density at the time as well. Today, unpaved road density ranges from 4 to 7 km·km^{-2} in the Carpathians [37]. In the late 19th century, unpaved roads were utilized more intensively and were an important source of both fine and coarse material as they were eroded as deep as bedrock [38]. If we were to consider channel evolution for periods prior to the period of high human impact (regulation works and large scale gravel mining), then we could argue that river channels in the Western Carpathians were multithreaded, wandering, or sinuous with lateral bars, and meandering in basins with a tendency for braiding [10,39,40].

Flood control levees have been built in Carpathian river valleys since the late 19th century. As a result, the channel migration zone has become smaller over time for rivers across the region. Major engineering works began in Carpathian river channels in the late 19th century. The aim was to regulate river channels in order to reduce flood risk and create commercial waterways. River regulation work proceeded in stages starting with reaches at lower elevations, and then shifting upriver [41]. Regulated river channels had gentler bends and longer straight sections [42]. River regulation work in the Carpathians can be divided into several distinct stages which are the following: (1) The period before World War II (described above), (2) the period from 1945 to 1989, and (3) the period after 1990. Each stage varied in terms of the method of regulation, extent of regulation, and intensity of regulation [4,8].

Stage 1 consisted of extensive work on long sections of river and some projects were never fully completed. Regulation work at this stage consisted of watercourse straightening, as well as a narrowing and shortening. Some Carpathian rivers were made shorter by as much as 10% [8,43]. Regulation projects were halted by the outbreak of World War II and resumed only in the late 1950s. In the beginning of Stage 2, since the 1940s to the 1960s, industrial-scale extraction of channel debris was introduced in Carpathian rivers [5]. In later decades, the local population continued to illegally extract debris from the river channels in the Carpathians [6,8]. The most advanced regulation works were performed in the 1960s and 1970s in the later part of Stage 2 [4]. This was a period of the most radical alteration of river channel patterns, longitudinal profiles, and cross-sections. This period produced mainly such river training structures such as groins, flow redirection piers, weirs, and debris dams, but few drop structures or steps.

In the 1970s, in Poland, river regulation works were planned at the local level [44]. The number of regulation projects decreased markedly in the 1980s, with drop structures being the main form of river regulation [6]. Some banks were also reinforced in this period of time. The main reason for this drought in investment was a major economic crisis in Poland, as well as the lack of major flood events. Regulation works became limited in scope when Poland made the transition to a market economy in the 1990s. Then, the river regulation Stage 3 began in the Polish Carpathians. New regulation works were initiated in the late 1990s, especially following a major regional flood in 1997. River engineering projects at the time were limited primarily to repairing damage produced by floods in 1997, 1998, and 2001. Work included the building of drop structures and bank reinforcements. Narrowed and straightened river channels lacked debris, which had been previously supplied by unreinforced banks, leading to an acceleration in downcutting erosion that undercut the base of existing hydrotechnical structures. The next stage would include the destruction of these structures. Given such outcomes, river engineers would reinforce the same banks two or three times [8].

River regulation systems built in the Carpathians turned out to be ineffective and harmful to the natural environment. Regulation works aimed to produce a single, narrow channel with no side channels. The large change in river channel geometry led to a lack of adjustment when paired with a river's actual hydrologic regime. Downcutting increased in river channels as a result [6,8,11,45,46]. In some cases, downcutting reached solid rock or damaged river engineering structures [4,11,47]. This forced new river regulation work to be performed. This repetition would eventually lead to extensive degradation in river channels, as much as 80% in the Raba river channel [8].

The gradient of regulated rivers would increase along straightened, shortened, and narrowed reaches, as would discharge and the rate of debris transport, which would lead to a deficit of debris in the river channel and, subsequently, a lowering of the channel floor [1,8]. Channel incision by as much as 1.3 to 3.8 m has occurred in the lower and middle reaches of Carpathian rivers since the early 20th century [9,11]. In the second half of the 20th century, channel floor lowering has been observed in upstream reaches of Carpathian rivers, ranging from 0.5 to 3.5 m [8,13,48,49].

In light of the general change in approach to river channel maintenance in the Carpathians, in Poland after 1989, human impact in river channel systems has been severely limited since. In place of extensive intervention, the strategy now is to pursue a minimal intervention focus, i.e., one that is close to what nature itself would pursue. The stoppage of river regulation work in the Carpathian region has led to spontaneous renaturalization of river channels [8].

5. Results and Discussion

5.1. Comparison of the Raba River Channel Structure in 1976 and 2015

Changes in the morphology and functioning of the Raba river channel is presented herein on the basis of a comparison between survey results from 1976 and 2016. Both surveys focused on changes in channel morphology and overall river patterns.

The Raba river channel had already been heavily regulated by the time of the survey study in 1976. However, the regulation structures had not been in place for very long and the channel was still alluvial almost in its entirety. Downcutting had not yet occurred. Narrowing had occurred to a substantial extent. In the 1970s, many semi-natural, multi-threaded reaches were still to be found in the middle section of the Raba River.

In the 1980s and early 1990s, these last semi-natural reaches were finally regulated or became submerged due to the construction of the Dobczyce Reservoir. In addition, the Raba channel became much deeper in the late 1970s and 1980s. This was a natural process that had been accelerated by the effects of river regulation in the Raba channel, gravel extraction from the river bed, limitation of the river free migration zone by railway and road infrastructure, and reduction in the arable land area in the Raba river catchment. This process prompted the Raba River to follow the regulated channel more permanently.

Over the next twenty or so years natural fluvial processes, including major floods, have led to major changes in the morphology of the Raba river channel in comparison with the year 1976.

In 1976, more than 90% of the length of the Raba river channel was alluvial type. The largest number of river reaches were shaped by deposition and redeposition processes as well as lateral erosion in some cases. These reaches occupied 49% of the total length of the Raba River and were characterized by the largest area of bars and large area of undercuts in relation to other reaches of the same river. Downcutting and lateral erosion were detected along 28% of the length of the Raba river channel. In the 1970s, as much as 23% of the river channel consisted of sections with transport function and with levees reinforcing banks (Figure 3).

In 2015, it is difficult to identify one predominant morphodynamic channel type for the Raba river channel in its longitudinal profile. The morphology of the channel is variable and adjusted to local conditions in mountain reaches, and much more homogeneous in basin reaches. The largest number of the river's reaches remain to be shaped by deposition, redeposition, and lateral erosion. Their number is, however, smaller than that in 1976, and they occupy only 36% of the length of the Raba river channel (Figure 3). It is these reaches that contain more than 70% of the area of bars in the entire Raba river channel system. In comparison with 1976, the share of straight sections of river channel has increased to 35% of the length of the channel. This type is present most often in urbanized areas, as well as in regulated reaches and those with levees. Reaches dominated by downcutting and lateral erosion are concentrated in the upper parts of the Raba river channel. These occupy about 30% of the length of the river channel, which is similar to their share from 1976. This type of river channel along with sinuous channels with lateral undercuts includes river reaches featuring rock formations such as rock steps, rock groins, and rock floor. Generally, a river channel of this type is currently rock alluvial in nature. This makes it different from channel reaches identified in 1976 dominated by downcutting and lateral erosion, as they had an alluvial channel floor. Bars present here are not thick and are often deposited directly atop the rocky channel floor.

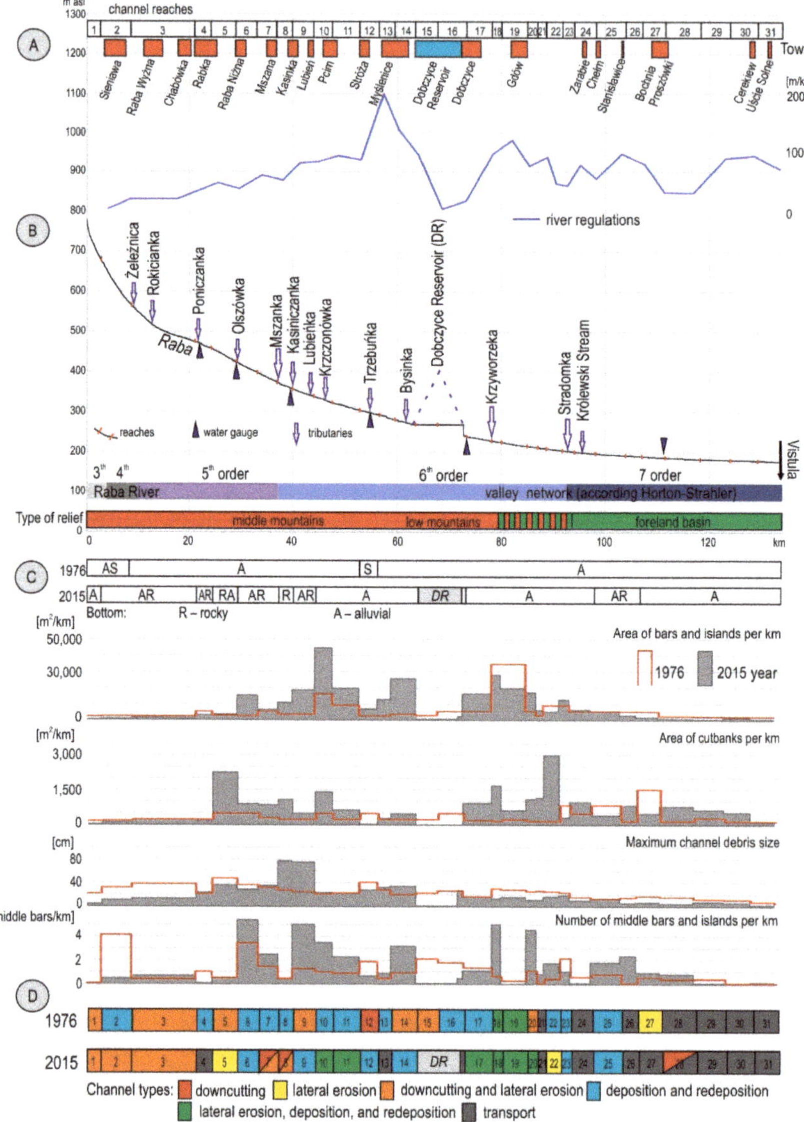

Figure 3. Raba river channel in 1976 and 2015: (**A**) Human impact in the Raba River valley, (**B**) longitudinal profile of the Raba River, (**C**) morphology of the Raba channel in 1976 and 2015, and (**D**) channel types of the Raba River in 1976 and 2015.

This study also examines differences in channel morphologic characteristics for analyzed Raba river channel reaches between 1976 and 2015.

In 2015, the width of the Raba river channel was larger than that in 1976 in more than 75% of the reaches examined. Channel width increased by an average of 14 m, with a maximum increase of 102 m in Channel Reach no. 10 (Figure 4). River bed width increased by 185 m at this site. The width was 45 m after the regulation of the Raba river channel in the 1970s and 1980s and increased to 230 m in 2015 as a result of a few major floods. Spontaneous renaturalization was supported by an artificial supply of gravel to the channel of the Raba River in this river section. This was part of a river restoration project called the "The Upper Raba River Spawning Grounds" [50]; however, the river bed width is still currently lower than the river bed width at the end of the 19th century, in the period before Raba River regulation. River width was ca. 380 m at the time [51].

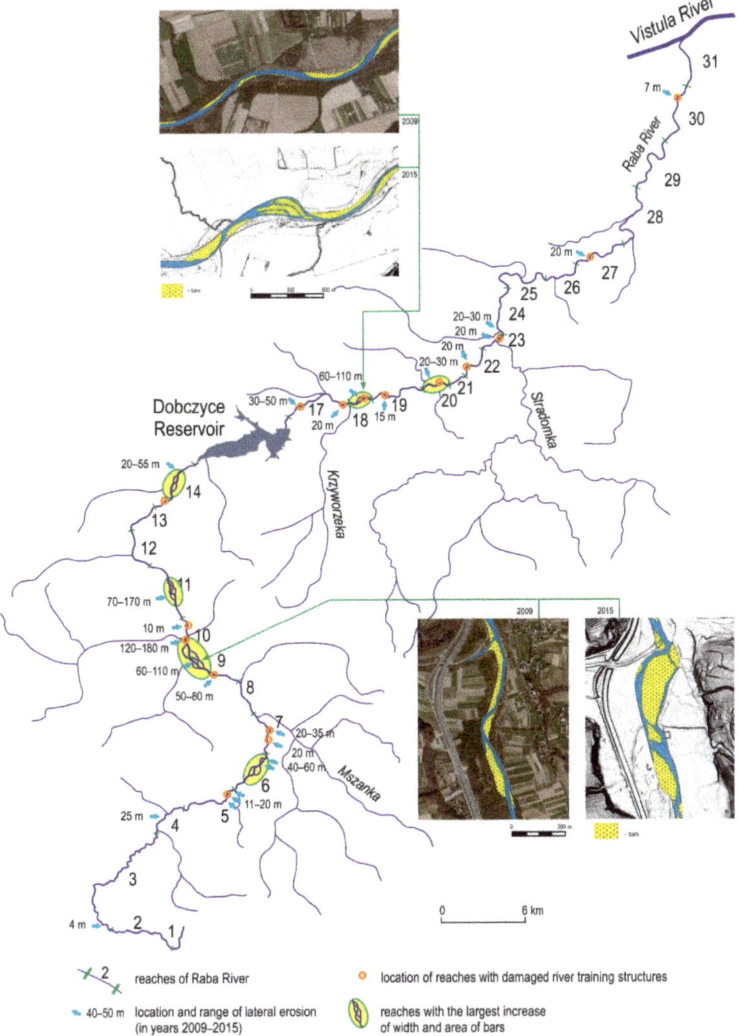

Figure 4. Location of sections undergoing a spontaneous renaturalization in the Raba river channel from 2009 to 2015.

There are some reaches where the channel is narrower than in 1976. In one case, it is narrower by 18 m (Figure 4).

Increases in channel depth were noted in all the examined reaches of the Raba River. Average increases equaled 1.8 m, with a maximum increase of 3.8 m in a downstream reach of the Raba river channel surrounded by levees (Figure 2). It is important to remember though that an increase in river channel depth in the Raba River is first and foremost caused by intense downcutting erosion occurring in the river following a period of intensive river regulation works.

Minimum annual water levels measured at gauging sites in the middle and lower course of the Raba River were used to assess the tendency towards channel deepening in the 20th century [10,52,53]. According to Wyżga [10,53] the Raba river channel became deeper by 2 or 3 m in its downstream section in the 20th century. Nonetheless, both periods of increased erosion and gradual debris accumulation occurred throughout the 20th century in the Raba river channel (Figure 2). Łapuszek and Ratomski [52] provide similar results. In the years from 1901 to 1996 the downstream reach of the river became deeper by 3.79 m in Gdów and 1.76 m in Proszówka. In the upstream reach of the river, the onset of erosion dates to 1971, and by 2002 the river channeled had become, on average, about 1.4 m deeper (Figure 2).

Additional information on channel deepening in lower reaches comes from an analysis of damage to regulation structures in the Raba river channel. Damage data were collected for 26 river training structures built in the 1960s and 1970s, most of which were located in the upper sections of the Raba River (Figure 2). River training structures damaged by downcutting and lateral erosion serve as unique types of reference sites for the rate of erosional change. River channel deepening at such sites ranged from 0.45 m to 2.5 m. Intensive channel deepening at more than 2 m was also observed upstream of the backwater area of the Dobczyce Reservoir. This reach is currently experiencing significant accumulation. A very high rate of downcutting erosion could have been associated with the construction of the Dobczyce Reservoir.

Accelerated downcutting and lateral erosion in the Raba river channel led to undercuts in the riverbank. Today more than 75% of the Raba River reaches feature a larger area of undercuts in relation to 1976 (Figure 3). The present-day area of undercuts is three times larger than in 1976. The average area of undercuts per reach increased by 515 m^2/km. Maximum increases in the area of undercuts per river reach equaled 3011 $m^2 \cdot km^{-1}$. An additional outcome of river regulation, although one not associated with downcutting erosion, is a rise in the length of sections of river dominated by water and sediment transport.

The most readily observable present-day change in the Raba river channel in comparison with 1976 is the increase in the area of bars (Table 1). Today the area of Raba bars is twice as large as that in the 1970s. Increases in bar and island areas, in 2015, relative to 1976 were noted in almost 75% of Raba River reaches and equal an average of 2190 $m^2 \cdot km^{-1}$. The largest increase in the area of accumulation landforms found in the Raba river channel reached almost 30,000 $m^2 \cdot km^{-1}$. It occurred at the point of confluence between the Raba River and one of its major tributaries, the Krzczonówka River (Figures 3 and 4). This reach was also characterized by the largest increase in river channel width. This reach has been viewed as a classic example of a bead of morphologic diversity strung on the string of the studied river channel [27]. Similar accumulation zones occur in most reaches of the mountain section of the Raba river channel. A total of six large accumulation zones and more than 25 rather small channel broadening zones with lateral bars were observed in the present study (Figure 4).

Table 1. Area of bars and undercuts in the Raba river channel.

Landform	Bars' Area (m^2)	Undercuts' Area (m^2)
1976	521,261	26,107
2009	712,080	no data
2015	1,057,447	83,222

The tendency described above is only seemingly inconsistent with the observed decline in the share of Raba river sections characterized by a predominance of deposition and redeposition processes. Deposition occurs along a shorter stretch of the Raba river channel, but it occurs with more intensity. In effect, reaches of channel where bar formation is a predominant morphodynamic process are characterized by channel widening. While the distribution of the area of bars along the Raba River remains uneven, it still differs from that in 1976, when, on the one hand, almost 50% of the area of bars in the studied channel was to be found in a several-kilometer-long stretch on the boundary between mountains and a foreland basin. On the other hand, accelerated deposition of river channel material today occurs at a much larger number of reaches along the Raba River.

In 1976, the gravel fraction of bed material (diameter 2 to 64 mm) dominated in nearly all reaches of the Raba River. In 2015, the most evident change in the surface grain size distribution was an increase in the share of the pebble fraction of bed material (diameter 64 to 256 mm). Today, this is the dominant bed-material fraction in eight Raba river reaches, with most being located upstream of the Dobczyce Reservoir. An increase in the mean grain size of bed material was a general rule in regulated Carpathian rivers in the second half of the 20th century [5,54]. Raba river channelization in the period from the 1960s to the 1980s increased the transport ability of the river. The sediment supply by Raba river tributaries also diminished as they too were regulated resulting in little or no sand and silt being deposited in the river beds, with the material becoming coarser and better sorted, and channel bed armoring developing in some reaches of the river [55]. A surprising discovery was the decline in the maximum bed-material grain size in the Raba river channel between 1976 and 2015, which occurred in more than 50% of the studied reaches (Figure 3). The average decline was 3 cm. The largest decline in the maximum bed-material grain size (24.5 cm) occurred in one of initial reaches of the studied river. A large decline also occurred in several reaches below the Dobczyce Reservoir, i.e., an average of 12 cm. This is likely due to the retention of debris in the area upstream of the Dobczyce Reservoir. On the other hand, marked growth (more than 50 cm) in the maximum bed-material grain size between 1976 and 2015 occurred in two reaches of the Raba River downstream of the point of confluence with the Mszanka, which is the most important tributary of the Raba River in its mountain section.

The river's braiding ratio [33] also increased to some degree, by an average of 0.43 middle landforms per kilometer of river, over the 40 year period between the first and second survey. An increase in the braiding ratio was noted in more than 50% of the reaches in the study. The largest increase in the braiding ratio (4.4 middle landforms per kilometer) was noted in a reach of the Raba River found downstream of the Dobczyce Reservoir where a tributary flowed into the Raba River (Figure 3).

5.2. Role of High Water Stages in the Evolution of a Post-Regulation River

Floodwaters are an extremely effective channel-shaping factor. Large floods, in particular, are effective in floodplain development [25,56–61]. Their significance is examined in this section in terms of how they affect spontaneous renaturalization in a gravel-bed river during the post-regulation period, in this case, the Raba River. The morphological features and functioning of the river channel in 2009 and 2015 are compared for this very purpose.

The state of the Raba river channel in 2009 can be described by hydrodynamic conditions that would be the equivalent of a medium-high water. Such water-stages occurred following an earlier period with large floods from 1996 to 2001. However, the river channel experienced some changes during this earlier period due to the occurrence of major floods. It has adapted to more dynamic flow conditions following a period of low water levels since the early 1980s.

After 2009, the morphology of the Raba river channel was strongly affected by the most recent three large floods, two floods in 2010 and one flood in 2014. The flood in May of 2010 was characterized as a catastrophic flood based on the Punzet classification [62]. The flood occurred along the Raba River both upstream of the Dobczyce Reservoir and downstream of it. A second flood occurred in early June. While it was smaller than the May flood, it occurred at a hydrologically disadvantageous moment when water levels wère still high in the river and groundwater levels were still high. The 2014 high-water stage was smaller than the first high-water stage in 2010, especially below the Dobczyce Reservoir (Q10%). However, in the upper reaches of the river it was higher than that in 2010. Water levels and discharge rates during the flood of 2014 were equivalent to a high-water stage that occurs once per 100 years (Q1%).

All the floods discussed above accelerated erosion and accumulation in the Raba river channel. The increase in the intensity of lateral erosion led to a widening of the river channel at about 50% of the studied reaches, although it did not affect the entire length of each reach. In areas where channel widening did occur, its width, in 2015, increased in most cases by several meters or several tens of meters relative to the year 2009. The river channel widened by more than 100 m relative to the year 2009 in several reaches, i.e., in the middle section of the river in the area where the Mszanka and Krzczonówka join the Raba River as well as in the section just below the Dobczyce Reservoir (Figure 4).

The area of bars also increased substantially throughout the Raba river channel after the floods of 2010 and 2014. By 2015 it was about 1.5 times larger than in 2009 (Table 1). The increase in the area of bars in the Raba channel, however, was not distributed evenly along the length of the river after the floods of 2010 and 2014.

In addition, increased channel sinuosity and the formation of multiple threads were observed in some reaches (Figure 4). The effects of the 2010 and 2014 flood events discussed herein are representative of the evolution of a river channel after a series of large floods, as described by the flood pulse model [21]. The widening of the channel and formation of a multithreaded river also suggest that connectivity between the river channel and floodplain has been regained for a period of time and the floodplain has reformed.

Today there are reaches of the Raba River that perform similar morphodynamic functions and repeat along the length of the river, although the morphodynamic functions of reaches also depend on the degree to which a river channel is regulated. Floods damage or destroy hydrotechnical structures, which is usually the first step in the shift from a channel's morphodynamic function, a transport function, or a downcutting function to a deposition function with lateral erosion.

The most dynamically evolving and morphologically diversified reaches of the Raba River were reaches that, already in 2009, were characterized by substantial sinuosity and large area of bars. These are mountain area reaches of the Raba River that were not "trapped" by road infrastructure and rural settlements.

The largest changes in the river channel were noted upstream of the Dobczyce Reservoir at sites where tributaries would supply large amounts of debris to the Raba River in the course of floods (Figure 4). Lateral erosion and deposition processes are very active at these sites, which has led to a rise in the Raba's riverbed due to the deposition of a portion of the material transported downstream of each point of confluence, as well as upstream of each point of confluence, due to limited discharge driven by abrupt increases in stream load in the confluence zone of the Raba and its tributaries. This evolution scenario for the Raba River was determined for reaches found downstream of the points of confluence of the following rivers: Olszówka, Kasiniczanka, Krzczonówka, and Bysinka (Figure 3).

Significant, although much smaller, changes in the morphology of the Raba river channel after foods were observed in the foothill section of the Raba channel downstream of the Dobczyce Reservoir. The retention of floodwaters and their sediment load by the Raba River inside the Dobczyce Reservoir limits the morphogenetic potential of flood events below the Dobczyce Dam. However, the floods of 2010 and 2014 have changed the morphodynamic function of these reaches from transport-dominated reaches to reaches with deposition with lateral erosion. A particularly large change in the morphology of the Raba river channel occurred downstream of the point of confluence of the Krzyworzeka River, which is also downstream of the Dobczyce Reservoir. In the flood period from 2010 to 2014 the channel of the Raba River made the transition from single-threaded to multi-threaded along this particular stretch of the river (Figure 4).

Straight and slightly sinuous reaches of the Raba River and reaches characterized by a small post-regulation width became only slightly altered during flood events in 2010 and 2014. As in the period prior to 2009, these reaches perform a mostly transport function, with small changes occurring in the channel due to the concentration of transported material along the main axis of the channel. These are reaches with mostly levees that are located in the lower course of the Raba, as well as two reaches located within the city limits of two largest towns in the Raba Valley, Rabka and Myślenice (Figures 1 and 3).

The last large floods period was followed by a gradual process of adjustment of the river channel to "average" hydrodynamic conditions in line with the flood pulse concept [21]. This period of adjustment could become interrupted by major floods in the future.

6. Conceptual Model of the Evolution of a Gravel-Bed River during the Post-Regulation Period

The fluvial system of a post-regulation river experiences a distorted debris supply pattern in its full longitudinal profile. A shortage of debris generates forces that aim to replenish the channel system with sediments. Erosion damages river training structures and liberates debris material into the channel system from the riverbanks and floodplain. A spatially uneven supply of debris from reaches experiencing less regulation pressure contributes to the formation of a number of wide accumulation zones similar to those discussed in the literature and designated sediment waves or sediment slugs [63,64]. The process of forming these zones is discontinuous. One additional characteristic of these zones in a post-regulation river is their limited mobility resulting from anthropogenic constraints in the river channel and the valley floor. A post-regulation river is characterized by long straight reaches and slightly sinuous reaches playing a primarily transport role where unit stream power values are high. Such reaches do not favor accumulation. Hence, the deposition of transported debris usually occurs in widened zones [65]. These zones can be classified as beads using the beads on a string conceptual model by Stanford and Ward [27].

The following four characteristic reach types were identified for post-regulation river channels based on the Raba River: (A) Regulated reaches, (B) reaches with destroyed regulation works but retained regulation channel, (C) sinuous reaches with bars, and (D) multithreaded reaches, i.e., deposition zones (Table 2 and Figure 5I,II). The first two types (A and B) are single-threaded, compact channels that often feature rock outcrops on the channel floor, characterized by low sinuosity and a W/D ratio implying a narrow, deep channel. The other two types (C and D) are channels at various stages of renaturalization. Type C is a single channel, sinuous, widened, and with bars. Type D is a multithreaded channel characterized by the highest W/D ratio, wide, with a number of threads and multiple mid-channel bars [8].

Table 2. Characteristics of the reaches of a post-regulation river.

Channel Reach	A	B	C	D
Channel	Single, compact	Single, compact	Single, expanded	Multi-threaded, wide
Sinuosity ratio	1.0–1.1	1.0–1.2	1.1–1.5	<1.5
W/D ratio	<12	<12	>12	>15
Dominant process	debris transport, downcutting	debris transport, downcutting	lateral erosion, accumulation	accumulation, lateral erosion

Types A and B represent a narrow string, which is consistent with the idea of beads on a string. Both small beads (Type C) and large beads (Type D) are scattered along this string at irregular intervals (Figure 5). These four types of reaches alternate along the longitudinal profile of the studied river. However, some river reaches are more homogeneous, where one type will exist over a significant distance, whereas the channel structure in other reaches will be fragmented to a greater extent. Longer homogeneous reaches are found in Type A and B channels. Type C and D reaches occur irregularly along the course of the river and are separated by Type A and B channel reaches. This sort of spatially variable system of channel reaches is found in gravel-bed river valleys in the Carpathians and is accurately described by the beads on a string idea proposed by Stanford and Ward [27].

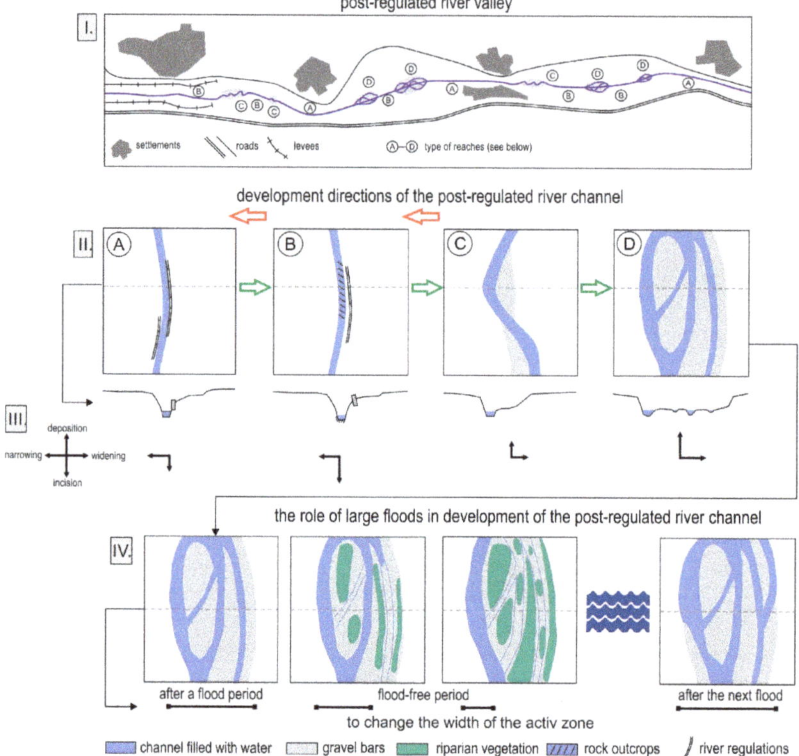

Figure 5. Conceptual model of the evolution of a gravel-bed river in a mountain area during a post-regulation period.

The location of the above-mentioned reaches, types A through D, in the river system is due to both natural determinants such as geologic structure, relief, and climate, as well as due to human impact associated with river channel regulation, road infrastructure, land use, and human settlement. This is why it is difficult to unequivocally determine the reasons for the occurrence of reaches of type C and D where the pattern of occurrence is the result of a combination of fluvial processes present in a post-regulation river (Figure 5I,II).

Spontaneous renaturalization occurs most intensively in reaches of type C and D, at locations where fluvial processes can proceed freely. Hence, spontaneous renaturalization along a post-regulation river does not occur in a continuous manner, but in a fragmented manner.

Type D reaches constitute initial sites of channel debris restoration in the river system. Debris accumulated in these zones comes primarily from bank erosion and is supplied by tributaries. These wide accumulation zones also serve as sites for the evolution of multithreaded systems and make it possible for the river channel to migrate freely (Figure 5II). The presence of type D reaches, and to a lesser degree type C reaches, is a sign of spontaneous renaturalization in a post-regulation river system. This process begins with the destruction of regulation structures and changes in principal fluvial processes along with a reduction in the share of downcutting and an increase in the share of lateral erosion, where type A and B reaches transition into type C reaches. In the next stage, type C reaches experience accumulation, expansion of the active zone, and expansion of the channel free migration zone or the formation of several channels. This process can occur gradually or abruptly in the course of a single large flood event where type C reaches transition into type D reaches. Type A and B reaches can also develop into type D reaches directly in the event of a catastrophic flood. However, type A and B reaches can suffer further decline as a result of increased downcutting, which makes it impossible for them to transition into type C. In the event of repeated river regulation, type B, C, and D reaches can transition back to type A (Figure 5II).

These wide alluvial reaches (type D) represent a source of channel material, and their evolution follows three different directions as follows: (1) They can become wider, (2) they can become longer, and (3) they can become thicker. The expansion of this zone in a post-regulation river is associated with lateral erosion. This is a necessary precondition for the river renaturalization, as the river is "trapped" in a deep-cut channel and a narrow, regulated flow pathway. Lateral erosion and accumulation are linked processes in this case. Extension of the accumulation reach or its migration along the length of the river is often made difficult by various forms of human impact in the valley. Difficulties include road infrastructure, close proximity to buildings, and river training structures.

The thickness of debris layers increases, as does the width and length of sedimentation zones in the course of flood events. These are zones wherein the river channel can freely migrate and consist of one or more threads and wide bars not covered with perennial vegetation. A period of several years without floods leads to changes in the functioning of the widened channel migration zone, where a river can freely choose its flow pathway (Figure 4III).

In periods without major floods, fluvial processes usually become limited to a narrower zone, i.e., the encroachment of vegetation onto bars, the decline of discharge inside the channels, and the limiting of discharge into the most deeply incised channel narrows down the active zone of the free migration zone. What remains in the active zone is a channel with a low level of water and bars not covered with vegetation [66,67]. However, this remains a zone for accumulation, and it is readily available to fluvial processes during subsequent floods. While this depends on the magnitude of the next flood, this zone can become only refreshed or it can become expanded.

This expansion can occur via the undercutting of floodplain terraces, which incorporates more land into the migration zone. In a post-regulation river, this process is limited in scope. A river that is "trapped" between road embankments, various other structures, or levees has only so much leeway to migrate. This pulse-type pattern of the evolution of accumulation zones with the capability for free migration during flood events and a narrowed down active zone between floods can be conceptually associated with the idea of a flood pulse [21,22] in a geomorphologic sense.

The final research question concerns the direction that mountain river systems are headed during the post-regulation period having been subjected to significant human impact in the last 100 years. Is it possible for river systems to return to their functioning from the 19th century? Or, will completely new channel patterns emerge as a way of adjusting to contemporary factors? A complete answer to these questions is not possible as of today. Currently river systems during the post-regulation period are in their initial stage of spontaneous renaturalization. In order to produce forecasts, it is necessary to conduct more research in such systems over the next few decades or at least over the next decade or so.

7. Conclusions

Compared with the 19th century, present-day river channels in the Carpathians possess more transport capacity, receive a smaller supply of catchment debris due to changes in land use, have a narrower free migration zone, experience less lateral erosion due to reinforced banks, and follow a distorted (i.e., narrowed and straightened) flow path that is not adapted to natural conditions.

Major changes in the morphology of post-regulation river channels occur mostly in the course of flood events. The initial stage of spontaneous renaturalization includes at least two potential channel evolution scenarios. A river channel can evolve in the direction of an accumulation channel with lateral erosion. In the second scenario, a river channel experiences erosion that helps make permanent its regulated structure. Both scenarios can play out in the same river channel, but in adjacent reaches. This indicates great complexity and fragmentation of the structure of a post-regulation river channel. In initial spontaneous renaturalization stages, this process can occur in limited sections and does not need to affect the entire channel system. In the case of river channels situated in urbanized areas, this process could never materialize. The same is true of fully regulated river reaches.

New patterns of channel evolution observed in river channels now affected by spontaneous renaturalization could be indicative of how gravel-bed rivers in mountains areas will evolve in the 21st century if human impact is reduced.

Author Contributions: Conceptualization, E.G. and K.K.; data curation, E.G.; investigation, E.G. and K.K.; methodology, E.G. and K.K.; software, E.G. and K.J.; visualization, E.G.; writing—original draft, E.G., K.J. and K.K.; writing—review and editing, K.J. All authors have read and agreed to the published version of the manuscript.

Funding: This research received no external funding.

Conflicts of Interest: The authors declare no conflict of interest.

References

1. Bravard, J.P.; Petts, G.E. Human Impacts on Fluvial Hydrosystems. In *The Fluvial Hydrosystems, Chapman and Hall*; Petts, G.E., Amoros, C., Eds.; Springer: Dordrecht, The Netherlands, 1996; pp. 242–262.
2. Wohl, E. Human Impacts to Mountain Streams. *Geomorphology* **2006**, *79*, 217–248. [CrossRef]
3. Rinaldi, M.; Surian, N.; Comiti, F.; Bussettini, M. A Method for the Assessment and Analysis of the Hydromorphological Condition of Italian Streams: The Morphological Quality Index (MQI). *Geomorphology* **2013**, *180*, 96–108. [CrossRef]
4. Korpak, J. The Influence of River Training on Mountain Channel Changes (Polish Carpathian Mountains). *Geomorphology* **2007**, *92*, 166–181. [CrossRef]
5. Wyżga, B.; Zawiejska, J.; Radecki-Pawlik, A.; Hajdukiewicz, H. Environmental Change, Hydromorphological Reference Conditions and the Restoration of Polish Carpathian Rivers. *Earth Surf. Process. Landf.* **2012**, *37*, 1213–1226. [CrossRef]
6. Korpak, J.; Krzemień, K.; Radecki-Pawlik, A. Wpływ czynników antropogenicznych na zmiany koryt cieków karpackich. *Infrastruktura I Ekologia Terenów Wiejskich* **2008**, *4*, 1–88.
7. Wyżga, B.; Zawiejska, J.; Radecki-Pawlik, A. Impact of Channel Incision on the Hydraulics of Flood Flows: Examples from Polish Carpathian Rivers. *Geomorphology* **2016**, *272*, 10–20. [CrossRef]
8. Gorczyca, E. *Rozwój Górskich Żwirodennych Koryt Rzecznych w Warunkach Antropopresji*; Instytut Geografii i Gospodarki Przestrzennej Uniwersytetu Jagiellońskiego: Kraków, Poland, 2016; pp. 1–240.
9. Wyżga, B. Present-day Downcutting of the Raba River Channel (Western Carpathians, Poland) and its Environmental Effects. *Catena* **1991**, *18*, 551–566. [CrossRef]
10. Wyżga, B. River Response to Channel Regulation: Case Study of the Raba River, Carpathians, Poland. *Earth Surf. Process. Landf.* **1993**, *18*, 541–556. [CrossRef]
11. Wyżga, B. Impact of the Channelization—Induced Incision of the Skawa and Wisłoka Rivers, Southern Poland, on the Conditions of Overbank Deposition. Regulated Rivers. *Res. Manag.* **2001**, *17*, 85–100.
12. Perzanowska, J. Pionierska roślinność na kamieńcach górskich potoków. In *Monitoring Siedlisk Przyrodniczych. Przewodnik Metodyczny. Część II*; Mróz, W., Ed.; Główny Inspektorat Ochrony Środowiska: Warszawa, Poland, 2012; pp. 170–180.
13. Rinaldi, R.; Wyżga, B.; Surian, N. Sediment Mining in Alluvial Channels: Physical Effects and Management Perspectives. *River Res. Appl.* **2005**, *21*, 805–828. [CrossRef]
14. Gorczyca, E.; Krzemień, K.; Sobucki, M.; Jarzyna, K. Can beaver impact promote river renaturalization? The example of the Raba River, southern Poland. *Sci. Total Environ.* **2018**, *615*, 1048–1060. [CrossRef]
15. Kondolf, G.M. Setting goals in river restoration: When and where can the river "heal itself". *Stream Restor. Dyn. Fluv. Syst.* **2011**, 29–43.
16. Rinaldi, M.; Simoncini, C.; Piégay, H. Scientific design strategy for promoting sustainable sediment management: The case of the Magra River (Central-Northern Italy). *River Res. Appl.* **2009**, *25*, 607–625. [CrossRef]
17. Comiti, F.; Da Canal, M.; Surian, N.; Mao, L.; Picco, L.; Lenzi, M.A. Channel adjustments and vegetation cover dynamics in a large gravel bed river over the last 200 years. *Geomorphology* **2011**, *125*, 147–159. [CrossRef]
18. Bańkowska, A.; Sawa, K.; Popek, Z.; Wasilewicz, M.; Żelazo, J. Studia wybranych przykładów renaturyzacji rzek. *Infrastruktura i Ekologia Terenów Wiejskich* **2010**, *9*, 181–196.

19. Duszyński, R. Ekologiczne techniki ochrony brzegów i rewitalizacja rzek. *Inżynieria Morska i Geotechnika* **2007**, *6*, 341–351.
20. Pedersen, M.L.; Andersen, J.M.; Nielsen, K.; Linnemann, M. Restoration of Skjern River and its valley: Project description and general ecological changes in the project area. *Ecol. Eng.* **2007**, *30*, 131–144. [CrossRef]
21. Junk, W.J.; Bayley, P.B.; Sparks, R.E. The flood pulse concept in river-floodplain systems. *Can. Spec. Publ. Fish. Aquat. Sci.* **1989**, *106*, 110–127.
22. Tockner, K.; Malard, F.; Ward, J.V. An extension of the flood pulse concept. *Hydrol. Process.* **2000**, *14*, 2861–2883. [CrossRef]
23. Ward, J.V. The four-dimensional nature of lotic ecosystems. *J. N. Am. Benthol. Soc.* **1989**, *8*, 2–8. [CrossRef]
24. Ward, J.V.; Tockner, K. Biodiversity: Towards a unifying theme for river ecology. *Freshw. Biol.* **2001**, *46*, 807–819. [CrossRef]
25. Benjankar, R.; Yager, E.M. The impact of different sediment concentrations and sediment transport formulas on the simulated floodplain processes. *J. Hydrol.* **2012**, *450*, 230–243. [CrossRef]
26. Juez, C.; Schärer, C.; Jenny, H.; Schleiss, A.J.; Franca, M.J. Floodplain Land Cover and Flow Hydrodynamic Control of Overbank Sedimentation in Compound Channel Flows. *Water Resour. Res.* **2019**, *55*, 9072–9091. [CrossRef]
27. Stanford, J.A.; Ward, J.V. An ecosystem perspective of alluvial rivers: Connectivity and the hyporheic corridor. *J. N. Am. Benthol. Soc.* **1993**, *12*, 48–60. [CrossRef]
28. Ward, J.V.; Tockner, K.; Arscott, D.B.; Claret, C. Riverine landscape diversity. *Freshw. Biol.* **2002**, *47*, 517–539. [CrossRef]
29. Stupnicka, E. *Geologia Regionalna Polski*; Wydawnictwa Uniwersytetu Warszawskiego: Warszawa, Poland, 1997; pp. 1–332.
30. Punzet, J. Przepływy charakterystyczne. In *Dorzecze Górnej Wisły*; cz. I, PWN, Warszawa: Kraków, Poland, 1991; pp. 167–215.
31. Soja, M.; Zborowski, A. Wybrane zagadnienia zagospodarowania przestrzennego zlewni Raby. *Pr. Geogr. IG UJ* **2000**, *106*, 119–140.
32. Soja, R. Hydrologiczne aspekty antropopresji w polskich Karpatach. *Pr. Geogr. IG UJ* **2002**, *186*, 1–130.
33. Kamykowska, M.; Kaszowski, L.; Krzemień, K. River Channel Mapping Instruction. Key to the River Bed description. *Pr. Geogr. IG UJ* **1999**, *104*, 9–25.
34. Krzemień, K. (Ed.) *Struktura Koryt Rzek i Potoków (Studium Metodyczne)*; Instytut Geografii i Gospodarki Przestrzennej Uniwersytetu Jagiellońskiego: Kraków, Poland, 2012; pp. 1–144.
35. Wolman, M.G. A method of sampling coarse river-bed material. *EOS Trans. Am. Geophys. Union* **1954**, *35*, 951–956. [CrossRef]
36. Kaszowski, L. Struktura i typy koryt rzecznych w dorzeczu Raby. *Sprawozdania z Posiedzeń Komisji Naukowych PAN* **1980**, *22*, 162–163.
37. Froehlich, W.; Walling, D.E. The role of unmetalled roads as a sediment source in the fluvial systems of the Polish Flysch Carpathians. *IAHS Publ.* **1997**, *245*, 159–168.
38. Froehlich, W.; Słupik, J. Rola dróg w kształtowaniu spływu i erozji w karpackich zlewniach fliszowych. *Przegląd Geogr.* **1986**, *58*, 67–87.
39. Starkel, L. Rozwój rzeźby Polskich Karpat fliszowych w holocenie. *Pr. Geogr. IG PAN* **1960**, *22*, 1–239.
40. Klimek, K.; Trafas, K. Young-Holocene Changes in the Course of the Dunajec River in the Beskid Sadecki Mts. (Western Carpathians). *Studia Geomorphol. Carpatho Balc.* **1972**, *6*, 85–103.
41. Starkel, L.; Łajczak, A. Kształtowanie rzeźby den dolin w Karpatach (koryt i równin zalewowych). In *Współczesne Przemiany Rzeźby Polski*; Starkel, L., Kostrzewski, A., Kotarba, A., Krzemień, K., Eds.; Instytut Geografii i Gospodarki Przestrzennej Uniwersytetu Jagiellońskiego: Kraków, Poland, 2008; pp. 95–108.
42. Kościelniak, J. Influence of River Training on Functioning of the Biały Dunajec River Channel System. *Geomorphol. Slovaca* **2004**, *1*, 62–67.
43. Kędzior, A. Roboty wodne i melioracyjne w południowej Małopolsce. cz. III. In *Regulacja Rzek Górskich, Zbiorniki Wody I Zabudowanie Potoków Górskich*; Ministerstwo Robót Publicznych: Lwów, Poland, 1931.
44. Łapuszek, M.; Lenar-Matyas, A. *Utrzymanie I Zagospodarowanie Rzek Górskich*; Wydawnictwo Politechniki Krakowskiej: Kraków, Poland, 2013; pp. 1–283.
45. Zawiejska, J.; Krzemień, K. Man—Induced Changes in the Structure and Dynamic of the Upper Dunajec River Channel. *Geogr. Časopis* **2004**, *56*, 111–124.

46. Gorczyca, E.; Krzemień, K.; Liro, M.; Sobucki, M. Changes of Mountain River Channels and their Environmental Effects. In *Open Channel Hydraulics, River Hydraulics Structures and Fluvial Geomorphology*; Radecki-Pawlik, A., Pagliara, S., Hradecký, J., Hendrickson, E., Eds.; Science Publishers, CRC Press, Taylor & Francis Group: Boca Raton, FL, USA, 2017; pp. 303–321.
47. Wharton, G. *Managing River Environments*; Cambridge University Press: Cambridge, UK, 2000; pp. 1–92.
48. Lach, J.; Wyżga, B. Channel Incision and Flow Increase of the Upper Wisłoka River, Southern Poland, Subsequent to the Reforestation of its Catchment. *Earth Surf. Process. Landf.* **2002**, *27*, 445–462. [CrossRef]
49. Krzemień, K. The Czarny Dunajec River, Poland, as an Example of Human-induced Development Tendencies in a Mountain River Channel. *Landf. Anal.* **2003**, *4*, 57–64.
50. Jeleński, J.; Wyżga, B. The Raba River at Lubień. Erodible river corridor as a restoration measure for mountains rivers. In Proceedings of the International Conference 'Towards the Best Practice of River Restoration and Maintenance', Kraków, Poland, 20–23 September 2016; Zawiejska, J., Wyżga, B., Eds.; Ab Ovo Association: Kraków, Poland, 2016; pp. 67–68.
51. Galicia and Bucowina (1861–1864) Second Military Survey of the Habsburg Empire. Available online: https://mapire.eu/en/map/secondsurvey-galicia/?layers=11&bbox=2648791.450423253%2C6410880.850074071%2C2700348.2260016045%2C6426168.255731106 (accessed on 13 December 2019).
52. Łapuszek, M.; Ratomski, J. Metodyka określania i charakterystyka przebiegu oraz prognoza erozji dennej rzek górskich dorzecza górnej Wisły. *Inżynieria Środowiska Monografia* **2006**, *332*, 1–122.
53. Wyżga, B. Wcinanie się rzek polskich Karpat w ciągu XX wieku. In *Stan Środowiska Rzek Południowej Polski I Możliwość Jego Poprawy—Wybrane Aspekty*; Wyżga, B., Ed.; Instytut Ochrony Przyrody PAN: Kraków, Poland, 2008; pp. 7–39.
54. Zawiejska, J.; Wyżga, B. Twentieth-century channel change on the Dunajec River, southern Poland: patterns, causes and controls. *Geomorphology* **2010**, *117*, 234–246. [CrossRef]
55. Wyżga, B. A geomorphologist's criticism of the engineering approach to channelization of gravel-bed rivers: Case study of the Raba River, Polish Carpathians. *Environ. Manag.* **2001**, *28*, 341–358. [CrossRef] [PubMed]
56. Erskine, W.D. Downstream Hydrogeomorphic Impacts of Eildon Reservoir on the Mid—Goulburn River. *Vic. Proc. Soc. Vic.* **1996**, *108*, 1–15.
57. Requena, P.; Weichert, R.B.; Minor, H.E. Self-widening by Lateral Erosion in Gravel Bed Rivers. In *River Flow*; Ferreira, A., Cardoso, L., Eds.; Taylor & Francis Group: Lisbon, Portugal, 2006; pp. 1801–1809.
58. Gorczyca, E.; Krzemień, K.; Wrońska-Wałach, D.; Boniecki, M. Significance of Extreme Hydro-geomorphological Events in the Transformation of Mountain Valleys (Northern Slopes of the Western Tatra Range), Carpathian Mountains, Poland. *Catena* **2014**, *121*, 127–141. [CrossRef]
59. Czech, W.; Radecki-Pawlik, A.; Wyżga, B.; Hajdukiewicz, H. Modelling the Flooding Capacity of a Polish Carpathian River: A Comparison of Constrained and Free Channel Conditions. *Geomorphology* **2016**, *272*, 32–42. [CrossRef]
60. Hajdukiewicz, H.; Wyżga, B.; Mikuś, P.; Zawiejska, J.; Radecki-Pawlik, A. Impact of a Large Flood on Mountain River Habitats, Channel Morphology and Valley Infrastructure. *Geomorphology* **2016**, *272*, 55–67. [CrossRef]
61. Lehotsky, M.; Rusnak, M.; Kidova, A. Application of Remote Sensing and the GIS in Interpretation of River Geomorphic Response to Floods. In *Open Channel Hydraulics, River Hydraulics Structures and Fluvial Geomorphology*; Radecki-Pawlik, A., Pagliara, S., Hradecký, J., Hendrickson, E., Eds.; Science Publishers, CRC Press, Taylor & Francis Group: Boca Raton, FL, USA, 2017; pp. 388–399.
62. Punzet, J. Zmiany w przebiegu stanów wody w dorzeczu górnej Wisły na przestrzeni 100 lat (1871–1970). *Folia Geogr. Ser. Geogr. Phys.* **1981**, *14*, 5–28.
63. Nicholas, A.P.; Ashworth, P.J.; Kirkby, M.J.; Macklin, M.G.; Murray, T. Sediment Slugs: Large-scale Fluctuations in Fluvial Transport Rates and Storage Volumes. *Prog. Phys. Geogr.* **1995**, *19*, 500–519. [CrossRef]
64. James, A.L. Secular sediment waves, channel bed waves and legacy sediment. *Geogr. Compass* **2010**, *4*, 576–598. [CrossRef]
65. Church, M.A.; Jones, D. Channel bars in gravel-bed rivers. In *Gravel-Bed Rivers: Fluvial Processes, Engineering and Management*; Hey, R.D., Bathurst, J.C., Theme, C.R., Eds.; John Wiley & Sons: Chichester, UK, 1982; pp. 291–338.

66. Gurnell, A.M.; Petts, G.E.; Harris, N.; Ward, J.V.; Tockner, K.; Edwards, P.J.; Kollmann, J. Large wood retention in river channels: The case of the Fiume Tagliamento, Italy. *Earth Surf. Process. Landf.* **2000**, *25*, 255–275. [CrossRef]
67. Hajdukiewicz, H.; Wyżga, B.; Amirowicz, A.; Oglęcki, P.; Radecki-Pawlik, A.; Zawiejska, J.; Mikuś, P. Ecological state of a mountain river before and after a large flood: Implications for river status assessment. *Sci. Total Environ.* **2018**, *610*, 244–257. [CrossRef]

Article

Hydrogeomorphic Impacts of Floods in a First-Order Catchment: Integrated Approach Based on Dendrogeomorphic Palaeostage Indicators, 2D Hydraulic Modelling and Sedimentological Parameters

Radek Tichavský, Stanislav Ruman and Tomáš Galia *

Department of Physical Geography and Geoecology, Faculty of Science, University of Ostrava, Chittussiho 10, 710 00 Ostrava, Czech Republic; radek.tichavsky@osu.cz (R.T.); stanislav.ruman@osu.cz (S.R.)

* Correspondence: tomas.galia@osu.cz; Tel.: +420-553-46-2307

Received: 22 November 2019; Accepted: 10 January 2020; Published: 12 January 2020

Abstract: Floods represent frequent hazards in both low- and first-order catchments; however, to date, the investigation of peak flow discharges in the latter catchments has been omitted due to the absence of gauging stations. The quantification of flood parameters in a first-order catchment (1.8 km^2) was realised in the moderate relief of NE Czechia, where the last flash flood event in 2014 caused considerable damage to the infrastructure. We used an integrated approach that included the dendrogeomorphic reconstruction of past flood activity, hydraulic modelling of the 2014 flash flood parameters using a two-dimensional IBER model, and evaluation of the channel stability using sedimentological parameters. Based on 115 flood scars, we identified 13 flood events during the period of 1955 to 2018, with the strongest signals recorded in 2014, 2009 and 1977. The modelled peak flow discharge of the last 2014 flood was equal to 4.5 m^3·s^{-1} (RMSE = 0.32 m) using 26 scars as palaeostage indicators. The excess critical unit stream power was observed at only 24.2% of the reaches, representing predominantly bedrock and fine sediments. Despite local damage during the last flood, our results suggest relatively stable geomorphic conditions and gradual development of stream channels under discharges similar to that in 2014.

Keywords: flood; dendrogeomorphology; palaeostage indicator; hydraulic modelling; peak flow discharge; unit stream power; bed shear stress

---

1. Introduction

Extreme rainfall resulting in flood events is a common phenomenon in different environments, including both mountain ranges and lowlands [1,2]. The systematic monitoring of flow stages at gauging stations, including precipitation and flow discharge prediction, is currently well applied and documented in many medium- and large-sized rivers [3–5]. In contrast, data from mountain steep headwater catchments and, in general, first-order catchments [6] are still poor, due to the insufficient network of stream gauges and the sporadic amount of processed documentary evidence [7]. Not only do mountain headwater streams generate sediment-laden flows with aftermaths within and along fans [8,9], but also streams and gullies of first-order catchments (up to 10 km^2) in moderate relief can be responsible for local damage to infrastructure. Ozturk et al. [10] analysed extraordinary flash flood events (140 mm per 2 h) in a small catchment (6 km^2) that resulted in damage to infrastructure and a high amount of suspended sediments (t/km^2) due to intense hillslope–channel coupling. In addition, Terti et al. [11] pointed to a short response time of small catchments to flash floods, thereby increasing the probability of trapping people during outdoor activities.

Despite the lack of gauging records from forested first-order catchments, several approaches exist to describe the hydrogeomorphic impacts of extremely high flow stages. The dating of flood scars on riparian vegetation using dendrogeomorphic methods [12,13] and recording their position and height above the channel bottom with a combination of hydraulic models is a well-established approach for the estimation of peak flood discharge, flow velocity, and unit stream power [14]. Using flood scars as a palaeostage indicator (PSI; maximum height of the scar above the channel bottom) allowed the interpretation of past flood events in medium- and large-scale rivers (catchment areas larger than 40 km^2), for example, in the Western Mediterranean [15], North America [16,17], the Carpathians [18], and the Himalayas [19]. The dendrogeomorphic response of flash floods in small catchments is generally considered lower [20] but may increase due to the presence of erodible sediments amplifying lateral bank erosion. In such conditions, while small streams may not generate flood waves as large as those of large rivers, the presence of exposed and scarred tree roots [21] may complete a relatively low number of scarred tree stems.

The alluvial streams draining first-order catchments often do not display clear relationships between channel geometry, bed substrate, unit stream power and drainage area, and their resulting form and evolution trajectory are unpredictable unless local conditions (e.g., bedrock resistance, intensity of hillslope-channel coupling processes, land use history, and presence of large instream wood) are constrained [22–26]. These streams are characterized by a more or less developed stepped-bed morphology with a wide range of sediment size, where individual steps controlling channel bed stability consist of interlocked boulders, bedrock outcrops or large wood pieces [27]. This implies the relative stability of their channel beds under relatively high discharges (up to floods of 20–50-year recurrence intervals) and thus only limited adjustments of the channel morphology and geometry to lower (e.g., bankfull) flows owing to the presence of generally shallow flows, particle-size interactions (hide/protrusion effect) and additional bed form resistance [25,28–31]. The correlations between the unit stream power, sediment calibre and prevailing fluvial process may exist at the reach scale when spatially limited depositional reaches can be accompanied by local bed sediment fining and an abrupt decrease in the unit stream power of a high-magnitude flood [23]. Nevertheless, our knowledge of the direct relationships between the transport capacity of a particular flood event, bed stability and the resulting fluvial processes in first-order catchments is still somewhat limited by the lack of detailed post-flood field surveys of geomorphic consequences.

As mentioned, two-dimensional hydraulic models have been successfully applied to peak discharge reconstructions in ungauged or poorly gauged catchments of various sizes and environments [18,32–35]. This peak discharge is mostly defined not as the single deterministic value but rather as a range of values due to uncertainties inherent in the process of the estimation [36]. Discharges predicted in such a way could be used in flood frequency analyses [18], where they serve as outliers to the measured discharges. Consequently, the predicted discharges could serve as a basis for the evaluation and mitigation of flood risk [34] or as information about the flood magnitude in the historical period [32]. Moreover, the results of hydraulic simulations can be used for the estimation of stream transport capacity and channel stability during a particular flood event at a very detailed scale. In this sense, the bed shear stress and unit stream power are relevant parameters for calculating the incipient motion of coarse bed particles and thus evaluating the stability of stepped-bed channels consisting of relatively stable cobble to boulder steps [30,37].

For our purposes, we selected a first-order catchment (1.8 km^2) in the moderate relief of the Eastern Sudetes (NE Czechia), where, during the last 15 years, the occurrence of several flash flood events caused substantial geomorphic imprints [38]. In particular, the last intense precipitation event (27 May 2014) resulted in a moderate flood risk in the case of medium-sized rivers (2-year recurrence interval), but a discharge of an approximately 100-year recurrence interval was estimated at ungauged small streams. The short-lived storm, with a total rainfall amount of 40 mm (locally up to 80 mm), had an intensity of between 40 and 60 mm/h. Its hydrogeomorphic response was primarily due to unprecedented rainfall intensity and large antecedent precipitation. The financial costs of the flash flood aftermaths in the affected region were calculated as approximately EUR 200,000 [39].

Our aims were to (i) create the chronology of the flash flood events in this small catchment using dendrogeomorphic approaches, (ii) estimate the parameters (peak flow discharge, flow velocity, bed shear stress, and unit stream power) of the last 2014 flash flood event using the combination of PSI and 2D hydraulic modelling, and (iii) describe the stream transport capacity and channel stability during the 2014 flash flood event based on the hydraulic simulation data and sedimentological parameters. Such a comprehensive approach may help to better quantify the flash flood parameters of ungauged streams and thus contribute to more reliable management of small streams in future extreme climate events.

2. Materials and Methods

2.1. Study Site

The hydrogeomorphic impacts of floods were studied on a small tributary of the Bělá River in the foreland of the Eastern Sudetes (NE part of Czechia, Central Europe; 50°17′ N, 17°17′ E; Figure 1). The area comprises a Proterozoic and Palaeozoic basement composed of orthogneiss, which is overlaid by glacifluvial deposits of the Saalian and Elsterian glaciations with different grain-size distributions of till sediments. These deposits are the result of deglaciation phases in proglacial areas on the ice margins and were recently covered by the Holocene sandy loam colluvium [40,41]. The study site is characterised by a temperate climate, with a mean annual precipitation between 850 and 900 mm. Most precipitation falls during the spring and summer months [42], with an occasional occurrence of extreme daily rainfall (more than 50 mm per day). Documentary evidence provides information about several flood events (e.g., 1903, 1921, 1971, 1977, 2007, 2009, and 2014), relating to both advective rains and short-term intense rainfall events that are currently responsible for intense gully incision and damage to infrastructure within the studied region [38,43].

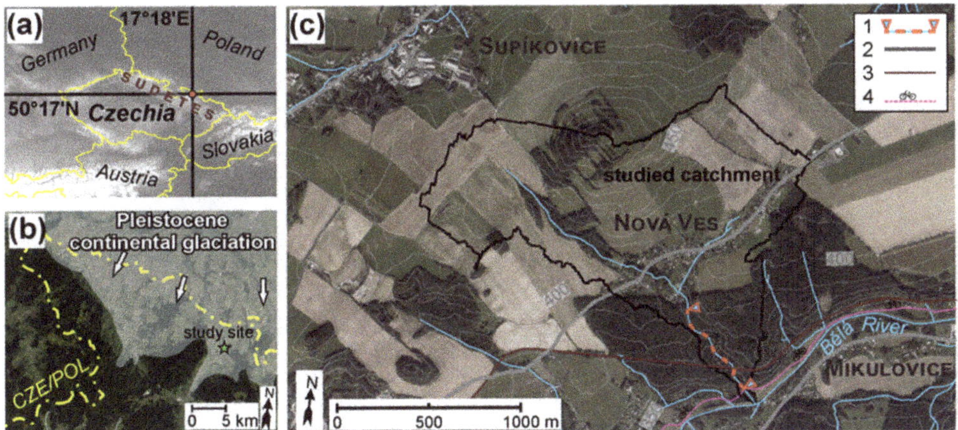

Figure 1. Location of the study site: position within Central Europe (**a**), within the Eastern Sudetes (**b**), and (**c**) orthophoto image of the studied catchment (1—study channel reach, 2—main road, 3—railway and 4—cycle path).

We focused on an approximate 700-m-long reach of an unnamed first-order stream draining a catchment area of 1.8 km^2 with a mean elevation of 417.5 m a.s.l. Fields and dwellings dominate the upper and middle catchment area, while the lower part (i.e., the study reach) is predominantly covered by a mixed forest (Figure 1c). The mean stream gradient of the study reach is 0.05 m/m, with a maximum up to 0.17 m/m. The channel cuts into the Pleistocene and Holocene deposits, with typical alternations of stepped-bed morphology and bedrock outcrops (Figure 2a) accompanied by poorly sorted bed sediments (Figure 2b,c). Fresh slope/bank failures (up to 50 m long and 10 m high) and generally unstable banks have caused channel widening during recent floods, resulting in the frequent occurrence of exposed tree root systems and flood scars on riparian vegetation along the entire reach (Figure 2b–d). In contrast, there is no evidence of deep channel incision due to the resistant gneiss bedrock and large interlocked boulders at several reaches within the stream. Slopes and floodplains surrounding the channel are overgrown by a mixed forest composed predominantly of *Pinus sylvestris* L., *Picea abies* (L.) Karst., *Alnus glutinosa* (L.) Gaertn, and *Tilia cordata* Mill. At 0.5 river km (r. km), the stream drains to a 50-m-long concrete trough within a viaduct under a railway. Moreover, culverts are presented under a cycle path in the lower part and under a road through a village in the middle part of the catchment.

Figure 2. Morphological features of the study channel reach: (**a**) channel bottom composed of gneiss bedrock; (**b**) channel widening evidenced by the system of exposed tree roots in boulder-step reach; (**c**) step-pool reach with the presence of channel widening and fresh, shallow slope failure; (**d**) flood scars (higher older, and lower younger) on the tree stem of *A. glutinosa* in the direction of flow (note the remnants of concrete parts of centring transported by past floods).

2.2. Dendrogeomorphic Fieldwork and Analyses

First, the terrain fieldwork focused on dendrogeomorphic sampling to create the chronology of past flood events and to select the flood scars caused by the 2014 flood event. The tree sampling followed the standard dendrogeomorphic procedure for flood reconstruction [13,44]. The sampling was focused on flood scars occurring either on tree stems or exposed tree roots. In the case of tree stems, increment cores were extracted from the edge of the scar and from the undisturbed part of the tree stem using the Pressler increment borer (40 × 0.5 mm). Tree stems of a small diameter (up to 10 cm) were cut by handsaw in the position of the flood scar to gain the stem disc or wedge while approximately 2-cm-wide cross-sections were sampled from exposed and scarred living roots. Only those scars oriented against the supposed direction of the flow path were considered for sampling. In addition, only scars that occurred on exposed but stabilized roots with the lowest tendency to flex during higher discharges were used to avoid the underestimation of modelled peak flow discharge [45]. All sampled trees were carefully described, and the height of the sampled scar was carefully noted, labelled with a visible object, photographed, and targeted using GPS. Reference trees growing near the study reach without any geomorphic influence on tree growth were sampled to cross-date with the disturbed samples and to eliminate false and missing rings.

Laboratory processing followed the standard dendrogeomorphic procedure [13,46]. All increment cores were glued into woody supports and—together with cross sections—dried and polished to be ready for dendrogeomorphic analysis. The tree rings of the increment cores and stem discs were counted and measured using TimeTable and PAST4 software [47], and their growth patterns were compared with the appropriate reference chronology (compiled in Arstan software [48] using a double detrending procedure) to ensure the reliability of dating. As root segments are more problematic regarding the occurrence of missing and wedging rings, the zig-zag segment tracing method [49] was applied to carefully count the years of each ring. In the case of problematic roots (i.e., dense growth increment and small roots), we used microslides cut by GLS-1 microtome and processed according to standard chemical procedures to precisely define the position of the flood scar within the root section [50,51].

In the next step, we identified the years with the occurrence of scars and onset of callus tissues (Figure 3) and compiled the chronology of past flood events. Scars represent an unequivocal signal of flood events and are considered the most reliable growth disturbance in dendrogeomorphic flood reconstructions [44]. The event identification was based on the event-response index (I_t index [52]), calculated as:

$$I_t = \frac{\sum R_t}{\sum A_t} \times 100\% \quad (1)$$

where R is the number of scars in a year t and A is the total number of sampled trees living in a year t. Then, a certain event was considered when $I_t \geq 10\%$ and the number of scars ≥ 3, while a probable event was determined when $10\% > I_t \geq 5\%$ and the number of scars ≥ 2. From the whole dataset of scars, we eventually selected the scars dated to 2014 as a PSI of the May 2014 flash flood event, whose parameters were modelled.

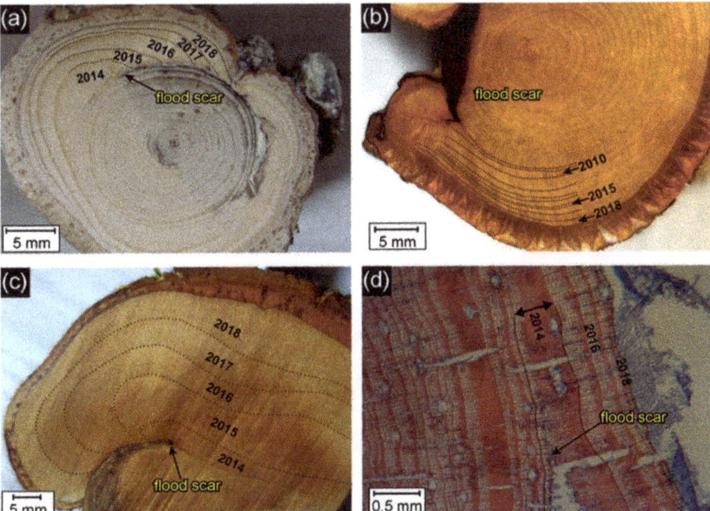

Figure 3. Identified flood scars: (**a**) the 2014 flood scar on the root section of *P. abies*; (**b**) the 2009 flood scar on the root section of *T. cordata*; (**c**) the 2014 flood scar on the stem wedge of *A. glutinosa*; and (**d**) the 2014 flood scar on the root microsection of *P. sylvestris*, with the obvious position within earlywood cells indicating the flood event on 27 May.

2.3. Channel Parameters and Channel Geometry

This part consisted of (i) the measurement of the longitudinal profiles (channel/floodplain geometry) to model the 2014 flood behaviour and (ii) the classification of erosive/depositional segments and the measurement of the largest clasts for the determination of channel stability during the 2014 floods:

(i) The channel/floodplain geometry and the height of the 2014 scars (PSI) were measured in cross sections using a total station (GTS-212) with an accuracy of ±2.5 cm and a GNSS geodetic receiver (Trimble R2) with an accuracy of ±4 cm. All measurements were taken in the spring of 2019. To register the complexity of the channel geometry, the density of cross sections was as follows: the position of the first cross section was located directly in the position of the scars. Afterwards, five cross sections were measured, with the distance of one meter among each other, followed by two cross sections at a distance of 2.5 m and one cross section at a distance of 5 m. Next, cross sections were measured at a distance of 10 m. In the case of close distances between the scars, no overall process was applied. In total, 341 cross sections were measured, resulting in 4300 surveyed points. The GNSS geodetic receiver was only applied to place the relative position of points to the absolute location based on the local geographical reference system. A digital terrain model (DTM) in raster format was created from these cross sections with a grid size equal to 0.1 m and was used as a source for the hydraulic modelling.

(ii) To reveal the channel stability during the last flood event, the middle axes of the five largest bed particles were measured by a tape (with ±0.01 m accuracy; with a less-precise ±0.05 m accuracy only in a few cases of partially buried boulders) in 10 ± 1 m intervals along the stream's longitudinal profile, except for the reach located in the culvert (0.51–0.54 r. km). Only the particles within the bankfull channel were accounted for. Consequently, the representative mean boulder diameter MBD (mm) for each of the channel cross sections was calculated as the arithmetical mean of these five measurements [23]. We classified each cross section as erosional, stable, or depositional by the observed signs of the present stability of the adjacent channel reach. The erosional reaches indicated trends of incision together with the frequent presence of exposed roots or bedrock outcrops in the channel banks, whereas the depositional reaches were typified by the occurrence of locally widened channels with developed unvegetated bars. The stable reaches represented the transport-balanced segments without evident signs of recent incision or bed aggradation. We observed no wood obstructions in the channel that would influence the channel bed stability and sediment transport processes.

2.4. Hydraulic Modelling

This phase comprised three steps: (i) hydraulic model creation, setup and calibration; (ii) estimation of scar peak discharges (SPD) and reach peak discharge; and (iii) scenario modelling of the 2014 flood event. The two-dimensional IBER model (version 2.5.1) was applied to the hydraulic modelling of the selected reach [53]. This is an established software that was applied to the estimation of palaeoflood discharges of small [18] to large rivers [35,45]. An unstructured mesh, which comprised almost 200,000 elements with an average size of 0.3 m, was developed over the reach. As an initial condition at $t = 0$, the river was set dry. The flow was subcritical throughout the whole domain for all used discharges. We imposed inlet boundary conditions based on the uniform discharge and the critical depth at the outlet of the studied reach. The first PSI was located 20 m from the outlet of the reach, which allowed the model to overcome inaccuracy in the selected boundary conditions. A wet–dry threshold of 0.01 m and 2nd order roe scheme was chosen. The Courant–Friedrichs–Lewy was set to 0.9 and the mixed length turbulence model was selected. Although the erosion and accumulation of alluvium could occur during the 2014 flood, we considered the stable riverbed during the modelling [54]. The model was calibrated to a single value of water stage in the cross section where we measured the discharge (equal to 25.23 cubic litres per second), using the velocity meter. The cross section was located in the downstream part of the selected reach and the measurement took place at the end of May 2019. Based on the calibration results, a uniform value for Manning's roughness coefficient (n) equal to 0.08 was applied to the overall reach. The selection of the roughness value was based on the studies dealing with high-gradient streams [55,56]. In the second step, an iterative process was run to find the SPD [57] that produced the best fit between the scars and modelled water depths [58] and to select the reach peak discharge that produced the best RMSE value [32]. Finally, the three scenarios were established to model the characteristics (unit stream power and bed shear stress) of the 2014 flood event: Q_{min} (discharge equal to the 1st quartile of the SPD), $Q_{optimum}$ (reach peak discharge which produced the optimum/lowest value of RMSE), and Q_{max} (the 3rd quartile of the SPD). Although the main results were created for the Qoptimum, the remaining scenarios allowed the evaluation of the uncertainty inherent in palaeoflood discharge estimation [36].

2.5. Relations between the Hydraulic and Sedimentologic Parameters and Calculation of Channel Stability

To evaluate the flow competence of the 2014 flood (i.e., the ability of this flood to transport coarse bed material and destabilize the channel bed), we used the critical unit stream power ω_{ci}–transported particle diameter D_i (mm) relationship. To approximate the local conditions of the frequent occurrence of a stepped-bed morphology, relatively high channel gradients and a small catchment area, we applied the relationship developed by field observations in similar steep channels ($0.06 \leq S \leq 0.14$ m/m) draining small catchments ($0.2 \leq A \leq 2.2$ km^2) in Central European medium–high mountain settings [23,59]:

$$\omega_{ci} = 0.72 D_i^{1.02} \tag{2}$$

This relationship (2) was derived from direct observations of the largest boulders that were transported by a high-magnitude flood event and from the displacements of marked particles during lower discharges (covering grain-sizes 20–400 mm) in stepped-bed streams with poorly sorted bed sediments. As D_i, we substituted MBD and calculated the potential critical unit stream power ω_{cMBD}, which will lead to the incipient motion of MBD in a given cross section. The resulting value was compared with the unit stream power ω simulated for the 2014 event, and the excess critical unit stream power ω_E was expressed in the following form:

$$\omega_E = \frac{\omega}{\omega_{cMBD}} \tag{3}$$

This implies that $\omega_E \geq 1$ indicates a potentially unstable bed structure consisting of coarse grains (i.e., rapids or individual step units) during the 2014 event. We simulated ω_E for all three scenarios (Q_{min}, $Q_{optimum}$ and Q_{max}).

In addition, we employed the peak bed shear stress of the 2014 event (τ_b) to find a possible relationship between the parameters of τ_b, ω and *MBD* and to assess potential differences between the groups of cross-sections by their present stability (erosional, vertically stable and depositional) during the $Q_{optimum}$ scenario. Due to the non-normality of the data, we used Spearman's correlation r_{sp} to test the potential relationships between τ_b, ω and *MBD*. A non-parametric Kruskal–Wallis test was used to compare τ_b, ω and *MBD* between the reaches with depositional, transport-balanced and erosional tendencies when the Z-value test with Bonferonni corrections was used to distinguish significantly different groups. We used a significance level of 0.05 for all the tested data.

3. Results

3.1. Chronology of Past Flood Events and Botanical Evidence of the May 2014 Flash Flood Event

In total, we successfully sampled and dated 75 scarred individuals (20 tree stems and 55 scarred roots) with a predominance of samples from *P. sylvestris* (37.3%), *P. abies* (18.7%), and *A. glutinosa* (17.3%), whereas five samples (root cross sections) had to be excluded from the chronology due to uncertainties during the dating procedure (Table 1). Eventually, it was possible to date 115 scars (1.5 scars per tree). Overall, we determined 13 flood events (seven certain and six probable events) during the period of 1955 to 2018 (limited by the minimum number of 10 sampled trees; Figure 4).

The oldest (probable) event was recorded in 1965. The strongest signals (i.e., the highest value of the It index) were identified during 2014 (26 scars; I_t = 34.7%), 2009 (25 scars; I_t = 33.3%), and 1977 (5 scars; I_t = 15.2%).

Table 1. Overall number of sampled and dated trees and the distribution of samples containing the 2014 scar.

Tree Species	Nr. of Trees	Percentage of Trees (%)	Nr. of Root Sections	Nr. of Stem Sections/Cores	Number of Samples Containing the 2014 Scar	
					Root	Stem
Pinus sylvestris	28	37.3	28	0	5	0
Picea abies	14	18.7	13	1	6	0
Alnus incana	13	17.3	6	7	4	2
Tilia cordata	9	12.0	7	2	4	0
Acer pseudoplatanus	6	8.0	0	6	0	2
Sorbus aucuparia	5	6.7	1	4	1	2
Total	75	100.0	55	20	20	6

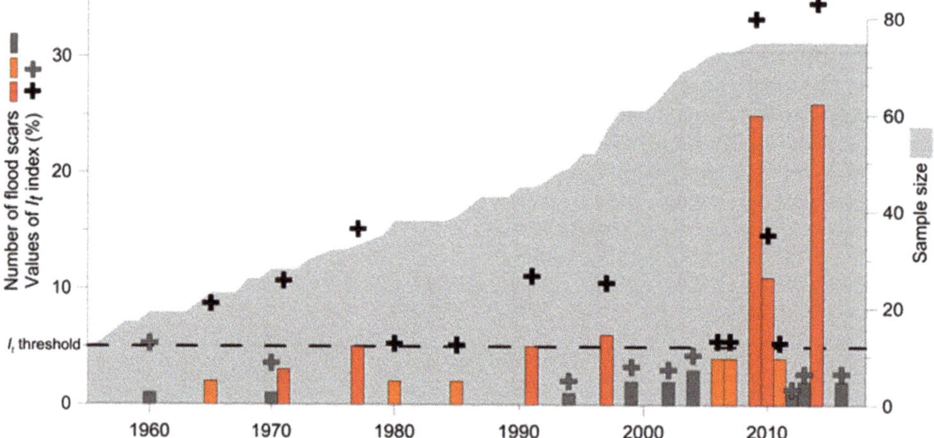

Figure 4. Number of identified flood scars and the final chronology of flood events (based on thresholds: $I_t \geq 5\%$ and number of scars ≥ 2). The red column indicates a certain event, and the orange column indicates a probable event. The grey column and grey cross indicate years that could not be considered flood events.

Focusing on the botanical evidence of the 2014 flood event, we identified 26 scars throughout the whole study reach (6 scars on tree stems and 20 scars on exposed roots; Table 1). The mean height of the scars above the thalweg was 96.6 ± 34.6 cm. The minimum height (41 cm) was observed at 0.33 r. km at a straight channel reach, while the maximum height (156 cm) was recorded at 0.07 r. km at the failure of a concave bank. The distribution of the 2014 scars (Figure 5) was zero at the reach between 0.35 and 0.5 r. km, where the total number of scarred trees was generally lower, and the scars were generally of older dates. In contrast, the highest abundance of the 2014 flood scars was recorded between 0.2 and 0.35 r. km (11 scars overall).

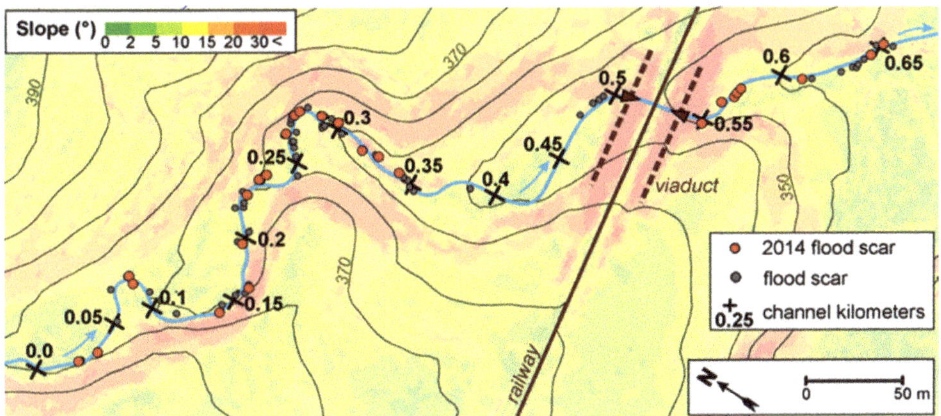

Figure 5. Localization of the 2014 flood scars within the studied channel reach.

3.2. Results of Hydraulic Modelling

The flow conditions during the simulations were subcritical in the major part of the reach (91–94% of the total area) for all simulated flow scenarios (Table 2). Other flow characteristics (depth, velocity, and Froude number) resulting from the raster of hydraulic modelling are described in Table 2.

Table 2. Selected flow characteristics for flow scenarios applied in this study.

Flow Characteristics/Scenario	Depth (m) Mean	Depth (m) Max	Velocity ($m \cdot s^{-1}$) Mean	Velocity ($m \cdot s^{-1}$) Max	Froude Mean	Froude Max	Area of Supercritical Flow (%)
Q_{min}	0.35	1.41	1.05	9.43	0.59	3.67	6.43
$Q_{optimum}$	0.43	1.50	1.17	11.97	0.61	3.79	7.37
Q_{max}	0.59	1.95	1.44	12.02	0.64	3.88	9.01

The characteristics of the RMSE curve of $Q_{optimum}$ during the 2014 flood and the SPD characteristics are shown in Figure 6. The value of the reach peak discharge ($Q_{optimum}$) of the 2014 flood estimated by the RMSE (0.32 m) was equal to 4.5 $m^3 \cdot s^{-1}$. The values of the SPD showed high variability (coefficient of variability = 0.79) when the minimal SPD was 1.5 $m^3 \cdot s^{-1}$ and the maximal SPD was equal to 16.5 $m^3 \cdot s^{-1}$. The values of the Q_{min} and Q_{max} scenarios were calculated from all estimated SPD and were equal to 2.63 and 9.38 $m^3 \cdot s^{-1}$, respectively. The deviation between the measured PSI elevations and the simulated water level for $Q_{optimum}$ ranged from −0.60 to 0.63 m with a median equal to −0.05 m. Similarly, the Q_{min} deviations ranged from −0.30 to 0.75 m, and the median was equal to 0.08 m. For Q_{max}, we registered deviations from −1.07 to 0.42 m, and the median was equal to −0.31 m. The highest magnitudes of velocity (up to 11.97 $m \cdot s^{-1}$; Figure 7) and bed shear stress (up to 1787.10 $N.m^{-2}$) were registered at the positions of the highest changes in riverbed topography.

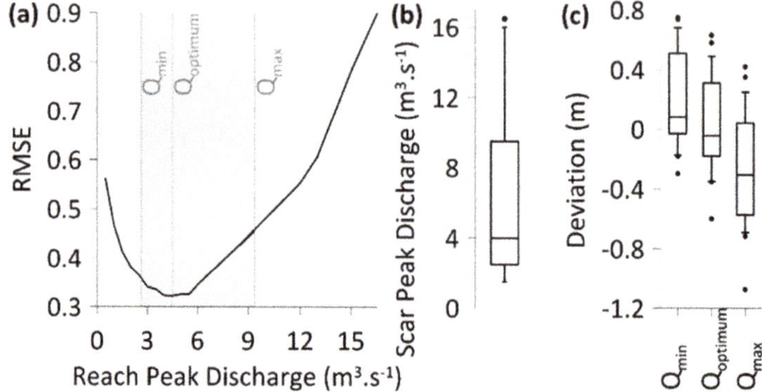

Figure 6. Parameters of the 2014 modelled flood discharge: (**a**) RMSE curve for the reach peak discharges where three scenarios are visualized with the uncertainty defined by Q_{min} and Q_{max} (filled rectangle); (**b**) the box plot of scar peak discharges; (**c**) the box plots of deviation between the measured PSI elevation and simulated water level. Whiskers in (**b**,**c**) are equal to the 10th and 90th percentiles (n = 26).

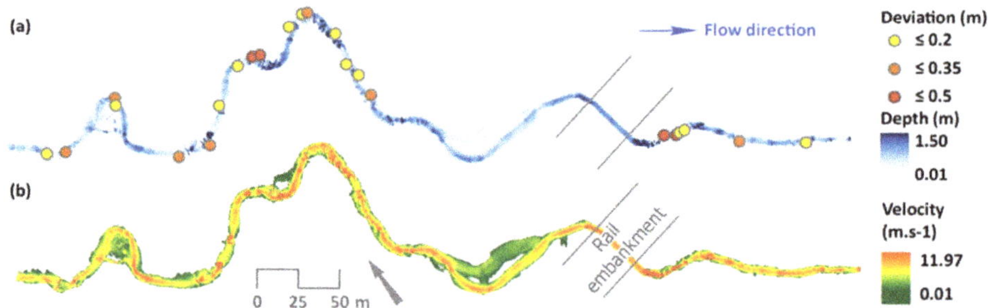

Figure 7. Raster outputs of hydraulic modelling for the $Q_{optimum}$ scenario: (**a**) water depth and (**b**) velocity. Colored circles illustrate the position of the 2014 PSI (scars) and deviations of their heights from the modelled flow.

We further investigated the effect of the hydraulic conditions in the positions of the PSIs using non-parametric Spearman's correlation. However, we did not observe any significant relationship between the PSI deviations and the velocity magnitudes for all three scenarios: $Q_{optimum}$ (r_{sp} = 0.19, p = 0.36), Q_{min} (r_{sp} = 0.08; p = 0.71), and Q_{max} (r_{sp} = 0.004; p = 0.98; Figure 8).

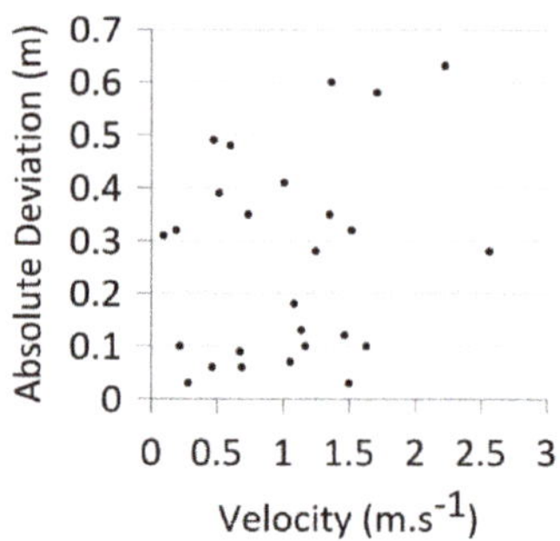

Figure 8. Plot describing the relationships between the flow velocity and absolute deviation of the PSI for $Q_{optimum}$ scenario.

3.3. Channel Stability and Relations between the Hydraulic and Sedimentologic Parameters

A positive significant relationship exists between the bed shear stress and the unit stream power calculated for all cross sections ($n = 62$) and the $Q_{optimum}$ scenario during the 2014 event ($r_{sp} = 0.65$, $p < 0.0001$) (Figure 9a). In contrast, there were no significant correlations between the parameters of τ_b and MBD ($r_{sp} = 0.05$, $p = 0.68$) or ω and MBD ($r_{sp} = -0.04$, $p = 0.74$) (Figure 9b,c). After the removal of 15 cross-sections assigned to deep pools or bedrock sections containing a limited number of coarse grains, the final relationship was significant for τ_b and MBD ($r_{sp} = 0.33$, $p = 0.024$) and remained insignificant for ω and MBD ($r_{sp} = 0.13$, $p = 0.38$).

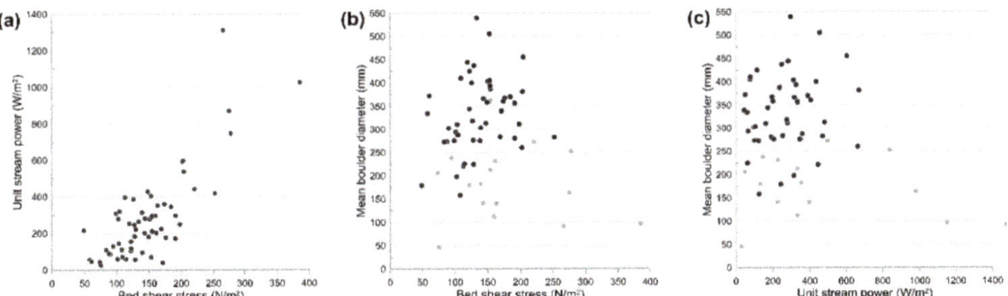

Figure 9. Plots between (**a**) the bed shear stress and unit stream power, (**b**) the bed shear stress and mean boulder diameter, and (**c**) the unit stream power and mean boulder diameter. The grey points in plots (**b**,**c**) indicate cross sections located in bedrock reaches or deep pools.

The mutual comparison of the MBD of cross sections located in depositional ($n = 8$), stable ($n = 30$) and erosional reaches ($n = 24$) showed significant differences between the groups ($p = 0.002$), when the cross sections of stable reaches indicated significantly higher values of MBD than those of locations with contemporary signs of prevailing depositional or erosional processes (Figure 10a). The analysis of the bed shear stress revealed significant differences between these groups ($p = 0.012$) when the cross sections of erosional reaches indicated significantly higher values of τ_b than those of the other groups (Figure 10b). On the other hand, we observed no significant differences among these groups in terms of ω ($p = 0.095$), although the erosional reaches were again characterised by somewhat higher values of ω than the other reaches (Figure 10c).

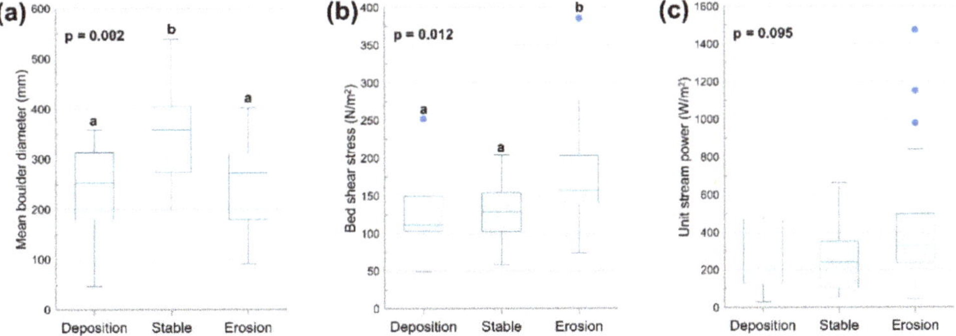

Figure 10. Boxplots of mean boulder diameter, bed shear stress and unit stream power calculated for the 2014 event ($Q_{optimum}$ scenario) in erosional, stable and depositional cross sections. The letters above the boxes show significantly different groups by Z-value test with Bonferonni corrections.

The evaluation of flow competence and the potential stability of coarse bed material in individual cross sections during the 2014 event showed that 15 of 62 cross sections (24.2%) indicated $\omega_E \geq 1$

for the $Q_{optimum}$ scenario (Figure 11). These cross sections were frequently representative of bedrock reaches with erosional tendencies with relatively fine grain-size character of the limited alluvial cover. An extraordinarily high excess was observed immediately downstream of the culvert (0.56–0.57 r. km). For the Q_{max} scenario, half of the cross sections (31 of 62; 50.0%) were perceived as those with unstable alluvial cover, still leaving the part upstream of the culvert (0.38–0.48 r. km) as relatively stable. On the other hand, the Q_{min} scenario predicted only seven unstable cross sections (11.3%) with a notable excess of critical unit stream power immediately downstream of the culvert (0.56–0.57 r. km).

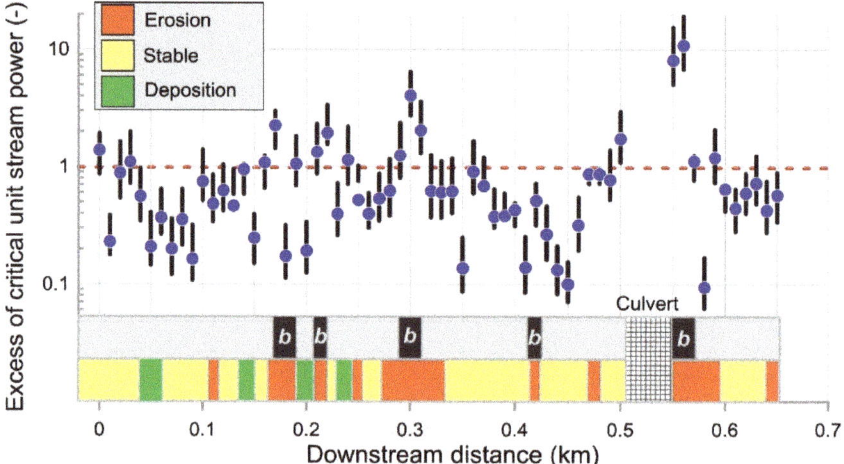

Figure 11. Downstream variations in the excess of critical unit stream power for the 2014 event (critical threshold 1 is indicated by the red dashed line) and dominant fluvial processes. The individual blue points represent values calculated for the $Q_{optimum}$ scenario for individual cross sections; the scatter lines around these points indicate the values of the Q_{min} and Q_{max} scenarios. The occurrence of bedrock reaches is indicated by black rectangles marked by 'b' letters.

4. Discussion

The dating of flood events, together with the determination of the 2014 flash flood parameters in the moderate relief of Central Europe, were introduced using the combination of dendrogeomorphic methods, 2D hydraulic modelling, and sedimentological parameters. Based on these approaches, we provide a quantitative estimation of the last flash flood event and a determination of channel reaches that tended to be (un)stable. Unlike the palaeoflood reconstructions from larger catchments, we included scarred roots as a possible PSI in the first-order catchments due to the generally lower peak flow discharges and lower volume of transported sediments. If the scarred root is a part of a stable root system and does not evidence a high rate of flexibility, it can be used not only as a helpful tool to date the time of (flood) erosion [21,60,61], but also as a PSI of recent floods.

4.1. Hydrogeomorphic Response of Flash Floods in the First-Order Catchment

The dendrogeomorphic results confirm that, despite the strong geomorphic impact of the last 2014 flash flood, there is evidence of former flood events within this catchment. We recorded the hydrogeomorphic impacts of floods in years with the occurrence of debris flows and rockfalls in the surrounding mountains (e.g., in 1991, 2006 and 2010 [62,63]). In addition, several identified years coincided with documentary data about local and/or regional flooding there (e.g., in 1971, 1977, 1997 and 2009). Polách and Gába [43] described a spatially limited downpour in July 1971 after strong antecedent precipitation, resulting in local damage to small streams within this region. This situation was likely similar to that of the last 2014 flash flood, which was considered unprecedented regarding the rainfall intensity, but the hydrogeomorphic response seems to be comparable to that of the 1971 event. Moreover, the transported parts of damaged culverts within the channel and several older scars on tree stems (Figure 2d) suggest an even higher hydrogeomorphic impact in the past. Furthermore, we identified that more than 40% of scars can be dated to the period 1951–2000, pointing to slow and progressive channel bank erosion during several flood events rather than to abrupt changes in channel geometry and the fast decay of exposed roots, as is typical for mountain regions [64].

We found no significant correlation by plotting the MBD and τ_b for all 62 cross sections, but a significant positive relationship existed after removal of fifteen cross sections assigned to deep pools and bedrock segments. This increase in statistical significance reflected the presence of the fine-grained character of the alluvial cover in these cross sections, the calibre of which did not necessarily correspond to the stream transport capacity (expressed by τ_b) during the examined flood event. Thus, we assumed more frequent transport of these fine particles during lower-magnitude events (e.g., bankfull or even lower flows). On the other hand, the set of remaining 47 cross sections had adjusted the calibre of the MBD to the calculated bed shear stresses. This positive relationship clearly pointed to the adjustment of the stepped-bed architecture during high-magnitude events towards a condition that provides the maximum possible bed stability, as was documented by previous flume experiments [65].

In general, the highest values of τ_b and ω were calculated for cross sections with erosional tendencies when compared to those of stable or depositional cross sections, although only the parameter of τ_b indicated statistically significant differences. These high values of τ_b and ω typically reflected a low width–depth ratio in erosional reaches with limited inundation ability together with the concentrated energy of flood flows. The presence of alluvial pockets of relatively fine sediments in the erosional reaches with exposed bedrock (resulting in very low values of MBD) explained our observations of the coarser bed particles in the stable reaches when one may expect local coarsening in erosional reaches with the highest calculated τ_b and, contrarily, fining in depositional reaches [23,66,67]. In addition, no differences in the MBD were observed between the erosional and stable reaches in the case of stepped-bed headwater streams based in flysch rocks, despite higher unit stream power being calculated for the erosional reaches [23]. These observations from small streams contradict the relation between the sediment coarsening and channel incision that was reported for gravel-bed rivers [68] by taking into consideration the specifics of steep headwater streams (e.g., mixture of alluvial and semi-alluvial reaches or strong effect of local lithology predicting patterns of sediment supply).

The parameter ω is frequently used to determine the capacity of a stream to mobilize specific bed grain sizes [37,69,70]. In our case, the excess critical unit stream power ω_E identified the segments with unstable alluvial cover during the examined flood event. Only one-quarter of the cross sections indicated transport of the MBD under the $Q_{optimum}$ scenario, when these cross sections were often representative of bedrock reaches with limited alluvial cover consisting of relatively fine grains. Even by applying the Q_{max} scenario, approximately 50% of the cross sections remained stable. Despite the large number of scars possibly related to intensive sediment transport and bank erosion during the 2014 flood, this event likely did not reach the critical threshold discharge to completely rework the stepped-bed character of the studied stream. Previous field measurements in steep mountain streams have perceived high-magnitude floods of up to a 50-year recurrence interval as those that mobilize most step-forming particles [71,72]. This suggests either a lower-than-expected magnitude of the 2014 flood described by local authorities or high channel bed stability of this first-order stream owing to the presence of large interlocked boulders.

4.2. Benefits and Limits of the Approach Used in the First-Order Catchment

Introducing the scarred roots as a PSI in a first-order catchment entails uncertainties similar to those in the case of scars on tree stems in larger rivers. As noted by several authors [18,36], PSIs located in straight channel reaches or on the inner side of channel bends are more suitable for peak discharge reconstructions than those of trees located on the outer side of channel bends or growing in overbank sections with dense vegetation cover. In our study, the majority of the PSIs (20) were located in the straight channel and/or inner side, reducing the overestimation of peak flow discharge. Moreover, the avoidance of root sampling in concave banks and slope failures for the PSI estimation reduces the probability of dating scars caused by bank/slope erosion. This situation may frequently occur, even during floods [73], but cannot be used as information about peak flow discharge.

A tricky situation may occur when several consecutive intense rainfalls affect the catchment. Therefore, more than one peak flow discharge may occur and thus enter as the uncertainty in hydraulic modelling. In our case, several storms generated during May 2014 could result in higher-than-average discharges [38], so we cannot exclude the possibility that several of the dated scars with lower PSI belong to some lower spring peak flows rather than to the surveyed flood of 27 May 2014. In contrast, anatomical analysis of some scarred roots revealed the position of the 2014 scar within the earlywood cells, pointing to the spring floods (Figure 3d) and thus eliminating the possible inclusion of scars during intense summer or autumn rainfall in 2014.

The results of the peak flow estimations show a high variation in SPD and high deviation in differences between the predicted water level and the observed PSI height. We hypothesize that this could be related to the small size of the catchment, where the role of the input data (PSI, DTM and roughness coefficient) is more crucial than in a large catchment (which produces large peak discharges). Similarly, Ballesteros-Cánovas et al. [18] reported the highest deviation of the estimated peak discharge in a catchment of the lowest size, while the opposite conclusions are described by Ballesteros-Cánovas et al. [57], where the authors reported much larger uncertainties for larger catchments than for smaller catchments. Following the approach of Bohorquez et al. [32], who defined the critical value of deviation between the simulated water level and the observed height of PSI as 0.2 m, for the Q_{min}, fifteen scars were within this range, followed by $Q_{optimum}$ (12 PSI) and Q_{max} (10 PSI).

Further uncertainties associated with peak discharge reconstruction using the hydraulic model are related to the DTM accuracy [58] and model calibration. Although we tried to create an accurate DTM, especially at the position and close surroundings of the PSIs, our DTM is not error free, particularly between the cross sections where the interpolation took place. Victoriano et al. [34] found a 7% difference in the estimated peak discharge of 316 $m^3 \cdot s^{-1}$ (catchment area: 5.72 km^2) when two different DTMs were compared. We call for further research to assess how the level of DTM accuracy influences the results of peak discharge reconstructions, even in small catchments. The uniform value of Manning's roughness coefficient was applied over the overall reach, and the final value was a product of the calibration exercise to single "low-flow" discharge. Indeed, differences in the calibration for the low- and high-flow discharges are well known [56] and the influence of various values of Manning's roughness coefficient on the results of the hydraulic model are described by Ballesteros-Cánovas et al. [57]. In fact, the hydraulic modelling of flood waves in large, low gradient rivers could cause serious problems with the approach used in this study (i.e., calibration to the discharge several orders of magnitude lower than simulated flood). However, in the case of small, high-gradient streams, this approach is better than the selection of the roughness value from the literature. Overall, we believe our approach reduced the uncertainty in the roughness coefficient to an acceptable level.

Although we considered the channel bed as a stable component with limited erosion and/or deposition during the 2014 flood, this does not reflect the natural condition in the modelled reach, because the majority of the riverbed is formed by poorly sorted alluvium with stripes of bedrock in a few places. This simplification could further influence the height of the PSI and the estimated discharge [14]. In addition, we did not observe any significant relationship between the velocity of the flood wave and the absolute deviation of the PSI from the modelled discharges (Q_{min}, $Q_{optimum}$, and Q_{max}), which is in agreement with the study of Ballesteros-Cánovas et al. [14]. We assume that the position of the PSI regarding the channel geometry (convex/concave/straight reach) is more important in small streams where the velocity is primarily changing with the stream gradient and channel width. Unfortunately, we were not able to test the influence of PSI deviations among convex, concave, and straight reaches due to the limited number of PSIs in each group.

We simulated subcritical flow over the overall reach, although supercritical flow occurred during the modelling in some parts. Flow conditions in steep channels can be transitional, with supercritical flow over the steps and subcritical flow in the pools. However, flow will never be supercritical throughout the simulated domain, because such flow would destabilize most channels [74]. This fact is supported by other studies [75,76], which described that, despite high velocity and extreme turbulence, flow in mountain streams is critical or subcritical, leading to the assumption that supercritical flow in natural streams does not exist for any extended length.

5. Conclusions

The estimations of peak flood discharges and the quantifications of related morphological changes and bedload transport in small streams remain a challenge, especially in the case of poorly gauged first-order catchments, despite their ability to transfer high amounts of water and sediment into low-order catchments. In this study, we presented an integrated approach based on palaeoflood reconstruction, which was performed in a first-order catchment in the moderate relief of Central Europe, to provide quantitative data about the last 2014 flash flood event and to create the chronology of former floods. The dendrogeomorphic results show the regular occurrence of flood events during at least the last 60 years, although the intensity of the last 2014 flash flood was considered to be unprecedented according to local authorities. The results of 2D hydraulic modelling are in favour of the optimum scenario of the 2014 peak flow discharge of 4.5 $m^3 \cdot s^{-1}$, which resulted in the formation of root and stem scars. Nevertheless, a sedimentological survey confirms the rather progressive development of the studied channel reach with limited ability to transport step-forming boulders during peak flow discharges similar to that in 2014 and thus relative bed stability during events of such magnitude. Despite notable geomorphic imprints and ongoing lateral erosion caused by the most recent flash flood events, with damage to infrastructure (especially in 2009 and 2014), we conclude that substantial geomorphic transition is practically excluded during the similar rainfall episodes that occurred within the last 60 years. The limited precision of the results of hydraulic modelling lead to the relatively high variability in possible peak flow in 2014 (from modelled Q_{min} to Q_{max}). This level of accuracy is still comparable to larger rivers, but may slightly change the interpretation of suggested event magnitudes in the case of smaller catchments. Therefore, further research dealing with the quality of input data into palaeoflood reconstruction (e.g., DTM accuracy and variable channel roughness) is crucial to investigation into first-order catchments.

Author Contributions: Conceptualization, R.T., S.R. and T.G.; methodology, R.T., S.R. and T.G.; validation, R.T., S.R. and T.G.; formal analysis, R.T., S.R. and T.G.; investigation, R.T., S.R. and T.G.; writing—original draft preparation, R.T., S.R. and T.G.; writing—review and editing, R.T.; visualization, R.T., S.R. and T.G. All authors have read and agreed to the published version of the manuscript.

Funding: This research received no external funding.

Acknowledgments: This research was supported by the University of Ostrava grant no. SGS02/PřF/2019–2020 "Quaternary development and contemporary state of Central European landscape in the context of geohazard action, effects of anthropopressure, and climate changes". Jakub Hrubý and Ondřej Vala are acknowledged for their help during fieldworks. English language was edited by American Journal Experts.

Conflicts of Interest: The authors declare no conflict of interest.

References

1. Baker, V.R. Geomorphological understanding of floods. In *Geomorphology and Natural Hazards*, 1st ed.; Morisawa, M., Ed.; Elsevier: Amsterdam, The Netherlands, 1994; pp. 139–156. [CrossRef]
2. Jakob, M.; Hungr, O. *Debris-Flow Hazards and Related Phenomena*; Springer: Berlin, Germany, 2005; p. 781. [CrossRef]
3. Petrow, T.; Merz, B. Trends in flood magnitude, frequency and seasonality in Germany in the period 1951–2002. *J. Hydrol.* **2009**, *371*, 129–141. [CrossRef]

4. Amponsah, W.; Marchi, L.; Zoccatelli, D.; Boni, G.; Cavalli, M.; Comiti, F.; Crema, S.; Lucia, A.; Marra, F.; Borga, M. Hydrometeorological characterization of a flash flood associated with major geomorphic effects: Assessment of peak discharge uncertainties and analysis of the runoff response. *J. Hydrometeorol.* **2016**, *17*, 3063–3077. [CrossRef]
5. Saharia, M.; Kirstetter, P.E.; Vergara, H.; Gourley, J.J.; Hong, Y. Characterization of floods in the United States. *J. Hydrol.* **2017**, *548*, 524–535. [CrossRef]
6. Strahler, A.N. Quantitative analysis of watershed geomorphology. *Trans. Am. Geophys. Union* **1957**, *38*, 913–920. [CrossRef]
7. Raška, P.; Brázdil, R. Participatory responses to historical flash floods and their relevance for current risk reduction: A view from a post-communist country. *Area* **2015**, *47*, 166–178. [CrossRef]
8. Fuchs, S.; Heiss, K.; Hübl, J. Towards an empirical vulnerability function for use in debris flow risk assessment. *Nat. Hazards Earth Syst. Sci.* **2007**, *7*, 495–506. [CrossRef]
9. Ballesteros-Cánovas, J.A.; Stoffel, M.; Corona, C.; Schraml, K.; Gobiet, A.; Tani, S.; Sinabell, F.; Fuchs, S.; Kaitna, R. Debris-flow risk analysis in a managed torrent based on a stochastic life-cycle performance. *Sci. Total Environ.* **2016**, *557*, 142–153. [CrossRef]
10. Ozturk, U.; Wendi, D.; Crisologo, I.; Riemer, A.; Agarwal, A.; Vogel, K.; López-Tarazón, J.A.; Korup, O. Rare flash floods and debris flows in southern Germany. *Sci. Total Environ.* **2018**, *626*, 941–952. [CrossRef]
11. Terti, G.; Ruin, I.; Anquetin, S.; Gourley, J.J. Dynamic vulnerability factors for impact-based flash flood prediction. *Nat. Hazards* **2015**, *79*, 1481–1497. [CrossRef]
12. Alestalo, J. Dendrochronological interpretation of geomorphic processes. *Fennia* **1971**, *105*, 1–140.
13. Stoffel, M.; Bollschweiler, M. Tree-ring analysis in natural hazards research? An overview. *Nat. Hazards Earth Syst. Sci.* **2008**, *8*, 187–202. [CrossRef]
14. Ballesteros-Cánovas, J.A.; Eguibar, M.; Bodoque, J.M.; Díez-Herrero, A.; Stoffel, M.; Gutiérrez-Pérez, I. Estimating flash flood discharge in an ungauged mountain catchment with 2D hydraulic models and dendrogeomorphic palaeostage indicators. *Hydrol. Process.* **2011**, *25*, 970–979. [CrossRef]
15. Garrote, J.; Díez-Herrero, A.; Bodoque, J.M.; Perucha, M.A.; Mayer, P.L.; Génova, M. Flood hazard management in public mountain recreation areas vs. ungauged fluvial basins. Case study of the Caldera de Taburiente National Park, Canary Islands (Spain). *Geosciences* **2018**, *8*, 6. [CrossRef]
16. Gottesfeld, A.S. British Columbia flood scars: maximum flood-stage indicators. *Geomorphology* **1996**, *14*, 319–325. [CrossRef]
17. Yanosky, T.M.; Jarrett, R.D. Dendrochronologic evidence for the frequency and magnitude of paleofloods. In *Ancient Floods, Modern Hazards: Principles and Applications of Paleoflood Hydrology*; House, P.K., Webb, R.H., Baker, V.R., Levish, D.R., Eds.; American Geophysical Union: Washington, DC, USA, 2002; Volume 5, pp. 77–89.
18. Ballesteros-Cánovas, J.A.; Stoffel, M.; Spyt, B.; Janecka, K.; Kaczka, R.J.; Lempa, M. Paleoflood discharge reconstruction in Tatra Mountain streams. *Geomorphology* **2016**, *272*, 92–101. [CrossRef]
19. Ballesteros-Cánovas, J.A.; Trappmann, D.; Shekhar, M.; Bhattacharyya, A.; Stoffel, M. Regional flood-frequency reconstruction for Kullu district, Western Indian Himalayas. *J. Hydrol.* **2017**, *546*, 140–149. [CrossRef]
20. McCord, V.A. Fluvial process dendrogeomorphology: Reconstruction of flood events from the southwestern United States using flood-scarred trees. In *Tree Rings, Environment and Humanity*; Dean, J.S., Meko, D.M., Swetnam, T.W., Eds.; Radiocarbon Department of Geosciences University of Arizona: Tucson, AZ, USA, 1996; pp. 689–699.
21. Stoffel, M.; Casteller, A.; Luckman, B.H.; Villalba, R. Spatiotemporal analysis of channel wall erosion in ephemeral torrents using tree roots—An example from the Patagonian Andes. *Geology* **2012**, *40*, 247–250. [CrossRef]
22. Chappell, P.R.; Brierley, G.J. Multi-scalar controls on channel geometry of headwater streams in New Zealand hill country. *Catena* **2014**, *113*, 341–352. [CrossRef]
23. Galia, T.; Škarpich, V. Do the coarsest bed fractions and stream power record contemporary trends in steep headwater channels? *Geomorphology* **2016**, *272*, 115–126. [CrossRef]
24. Harvey, A.M. Coupling between hillslopes and channels in upland fluvial systems: Implications for landscape sensitivity, illustrated from the Howgill Fells, northwest England. *Catena* **2001**, *42*, 225–250. [CrossRef]

25. Kavage Adams, R.; Spotila, J.A. The form and function of headwater streams based on field and modeling investigations in the Southern Appalachian Mountains. *Earth Surf. Process. Landf.* **2005**, *30*, 1521–1546. [CrossRef]
26. May, C.L.; Gresswell, R.E. Processes and rates of sediment and wood accumulation in headwater streams of the Oregon Coast Range, USA. *Earth Surf. Process. Landf.* **2003**, *28*, 409–424. [CrossRef]
27. Comiti, F.; Mao, L. Recent advances in the dynamics of steep channels. In *Gravel-Bed Rivers: Processes, Tools, Environments*; Church, M., Biron, P.M., Roy, A.G., Eds.; John Wiley & Sons: Hoboken, NJ, USA, 2012; pp. 351–377. [CrossRef]
28. Bunte, K.; Abt, S.R.; Swingle, K.W.; Cenderelli, D.A. Effective discharge in Rocky Mountain headwater streams. *J. Hydrol.* **2014**, *519*, 2136–2147. [CrossRef]
29. Lenzi, M.A.; Mao, L.; Comiti, F. Effective discharge for sediment transport in a mountain river: Computational approaches and geomorphic effectiveness. *J. Hydrol.* **2006**, *326*, 257–276. [CrossRef]
30. Lenzi, M.A.; Mao, L.; Comiti, F. When does bedload transport begin in steep boulder-bed streams? *Hydrol. Process.* **2006**, *20*, 3517–3533. [CrossRef]
31. Chiari, M.; Rickenmann, D. Back-calculation of bedload transport in steep channels with a numerical model. *Earth Surf. Process. Landf.* **2011**, *36*, 805–815. [CrossRef]
32. Bohorquez, P.; Daby, S.E. The use of one- and two-dimensional hydraulic modelling to reconstruct a glacial outburst flood in a steep Alpine valley. *J. Hydrol.* **2008**, *341*, 240–261. [CrossRef]
33. Ruiz-Villanueva, V.; Díez-Herrero, A.; Bodoque, J.M.; Ballesteros-Cánovas, J.A.; Stoffel, M. Characterisation of flash floods in small ungauged mountain basins of Central Spain using an integrated approach. *Catena* **2013**, *110*, 32–43. [CrossRef]
34. Victoriano, A.; Díez-Herrero, A.; Génova, M.; Guinau, M.; Furdada, G.; Khazaradze, G.; Calvet, J. Four-topic correlation between flood dendrogeomorphological evidence and hydraulic parameters (the Portainé stream, Iberian Peninsula). *Catena* **2018**, *162*, 216–229. [CrossRef]
35. Balasch, J.C.; Pino, D.; Ruiz-Bellet, J.L.; Tuset, J.; Barriendos, M.; Castelltort, X.; Peña, J.C. The extreme floods in the Ebro River basin since 1600 CE. *Sci. Total Environ.* **2019**, *646*, 645–660. [CrossRef]
36. Bodoque, J.M.; Díez-Herrero, A.; Eguibar, M.A.; Benito, G.; Ruiz-Villanueva, V.; Ballesteros-Cánovas, J.A. Challenges in paleoflood hydrology applied to risk analysis in mountainous watersheds—A review. *J. Hydrol.* **2015**, *529*, 449–467. [CrossRef]
37. Mao, L.; Uyttendaele, G.P.; Iroumé, A.; Lenzi, M.A. Field based analysis of sediment entrainment in two high gradient streams located in Alpine and Andine environments. *Geomorphology* **2008**, *93*, 368–383. [CrossRef]
38. Tichavský, R.; Kluzová, O.; Břežný, M.; Ondráčková, L.; Krpec, P.; Tolasz, R.; Šilhán, K. Increased gully activity induced by short-term human interventions—Dendrogeomorphic research based on exposed tree roots. *Appl. Geogr.* **2018**, *98*, 66–77. [CrossRef]
39. Povodí Odry. *Zpráva o Povodni Květen 2014 v Dílčím Povodí Horní Odry*; Povodí Odry, Státní Podnik: Ostrava, Czech Republic, 2014; p. 34. (In Czech)
40. Cháb, J.; Čurda, J.; Kočandrle, J.; Manová, M.; Nývlt, D.; Pecina, V.; Skácelová, D.; Večeřa, J.; Žáček, V. *Základní Geologická Mapa České Republiky 1: 25000, List 14–224 Jeseník s Textovými Vysvětlivkami*; Česká Geologická Služba: Prague, Czech Republic, 2004; p. 75. (In Czech)
41. Hanáček, M.; Nývlt, D.; Skácelová, Z.; Nehyba, S.; Procházková, B.; Engel, Z. Sedimentary evidence for an ice-sheet dammed lake in a mountain valley of the Eastern Sudetes, Czechia. *Acta Geol. Pol.* **2018**, *68*, 107–134.
42. Tolasz, R.; Brázdil, R.; Bulíř, O.; Dobrovolný, P.; Hájková, L.; Halásová, O.; Hostýnek, J.; Janouch, M.; Kohut, M.; Krška, K.; et al. *Atlas Podnebí ČR*, 1st ed.; Czech Hydrometeorological Institute/Palacky University: Praha, Czech Republic; Olomouc, Czech Republic, 2007; p. 256. (In Czech)
43. Polách, D.; Gába, Z. Historie povodní na šumperském a jesenickém okrese. *Severní Morava* **1998**, *75*, 3–28. (In Czech)
44. Ballesteros-Cánovas, J.A.; Stoffel, M.; St George, S.; Hirschboeck, K. A review of flood records from tree rings. *Prog. Phys. Geogr.* **2015**, *39*, 794–816. [CrossRef]
45. Ballesteros-Cánovas, J.A.; Márquez-Peñaranda, J.F.; Sánchez-Silva, M.; Díez-Herrero, A.; Ruiz-Villanueva, V.; Bodoque, J.M.; Eguibar, M.A.; Stoffel, M. Can tree tilting be used for paleoflood discharge estimations? *J. Hydrol.* **2015**, *529*, 480–489. [CrossRef]
46. Bräker, O.U. Measuring and data processing in tree-ring research—A methodological introduction. *Dendrochronologia* **2002**, *20*, 203–216. [CrossRef]

47. Vienna Institute of Archaeological Science (VIAS). *Time Table. Installation and Instruction Manual*; Version 2.1; University of Vienna: Vienna, Austria, 2005.
48. Holmes, R.L.; Adams, R.K.; Fritts, H.C. *Tree-Ring Chronologies of Western North America: California, Eastern Oregon and Northern Great Basin with Procedures Used in the Chronology Development Work Including Users Manuals for Computer Programs COFECHA and ARSTAN*; Laboratory of Tree-Ring Research, University of Arizona: Tucson, AZ, USA, 1986; p. 182.
49. Wrońska-Wałach, D.; Sobucki, M.; Buchwał, A.; Gorczyca, E.; Korpak, J.; Wałdykowski, P.; Gärtner, H. Quantitative analysis of ring growth in spruce roots and its application towards a more precise dating. *Dendrochronologia* **2016**, *38*, 61–71. [CrossRef]
50. Gärtner, H.; Lucchinetti, S.; Schweingruber, F.H. A new sledge microtome to combine wood anatomy and tree-ring ecology. *IAWA J.* **2015**, *36*, 452–459. [CrossRef]
51. Gärtner, H.; Cherubini, P.; Fonti, P.; von Arx, G.; Schneider, L.; Nievergelt, D.; Verstege, A.; Bast, A.; Schweingruber, F.H.; Büntgen, U. A Technical Perspective in Modern Tree-ring Research—How to Overcome Dendroecological and Wood Anatomical Challenges. *J. Vis. Exp.* **2015**, *97*, e52337. [CrossRef]
52. Shroder, J.F. Dendrogeomorphological analysis of mass movement on Table Cliffs Plateau, Utah. *Quat. Res.* **1978**, *9*, 168–185. [CrossRef]
53. Bladé, E.; Cea, L.; Corestein, G.; Escolano, E.; Puertas, J.; Vázquez-Cendón, E.; Dolz, J.; Coll, A. Iber: Herramienta de simulación numérica del flujo en ríos. *Revista Internacional de Métodos Numéricos para Cálculo y Diseño en Ingeniería* **2014**, *30*, 1–10. [CrossRef]
54. Webb, R.H.; Jarrett, R.D. One-dimensional estimation techniques for discharges of paleofloods and historical floods. In *Ancient Floods, Modern Hazards: Principles and Applications of Paleoflood Hydrology*; House, P.K., Webb, R.H., Baker, V.R., Levish, D.R., Eds.; American Geophysical Union: Washington, DC, USA, 2002; Volume 5, pp. 111–126.
55. Yochum, S.E.; Bledsoe, B.P.; David, G.C.; Wohl, E. Velocity prediction in high-gradient channels. *J. Hydrol.* **2012**, *424*, 84–98. [CrossRef]
56. Ferguson, R.I.; Sharma, B.P.; Hardy, R.J.; Hodge, R.A.; Warburton, J. Flow resistance and hydraulic geometry in contrasting reaches of a bedrock channel. *Water Resour. Res.* **2017**, *53*, 2278–2293. [CrossRef]
57. Ballesteros-Cánovas, J.A.; Bodoque, J.M.; Díez-Herrero, A.; Sanchez-Silva, M.; Stoffel, M. Calibration of floodplain roughness and estimation of palaeoflood discharge based on tree-ring evidence and hydraulic modelling. *J. Hydrol.* **2011**, *403*, 103–115. [CrossRef]
58. Casas, A.; Benito, G.; Thorndycraft, V.R.; Rico, M. The topographic data source of digital terrain models as a key element in the accuracy of hydraulic flood modelling. *Earth Surf. Process. Landf. J. Br. Geomorphol. Res. Group* **2006**, *31*, 444–456. [CrossRef]
59. Galia, T.; Hradecký, J. Critical conditions for the beginning of coarse sediment transport in the torrents of the Moravskoslezské Beskydy Mts (Western Carpathians). *Carpath. J. Earth Environ. Sci.* **2012**, *7*, 5–14.
60. Malik, I.; Matyja, M. Bank erosion history of a mountain stream determined by means of anatomical changes in exposed tree roots over the last 100 years (Bílá Opava River—Czech Republic). *Geomorphology* **2008**, *98*, 126–142. [CrossRef]
61. Šilhán, K.; Galia, T. Sediment (un) balance budget in a high-gradient stream on flysch bedrock: A case study using dendrogeomorphic methods and bedload transport simulation. *Catena* **2015**, *124*, 18–27. [CrossRef]
62. Tichavský, R.; Šilhán, K.; Tolasz, R. Tree ring-based chronology of hydro-geomorphic processes as a fundament for identification of hydro-meteorological triggers in the Hrubý Jeseník Mountains (Central Europe). *Sci. Total Environ.* **2017**, *579*, 1904–1917. [CrossRef] [PubMed]
63. Lenart, J.; Tichavský, R.; Večeřa, J.; Kapustová, V.; Šilhán, K. Genesis and geomorphic evolution of the Velké pinky stopes in the Zlatohorská Highlands, Eastern Sudetes. *Geomorphology* **2017**, *296*, 91–103. [CrossRef]
64. Tichavský, R. Unravelling the recent dynamics of headwaters based on a combined dendrogeomorphic approach (a case study from the Sudetes Mts., Czech Republic). *Geogr. Ann. Ser. A Phys. Geogr.* **2019**, *101*, 16–33. [CrossRef]
65. Weichert, R.B.; Bezzola, G.R.; Minor, H.E. Bed morphology and generation of step-pool channels. *Earth Surf. Process. Landf. J. Br. Geomorphol. Res. Group* **2008**, *33*, 1678–1692. [CrossRef]
66. Addy, S.; Soulsby, C.; Hartley, A.J.; Tetzlaff, D. Characterisation of channel reach morphology and associated controls in deglaciated montane catchments in the Cairngorms, Scotland. *Geomorphology* **2011**, *132*, 176–186. [CrossRef]

67. Thompson, C.J.; Croke, J.; Ogden, R.; Wallbrink, P. A morpho-statistical classification of mountain stream reach types in southeastern Australia. *Geomorphology* **2006**, *81*, 43–65. [CrossRef]
68. Zawiejska, J.; Wyżga, B.; Radecki-Pawlik, A. Variation in surface bed material along a mountain river modified by gravel extraction and channelization, the Czarny Dunajec, Polish Carpathians. *Geomorphology* **2015**, *231*, 353–366. [CrossRef]
69. Gob, F.; Bravard, J.P.; Petit, F. The influence of sediment size, relative grain size and channel slope on initiation of sediment motion in boulder bed rivers. A lichenometric study. *Earth Surf. Process. Landf.* **2010**, *35*, 1535–1547. [CrossRef]
70. Petit, F.; Gob, F.; Houbrechts, G.; Assani, A.A. Critical specific stream power in gravel-bed rivers. *Geomorphology* **2005**, *69*, 92–101. [CrossRef]
71. Lenzi, M.A. Displacement and transport of marked pebbles, cobbles and boulders during floods in a steep mountain stream. *Hydrol. Process.* **2004**, *18*, 1899–1914. [CrossRef]
72. Molnar, P.; Densmore, A.L.; McArdell, B.W.; Turowski, J.M.; Burlando, P. Analysis of changes in the step-pool morphology and channel profile of a steep mountain stream following a large flood. *Geomorphology* **2010**, *124*, 85–94. [CrossRef]
73. Šilhán, K.; Galia, T.; Škarpich, V. Detailed spatio-temporal sediment supply reconstruction using tree roots data. *Hydrol. Process.* **2016**, *30*, 4139–4153. [CrossRef]
74. Church, M.; Zimmermann, A. Form and stability of step-pool channels: Research progress. *Water Resour. Res.* **2007**, *43*, W03415. [CrossRef]
75. Jarrett, R.D. Hydraulics of high-gradient streams. *J. Hydraul. Eng.* **1984**, *110*, 1519–1539. [CrossRef]
76. Trieste, D.J. Evaluation of supercritical/subcritical flows in high-gradient channel. *J. Hydraul. Eng.* **1992**, *118*, 1107–1118. [CrossRef]

Article

The Role of Gate Operation in Reducing Problems with Cohesive and Non-Cohesive Sediments in Irrigation Canals

Shaimaa A. Theol [1,*], Bert Jagers [2], F. X. Suryadi [1] and Charlotte de Fraiture [1,3]

1 Land and Water Development for Food Security, Water Science Engineering, IHE Delft Institute for Water Education, 2611 AX Delft, The Netherlands; f.suryadi@un-ihe.org (F.X.S.); c.defraiture@un-ihe.org or charlotte.defraiture@wur.nl (C.d.F.)
2 Numerical Simulation Software, Deltares, 2629 HV Delft, The Netherlands; Bert.Jagers@deltares.nl
3 Department of Environmental Sciences, Water Resources Management, Wageningen University and Research, 6708 PB Wageningen, The Netherlands
* Correspondence: s.theol@un-ihe.org

Received: 24 October 2019; Accepted: 3 December 2019; Published: 6 December 2019

Abstract: Sediments cause serious problems in irrigation systems, adversely affecting canal performance, driving up maintenance costs and, in extreme cases, threatening system sustainability. Multiple studies were done on the deposition of non-cohesive sediment and implications for canal design, the use of canal operation in handling sedimentation problems is relatively under-studied, particularly for cohesive sediments. In this manuscript, several scenarios regarding weirs and gate operation were tested, using the Delft3D model, applied to a case study from the Gezira scheme in Sudan. Findings show that weirs play a modest role in sedimentation patterns, where their location influences their effectiveness. On the contrary, gate operation plays a significant role in sedimentation patterns. Reduced gate openings may cause canal blockage while intermittently fully opening and closing of gates can reduce sediment deposition in the canal by 54% even under conditions of heavy sediment load. Proper location of weirs and proper adjusting of the branch canal's gate can substantially reduce sedimentation problems while ensuring sufficient water delivery to crops. The use of 2D/3D models provides useful insights into spatial and temporal patterns of deposition and erosion but has challenges related to running time imposing a rather coarse modelling resolution to keep running times acceptable.

Keywords: Gezira irrigation scheme; 2D/3D models; weirs; gate operation; Delft3D

1. Introduction

Improved irrigation water management plays a crucial role in enhancing crop production for food security. Sediment control in irrigation systems is of great concern for irrigation managers and farmers because sedimentation in canals and near structures often contributes to water management problems. Further, problems of heavy sedimentation loads may jeopardize the sustainability of irrigation systems due to the high costs of cleaning canals [1]. Therefore, understanding the mechanisms underlying sediment transport in irrigation canals received substantial scholarly attention [2–7]. However, most of these studies focus on system design and relatively few take into consideration the effects of irrigation structures and the operation of gates.

Crop water requirements are not constant but change throughout the season depending on the crop growth stage. Consequently, flows in canals that supply water to fields are variable depending on the use of control structures such as gates and weirs. Structures often cause unsteady flow in the canals, even where they are designed for steady or uniform flow. The change in flow affects the sediment transport which leads to sediment deposition and erosion in different locations of the canal. Even though canals are typically designed to keep the bed free from sediments and convey sediments to fields, the improper placement and operation of gates and weirs in the absence of optimal canal operation plans may lead to deposition and erosion of sediment in canals and reduce canal performance. The impact of canal operation on sedimentation problems in irrigation systems deserves more attention in modelling studies of irrigation systems.

Examples of studies simulating the effect of canal operation on sediment transport include Reference [8] in the Sunsari Morang system in Nepal and Reference [9] in the Machai Maira Branch Canals in Pakistan. However, both studies only considered non-cohesive sediment, mostly transported as bed material. In reality, many irrigation systems deal with a mix of coarse (non-cohesive) and fine (cohesive) sediment. Dealing with sedimentation in irrigation canals becomes more complex in case of cohesive sediments due to its physiochemical properties and inter-particle forces. Most studies regarding cohesive sediment behavior have been done in rivers and estuaries [10–15]. There is a great need to study the mechanism of cohesive sediment transport in irrigation canals [16], in particular under different scenarios of gate operation.

The impact of gate operation on the cohesive sediment in the Gezira Scheme in Sudan has been studied by Reference [17]. They considered two gate operation scenarios: (1) a system in which water allocation is based on water duty and the cropped area and water is given by a fixed discharge for one week. This so-called indent system has been followed for several ago in Gezira system and (2) a system in which water supply is reduced based on the crop water requirement when sediment concentrations reach its peak. They found that the latter scenario performs best, reducing sediment deposition to 48%, primarily because the intake of the amount of sediment-laden water is reduced. References [9,17] both used a 1D model while the behavior of cohesive sediments is best reflected in 2D/3D models [18].

The main objective of this paper is to investigate the role of gate operation in reducing the amount of cohesive and non-cohesive sediment in the canals using a 2D/3D model. This paper builds on the work by Reference [1] on the sediment deposition patterns in the Gezira irrigation scheme and uses the baseline data collected by her. However, we use a mix of cohesive and non-cohesive sediment and use Delft3D, a model that can be used in 2D or 3D mode [18,19], to test different scenarios of weir height and duration of gate openings. We consider the Gezira irrigation scheme in Sudan as illustrative for many irrigation systems in semi-arid areas suffering from high sediment loads originating from river intakes.

2. Materials and Methods

2.1. Modelling

2.1.1. Model Selection

Using 1D models to study hydrodynamics in irrigation canals computationally is efficient, however, these models may not be representative in morphologic simulations. 1D-models have a simple ability to present several basic phenomena exist in nature which is usually found in three-dimensional [20,21]. On other hand, 2D or 3D models can detect sediment movement and sediments patterns near offtakes and structures in more detail and simulate deposition and/or erosion locations within the cross-section in addition to those in the longitudinal direction. While beneficial from a morphological point of view, the biggest constraint of 2D and 3D models is the long simulation time.

To explain why we selected Delft3D, we compare three well-known 2D/3D models that are able to simulate sediment transport in canals (Table 1), namely Delft3D, Telemac [22] and Mike21 [20].

Table 1. Comparison between different models.

Features Available	Delft3D	Telemac	Mike 21
Grid construction	Structured with DD * Unstructured (FM) **	FM	FM
Simulating: Cohesive sediments	Yes	No	Yes
Simulating: Non-cohesive sediments	Yes	Yes	Yes
Open-source	Yes	Yes	No
RTC ***	Yes	No	Yes

* DD: domain decomposition, ** FM: flexible mesh, *** RTC: real-time control.

Structured grids with rectangular cells and areas are computationally efficient if aligned with long straight canals. However, in reality, a canal system consists of main canals and branches, with large 'empty' areas in between (Figure 1). These 'empty' inactive parts, which fall outside the area of interest, render the structured grids inefficient since the model domain includes large inactive parts taking up unnecessary computation time. Unstructured grids (or flexible mesh) can model irregular shapes that only include the active parts of the channel networks. However, these grids mostly consist of triangular cells which are not conducive for long canals, since they cannot be stretched in stream direction, leading to a higher number of triangular cells and hence longer simulation time. One possible solution is to use an unstructured grid with quadrilateral cells aligned with the flow direction along the canal (e.g., Delft3D FM or Mike FM). A more efficient solution is to use a structured grid with the domain decomposition (DD) tool available in Delft3D. This tool allows to divide the grid in separate parts that can be modelled and compiled. In this way inactive parts can be excluded, substantially reducing simulation time. The latter method, combining structured grids and domain decomposition, was used in this paper.

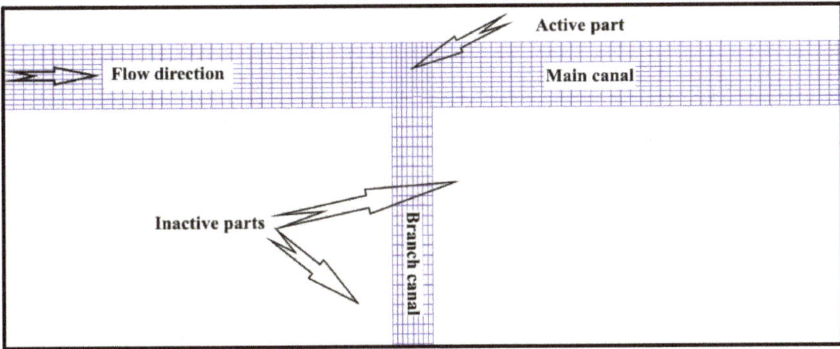

Figure 1. The active and inactive parts in the computational domain.

The Real-Time Control (RTC) tool enables changing weirs and gate settings during the simulation. This property can be activated in the Delft3D-FLOW input file by using the Rtcmod keyword. It allows simulating canal operation in which gates are opened and closed multiple times during the irrigation season. The morphological factor (Morfac or MF) feature in Delft3D further shortens the model running time and enables predictions of the morphologic developments in the medium term (months or seasons).

Comparing the three models Delft3D has all features necessary to simulate the effect of gate operation under scenarios of cohesive and non-cohesive sediment and their interaction. It is also open source and can handle non-steady flows.

Delft3D has been validated by Reference [21] for a series of simplified theoretical, laboratory and full-scale test cases. Furthermore, it was also validated against the results of prototype-scale measurements. A big advantage of numerical simulations is that there is no need to apply scale factors [21], unlike physical morphological models where sediment scaling is a major problem. Numerical morphological models can be tested directly against both the laboratory observations and prototype-scale observations.

So far the Delft3D model has been used primarily for rivers [23–26] and for estuaries [21,27] in 2D and/or 3D modes. The model has been used by Reference [16] for irrigation canals in both 2D and 3D. Running the model in 2D mode is to ensure better representation of the sediment processes and the large scale behavior with an acceptable simulation time period. Running the model in 3D mode provides information about the vertical and gives more details near structures.

2.1.2. Model Equations

Delft3D-FLOW solves the shallow water equations (derived from Reynolds-averaged Navier-Stokes equations for incompressible fluid under shallow water and Boussinesq assumptions). In the 3D simulations we have applied the k-ε turbulence closure model for vertical shear and a constant eddy viscosity model for horizontal shear. The sediment transport is split into both bedload and suspended load transport of non-cohesive sediments and suspended load of cohesive sediments; for the suspended load the advection-diffusion equation is solved with source and sink terms for sediment entrainment and deposition. The morphological change is computed based on an explicit integration of the bedload fluxes entering or leaving the grid cell plus the net deposition and entrainment.

For cohesive sediment fractions, the fluxes between the water phase and the bed are calculated with Partheniades-Krone formulations [28] for deposition and erosion [29]:

Erosion formula:

$$\mathbf{E^l} = \mathbf{M^l} \times \mathbf{S}\left(\boldsymbol{\tau}_{\mathbf{cw}}, \boldsymbol{\tau}^{\mathbf{l}}_{\mathbf{cr,e}}\right) \tag{1}$$

where:

$\mathbf{E^l}$: Erosion flux (kg $m^{-2}s^{-1}$);
$\mathbf{M^l}$: User-defined erosion parameter (kg $m^{-2}s^{-1}$);
$\mathbf{S}(\boldsymbol{\tau}_{\mathbf{cw}}, \boldsymbol{\tau}^{\mathbf{l}}_{\mathbf{cr,e}})$: Erosion step function;
$\boldsymbol{\tau}_{\mathbf{cw}}$: Maximum bed shear stress (N/m^2);
$\boldsymbol{\tau}^{\mathbf{l}}_{\mathbf{cr,e}}$: User-defined critical erosion shear stress (N/m^2).

$$\mathbf{S}(\boldsymbol{\tau}_{\mathbf{cw}}, \boldsymbol{\tau}^{\mathbf{l}}_{\mathbf{cr,e}}) = \begin{cases} \left(\frac{\tau_{cw}}{\tau^{l}_{cr,e}} - 1\right) \text{ when } \boldsymbol{\tau}_{\mathbf{cw}} > \boldsymbol{\tau}^{\mathbf{l}}_{\mathbf{cr,e}} \\ \mathbf{0} \quad \text{when } \boldsymbol{\tau}_{\mathbf{cw}} < \boldsymbol{\tau}^{\mathbf{l}}_{\mathbf{cr,e}} \end{cases} \tag{2}$$

Deposition formula:

$$\mathbf{D^l} = \mathbf{W^l_s} \times \mathbf{C^l_b} \times \mathbf{S}\left(\boldsymbol{\tau}_{\mathbf{cw}}, \boldsymbol{\tau}^{\mathbf{l}}_{\mathbf{cr,d}}\right) \tag{3}$$

$$\mathbf{C^l_b} = \mathbf{C^l}(\mathbf{Z} = \frac{\Delta \mathbf{Z_b}}{2}, \mathbf{t}) \tag{4}$$

where:

$\mathbf{C^l_b}$: Average sediment concentration in the near bottom computational layer (kg/m^3);
$\mathbf{D^l}$: Deposition flux (kg $m^{-2}s^{-1}$);
$\mathbf{S}(\boldsymbol{\tau}_{\mathbf{cw}}, \boldsymbol{\tau}^{\mathbf{l}}_{\mathbf{cr,d}})$: Deposition step function, $\boldsymbol{\tau}_{\mathbf{cw}}$ Maximum bed shear stress (N/m^2);
$\boldsymbol{\tau}^{\mathbf{l}}_{\mathbf{cr,d}}$: User-defined critical deposition shear stress (N/m^2);
$\mathbf{W^l_s}$: Fall velocity (hindered) (m/s);
$\mathbf{Z_b}$: Depth down to the bed from a reference height (positive down) (m);

ΔZ_b: Thickness of the bed layer (m).

$$S(\tau_{cw}, \tau^l_{cr,d}) = \begin{cases} (1 - \frac{\tau_{cw}}{\tau^l_{cr,d}}) \text{ when } \tau_{cw} > \tau^l_{cr,d} \\ 0 \text{ when } \tau_{cw} < \tau^l_{cr,d} \end{cases} \tag{5}$$

The high value for $t_{cr,\,d}$ causes $S(\tau_{cw}, \tau^l_{cr,d})$ to be effectively equal to 1, therefore we can neglect this term from Equation (4) and the equation will be as below:

$$D^l = W^l_s \times C^l_b. \tag{6}$$

For the computation of the behavior of non-cohesive sediments, we have applied the approach developed by Reference [30]. Van Rijn [30] predicts the sediment transport as bed load and suspended load. A reference height (a) is used to differentiate between these loads; sediments transport below this reference height is treated as bedload transport and above it as suspended-load transport. The layer situated directly above the Van Rijn reference height is called the kmx-layer. The sediments in this layer which move between bed and water flow are modelled using sink and source terms.

The advection-diffusion equation solves the sink term implicitly, whereas the source term is solved explicitly. The concentration and concentration gradient at the bottom of the kmx-layer needs to be approximated, in order to determine the sink and source terms. The model assumes a standard Rouse-profile between the reference height (a) and the center of the kmx-layer:

$$C^l = C_a{}^l \left[\frac{a(h-z)}{z(h-a)}\right]^{A^l}. \tag{7}$$

In the Delft3D model, the reference height can be represented as:

$$a = \min\left[\max\left\{f \times k_s, 0.01\,h\right\}, 0.2h\right], \tag{8}$$

where;

$A^{(l)}$ = Rouse number
a = Van Rijn's reference height (m);
$C^{(l)}$ = concentration of sediment fraction (l) (kg/m^3);
$Ca^{(l)}$ = reference concentration of sediment fraction (l) (kg/m^3);
f = user defined proportionality factor (–) (equals to 1);
h = water depth (m);
k_s = roughness height (m);
z = elevation above the bed (m).

Sink and source terms of the kmx-layer are subsequently calculated as follows:

Erosion formula:

$$E^{(l)} = \frac{\propto_2{}^l \varepsilon_s{}^l C^l_a}{\Delta z} - \frac{\propto_2{}^l \varepsilon_s{}^l C^l_{kmx}}{\Delta z} \tag{9}$$

where the first term is ($\mathbf{source}^{\,l}_{erosion}$) and the second term is ($\mathbf{sink}^{\,l}_{erosion}$).

Deposition formula:

$$D = \propto_1{}^l C^l_{kmx} w_s{}^l \tag{10}$$

$$(\text{source}^{\,l}_{deposition}) = \propto_2{}^l C^l_a \left(\frac{\varepsilon_s{}^l}{\Delta z}\right) \tag{11}$$

$$(\text{sink}^{\,l}_{deposition}) = \left[\propto_2{}^l \left(\frac{\varepsilon_s{}^l}{\Delta z}\right) + \propto_1{}^l w_s{}^l\right] C^l_{kmx} \tag{12}$$

where:

C_a^l: Reference concentration of sediment fraction (l) (kg/m^3);
C_{kmx}^l Average concentration of the kmx cell of sediment fraction (l) (kg/m^3);
$w_s{}^l$: Settling velocity (m/s);
Δz: Difference in elevation between the center of the kmx cell, Van Rijn's reference height: Δz = zkmx − a (m);
$\propto_1{}^l$: First correction factor for sediment concentration (−);
$\propto_2{}^l$: Second correction factor for sediment concentration (−);
$\varepsilon_s{}^l$: Sediment diffusion coefficient evaluated at the bottom of the kmx cell of sediment fraction (l) (−).

The reader is referred to Section 2.3.2 for the parameter settings used in this study.

2.2. Case Study

The Gezira Scheme is the largest irrigation scheme in Sudan, serving 8,800,000,000 m^2 and taking water from the Blue Nile River which carries large amounts of sediment. Since its construction in 1920, the scheme suffers from sediment accumulation in the canals, representing a big challenge for the operation and maintenance. The annual costs of desilting were estimated at aroundUS $12 million [31]. The irrigation system consists of a network of main, major, minor and field canals. Two canals were selected for this study—the Zananda Major canal and Toman Minor canal, fed by the Zananda Canal. The Zananda canal is the first canal that takes water from the Gezira Main Canal by gravity irrigation [1] (Figures 2 and 3).

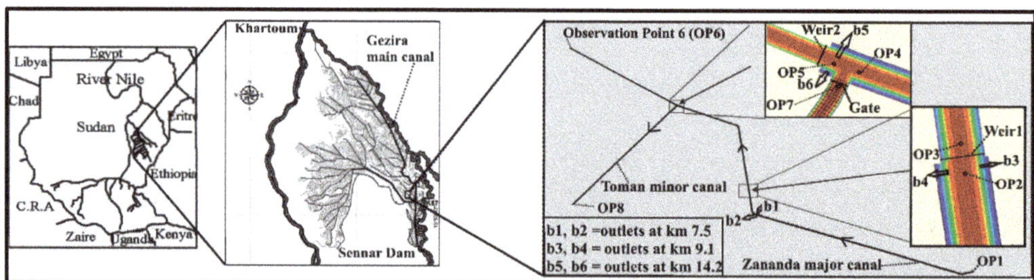

Figure 2. Scheme of the Zananda Major Canal and Toman minor canal, labels b1–b6 indicate the locations of the outlets while labels OP1–OP8 indicate the locations of the observation points.

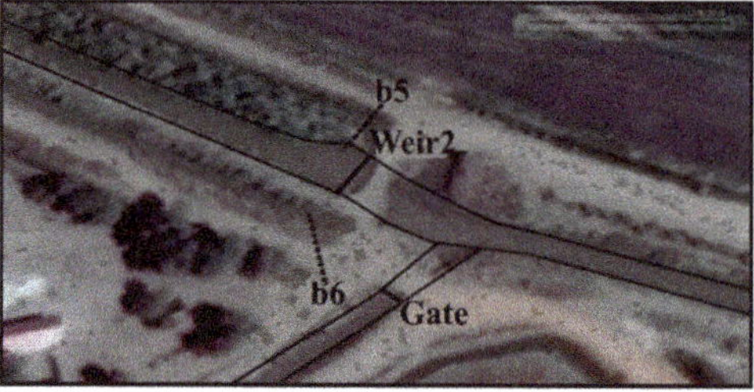

Figure 3. Location of Toman minor canal (Google Earth).

The location of the off-take is 14°01′42″ N and 33°32′33″ E. The Zananda canal is 17 km long and provides water to seven minor canals in which irrigate about 8520 ha, one of these minor canals is Toman Minor canal. In Figure 2 the other minor canals are presented as outlets named b1, b2, b3, b4, b5 and b6. In the selected area, 75% of the sediment is silt and clay with grain sizes less than 0.063 mm (considered as cohesive sediment); the remaining 25% is fine sand as mentioned in the analysis of the bed materials done by Reference [1]. Reference [32] concluded that sediment is transported in suspension, based on sediment analysis. More details regarding the canal geometry and hydraulic parameters are presented in Table 2.

Table 2. Geometric data.

Property	Major Canal	Minor Canal
Canal length	17 km	6.5 km
Canal width	4 m from (0–9.1) km 3 m from (9.1–17) km	2 m
Canal bank height	5 m	4 m
Roughness (n)	0.029	0.029
Slope	0.0001	0.00005
Side slope	1:2 from (0–14.2) km1:1 from (14.2–17) km	1:1.5
Structures	Weir 1 and weir2 with a height of 0.3 m, length of 3 m.	Gate fully opened

2.3. Model Setup

2.3.1. Grid Construction, Bathymetry and Other Parameters Assumptions

We constructed a grid for the Zananda major canal of 17 km long and 4 m wide from the inlet till the first contraction after the first weir where the width becomes 3 m till the end of the major canal. The grid for the Toman minor canal is 6.5 km long and 2 m wide with eight observation points as depicted in Figure 1. We followed the grid quality criteria of Delft3D with the orthogonality = 0.05 (i.e., cells are almost perpendicular to each other which proved the most suitable grid setting for reducing the Courant number that causes model instability in the course of the simulation) and smoothness = 1.2 for both M and N directions. The grid for the major canal contains 1125 and 14 cells in the M and N-direction respectively. The grid for the minor canal is 581 by 6 cells. For the 3D simulations, we have used five equidistant sigma-layers. The grid size for the long straight canal is 18 m. The more accurate mesh size of about 1 meter is used in important areas such as near the structure and near the minor canal. To reduce the computation time, the network domain is divided into the major grid domain and minor grid domain. The simulation results for both domains are compiled using the Domain Decomposition tool (DD), which reduces the simulation time to 40% of the total simulation time. Using the field data presented by Reference [1], we took the elevation of the upstream part of the Zananda major canal as a starting point and built the bathymetry of the remainder of the canal based on slope, length and canal geometry such as bed width, side slopes and roughness. We estimate the design discharge of the major canal as 5.5 m^3/s, based on available data and field reports [1]. When canals are free of sediment, the flow is assumed to be as steady non-uniform flow during the time step (i.e., flow rates of the outlets do not change with time but depth of water varies with the location along the canal). For hydrodynamic reasons, the model is first executed without sediment to get a steady-state flow condition and check some crucial flow parameters such as velocity, water levels and the bed shear stress which is important in calculating the sedimentation and erosion of cohesive sediments. The steady-state flow condition was validated with results from the DUFLOW model following the method described in Reference [16]. Due to the absence of the detailed field data regarding velocity and bed shear stress, we compare Delft3D results of water level to the DUFLOW model, where the DUFLOW model was previously calibrated by Reference [1] against field data. Our water levels match those of DUFLOW within 5 cm. Reference [1] validated the DUFLOW model against field data, the water level of DUFLOW match the field data within 3 cm.

After adjusting uniform bed roughness and wall roughness for hydrodynamic parameters, we test the scenarios assuming the entrance of sediment at a constant rate, evaluating the morphological changes in the canal bed after a simulation time of three months and comparing the results to the initial bed levels.

2.3.2. Model Runs

The model was run for a simulation time of three months using a time-step of 0.6 s and a morphological factor (MF) of 10 using both 2D and 3D modes. The results of the 2D and 3D simulations look identical. In this paper, the graphs are based on the 3D simulations. The small-time step is chosen to avoid the Courant number exceeding 1.0 which would destabilize the model. The MF enables the computation of the morphodynamics together with the hydrodynamics. This MF was used to speed up the changes of bed morphology by 10 times per time step, which reduces the time by a factor of 10. Thus, simulating the effective morphological changes over 3 months requires only a simulation period of 9 days. The MF approach simplifies the model setup and operation in comparison with other approaches [33,34], in this way, the Delft3D model was capable to predict the changes in canal morphology over a long time span within small simulation time.

Two different computers were used in this study, One has simple specifications (dual-core Hp ProBook 6570b) and the other is a higher-performance computer is (quad core Hp Z Book15 G3); the latter reduced the simulation time by 40%. The CPU time was 3.5 days and 2days for 3D and 2D modelling respectively.

In this study the maximum concentration is assumed to be (C_b^l = 3 kg/m^3 or 3000 ppm) for cohesive sediments, this concentration lies in the range of typical concentrations which are relevant for the Gezira Scheme [1]. As input data in Delft3D, the settling velocity (W_s) is set to 0.12 mm/s which corresponds to the Krone [35] formula for the aforementioned concentration. The value of the critical shear stress for erosion ($\tau_{cr,e}^l$) is set to 1 N/m^2. For the erosion parameter M ^{1} the default value of 0.0001 kg m^{-2}s^{-1} is used. For the critical shear stress for deposition, the authors used $t_{cr,d}$ = 1000 N/m^2.

The initial conditions are set as follows: water level = 34 m + (MSL) Mean Sea Level. The initial sediment concentration for each type of sediment equal to 0 kg/m^3, the canal bed is erodible (movable) limited by the available amount of sediment. The initial sediment layer assumed to be 20 cm consist of 50% sandy material and 50% muddy material. The boundary conditions are: discharge equals 5.5 m^3/s with sediment concentration is 3000 ppm and 100 ppm for cohesive and non-cohesive, respectively, as un upstream boundary condition. The downstream boundary condition for each canal was taken as Q-h relation which is based on the canal characteristics. For the other branches b1, b2, b6, they have been considered as only outflow, where each branch drags 0.5 m^3/s of water from the major canal.

Other parameter values regarding non-cohesive sediment are D_{50} = 100 μm (fine sand) with a specific density of 2650 kg/m^3. For the transport of non-cohesive sediment, we use the Van Rijn formula [30].

2.3.3. Morphological Comparison

Reference [1] collected sediment data in the Gezira irrigation system in Sudan for the years 2011 and 2012. Given the difficulty of getting actual field data, we used the data from Reference [1] to verify the Delft3D model.

Figure 4 shows that the results obtained from Delft3D are qualitatively comparable with the field data measured by Reference [1], giving confidence that the Delft3D model is able to replicate field conditions. Delft3D results' are also qualitatively comparable with the simulation results by Reference [1]. Observed differences in bed level are partly explained by differing modelling assumptions, the two simulation models use a different numerical technique. Reference [1] used a series of quasi-steady-state computations in her model, whereas Delft3D a dynamic model showing changes in time for flow characters like velocity, water depth and bed level changes due to sediment transport over time.

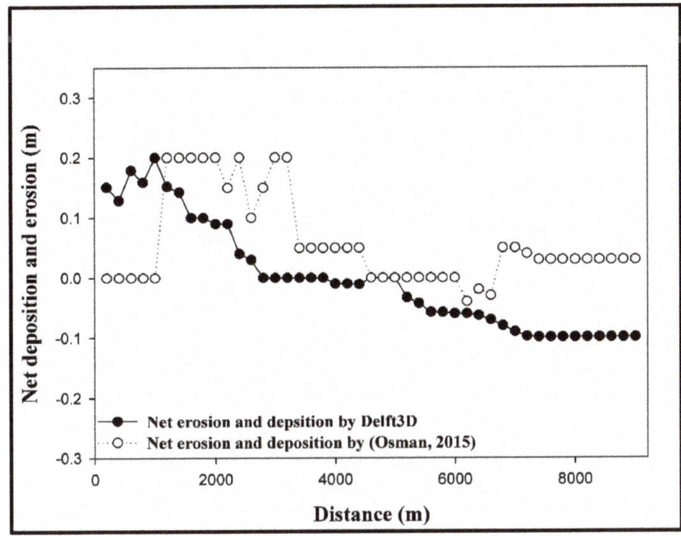

Figure 4. Comparing results from the Delft3D model with field and simulation data obtained from Reference [1].

2.4. *Scenarios*

To assess the effect of operation and structures, we tested different scenarios regarding gate opening and height of weirs; weirs and gates influence the Delft3D computation by changing the through-flow area (2D) and (partially) blocking specific layers in the 3D model. The structures do not directly affect the sediment transport; the sediment transports are affected indirectly by the changed flow patterns. Gate opening and weir heights vary according to the scenario (Table 3) while sediment concentration and other parameters are kept constant during the simulation. Regarding the gate operation, the Real-time Control (RTC) module is applied because this tool allows us to open the gate fully, partially or fully close during the simulation.

Table 3. Scenarios in the study.

Scenario	Description	Remarks
1. Reference case	Full open gate and fixed weirs' heights.	Gate fully opened; w1, w2 with fixed height at 0.3 m.
2. Effect of the weirs	a. Setting of the upstream weir height b. Setting of the downstream weir height c. Setting of both weirs.	Gate fully opened; lowering or removing, raising the weir (0 m, 0.6 m).
3. Effect of the gate	a. gate setting with constant openings	Lowering the gate (0.2 m–0.8 m); weir1 and weir2 with fixed height at 0.3 m.
	b. gate setting with variable openings	Operation plans for the gate; weir1 and weir2 with fixed height at 0.3 m.

3. Results

3.1. *Reference Scenario*

In the reference case, the gate in the Toman minor canal is fully opened while in the Zananda major canal, the height of both weirs is fixed at 0.3 m. During the simulation, sediments start depositing in the upstream part of the major canal (Figure 5). The cohesive sediment deposit gradually, distributed over the major canal while the non-cohesive sediments deposit mostly in the upstream of the canal. Because of the mixed sediment and interaction between cohesive and non-cohesive particles, the sedimentation pattern differs from the case of pure non-cohesive sediment. In the case of pure non-cohesive sediment, the heavy non-cohesive particles would rapidly deposit in the upstream the major canal. In the case of mixed sediment, some non-cohesive sediments are transported all the way to the downstream of the canal due to the interaction with the suspended cohesive particles.

In the canal stretch between 0 and 8 km, sediment deposition increases with time, with most accumulation (1.5 m) in the upstream of the major canal, the deposition in the first 8 km of the major canal consists mostly of non-cohesive sediments (Figure 5). With sufficient flow velocity, the transport capacity is sufficient to convey the sediments along the major canal. Just after 8 km in the vicinity of the first two outlets (b1 and b2) sediment locally accumulate. In the Delft3D model, we specify the amount of water drawn by the outlets (0.5 m^3/s); the amount of suspended sediment removed by the outlet cannot be specified—it equals the amount of water withdrawn times the locally computed sediment concentration. After the outlet with less water remaining in the major canal, velocity and hence sediment transport capacity reduces leading to sediment deposition.

The sediment deposition gradually decreases until 9.1 km where the first weir and two outlets (b3 and b4) are located. One would expect the deposition to increase again due to the low velocity. However, due to the canal contraction close to these outlets the flow velocity increases. These two opposite effects more or less even each other out and the velocity remains approximately equal. As a result, the sediments continue to be moving downstream to 14.2 km. Just after 14.2 km, there is a big canal contraction causing erosion in the canal section upstream of the second weir due to the acceleration of flow velocity. The Toman minor canal and the last two outlets (b5 and b6) are located upstream of the second weir at 14.2 km. The outlets should decrease the velocity since they draw water from the major canal but because of the big canal contraction, velocity increases and erosion occurs. Thereafter the sediments continue to be transported till the end of the major canal.

In the minor canal, the gate is fully opened so the minor canal gets water carrying mostly cohesive sediments which deposit in the upstream of the minor canal (deposition reaches to 0.7 m). Since there is no structure disturbing their movement, the sediments are transported along the minor canal till the end (Figure 5), where the profile of the bed level shown is along the centerline (which is typically the deepest point of the cross-section). (For more details, see the PowerPoint contains movies showing the updating of morphology within the cross-section at different locations in the major canal. The link for the supplementary data is: https://drive.google.com/drive/folders/1Wlw9SQSqGgRBLxyIoQ5FOjqdYAmxXFVV?usp=sharing).

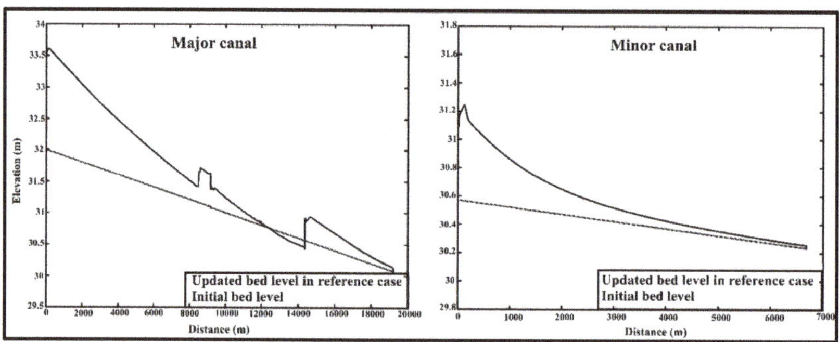

Figure 5. Sedimentation and erosion of sediments in the reference case.

The flow velocity gradually reduces along the minor and major canals except above the weirs explaining the sedimentation and erosion patterns along the canals and within the cross-section (Figure 6).

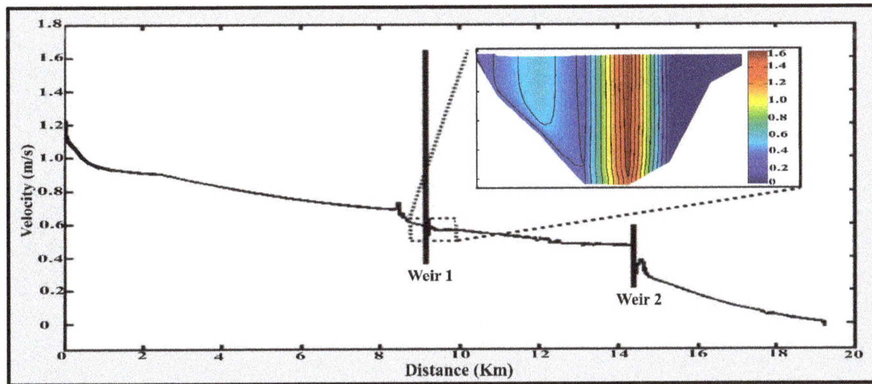

Figure 6. Flow velocity along the Zananda major canal and in the cross-section near the first weir.

The flow velocity (averaged over the cross-section) gradually reduces along the major canal (Figure 6). Within the cross-sections along the canal, the velocity distributions differ. For example, at the first weir, the average velocity is 0.6 m/s. The maximum velocity of 1.6 m/s occurs in the middle of the cross-section, while the velocity is at the sides is much less with 0 m/s close to both sidewalls. The left side has a higher velocity than the right side due to the asymmetric shape of the canal contraction and offtakes nearby. (For more details, see the PowerPoint contains other movies showing the behavior of velocity in the system near the diversion to minor canal. The link for the supplementary data is: https://drive.google.com/drive/folders/1Wlw9SQSqGgRBLxyIoQ5FOjqdYAmxXFVV?usp=sharing).

Also along the minor canal, the average flow velocity drops from 0.5 m/s upstream of the gate to 0.21 m/s at the downstream (Figure 7). Likewise, the flow velocity distributions within the cross-section vary with the highest velocities in the middle and lower velocities on both sides due to the roughness of the wall. In the downstream of the canal, the velocity distribution is logarithmic where higher velocities at the top layer of water and lower velocity near the bed. In the upstream near the gate, the water flows underneath the gate and the top layer velocity became less than the bottom layer velocity (Figure 7).

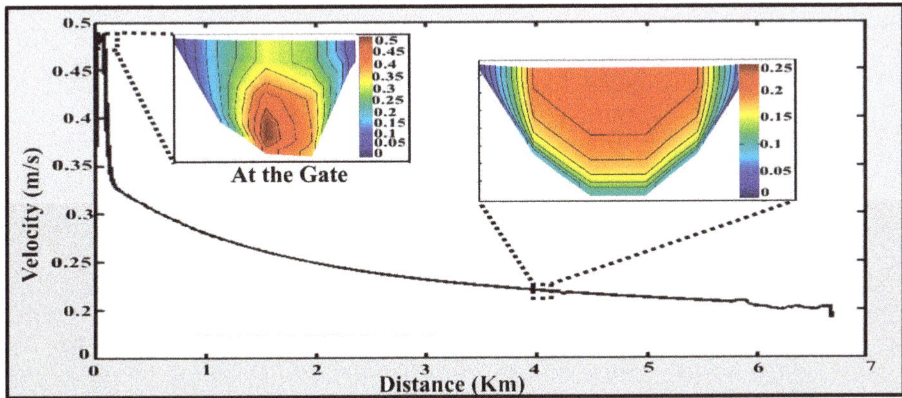

Figure 7. Velocity distribution in the Toman minor canal at different cross-sections.

Due to differences in velocity distribution, sediment is distributed in an asymmetric way within the cross-sections of the major and minor canal. The sediment behavior is influenced by multiple factors such as the velocity, widening and contractions of the canals and bed shear stress.

Figure 8 displays the difference in the deposition pattern between cohesive and non-cohesive sediments along the Zananda major canal. While cohesive sediments are gradually depositing along the major canal, the non-cohesive sediments are rapidly depositing in the first kilometers upstream of the canal with pronounced peaks and troughs in concentration near the weirs at 9 km and 14 km and canal contraction at 12.5 km. Non-cohesive sediments are deposited in the middle of the cross-section more than at both sides while the cohesive sediments are depositing almost equally in the middle and on both sides. In the case of pure cohesive sediments entering the irrigation system, most suspended sediments would be carried with the flow till the end of the major and minor canal. However, because in this case, the sediment is a mix of cohesive and non-cohesive, due to interaction with the heavier non-cohesive particles, some of the suspended cohesive particles start depositing in the upstream and middle of the canal stretches.

Figure 9 displays the difference in the deposition pattern between cohesive and non-cohesive sediments along the Toman minor canal. The same behavior will be there, where the cohesive sediments are gradually depositing along the minor canal, while the non-cohesive sediments are rapidly depositing at the beginning of the minor canal near the gate.

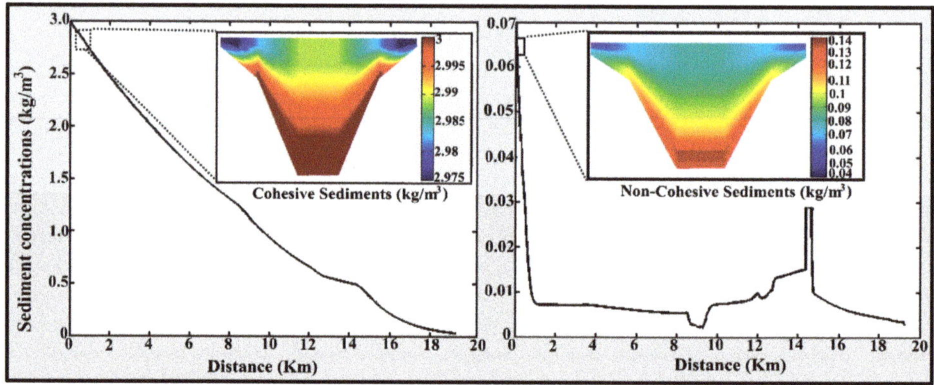

Figure 8. Distribution patterns of cohesive and non-cohesive sediments in Zananda major canal at different cross-sections.

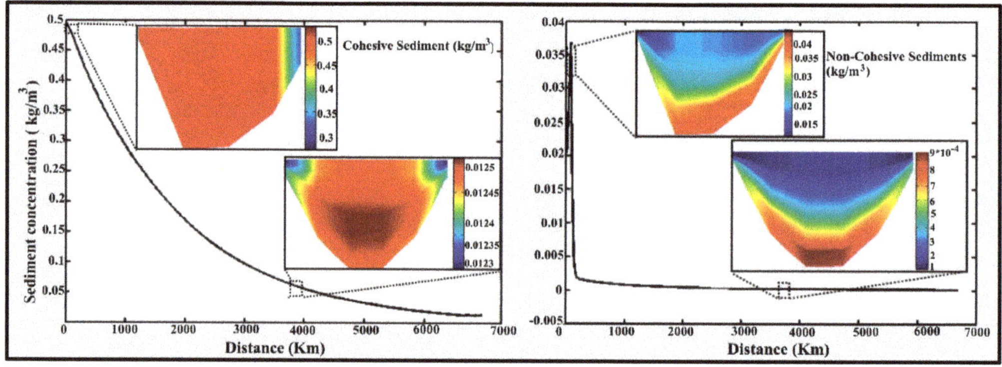

Figure 9. Sediment distribution patterns of cohesive and non-cohesive sediments in the Toman minor canal at different cross-sections.

The cohesive sediment concentration in the minor canal is much higher than the non-cohesive sediment concentrations. This is the opposite of the situation in the major canal.

The deposition pattern between cohesive and non-cohesive sediments is different in the minor canal. Where cohesive sediments are gradually depositing along the minor canal, vice versa for the non-cohesive sediments. The pattern of depositing of cohesive and non-cohesive sediments within the cross-section is the same as in the major canal, with the highest concentrations at the bottom and sides.

At the diversion to the minor canal, the velocity at the left side of the canal is reduced due to less water. Because of the subsequent reduction in velocity, a considerable amount of both non-cohesive and cohesive sediment is deposited, especially upstream the gate in the minor canal (Figure 10).

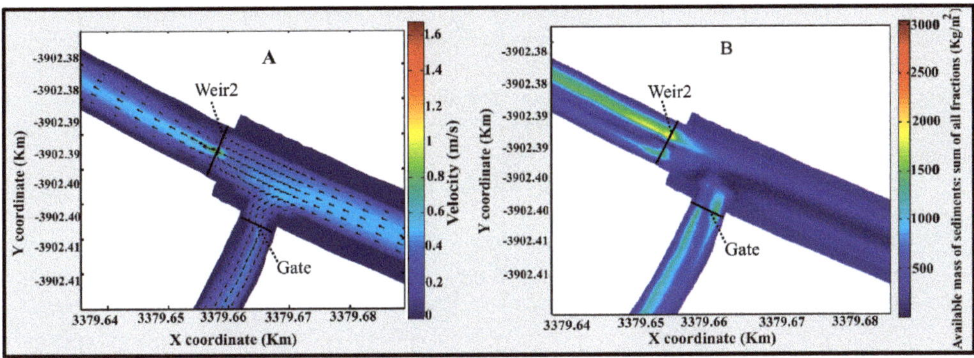

Figure 10. The relation between the velocity (**A**) and the amount of sediments (**B**) deposited at the diversion to the minor canal.

Figure 10 illustrates the effect of the velocity on the deposition and the transportation pattern of sediments. Panel (A) shows the reduced velocities at the right side of the canal after the diversion and the contraction. Panel (B) shows a higher deposition in these locations.

Acknowledging the asymmetric deposition patterns in the figures above, it can be noted the importance of using 2D/3D models to simulate sediment transport in the irrigation systems. Using Delft3D in this study proved useful in showing where the sediment is eroded or deposited and distributed along and within the cross-sections. Furthermore, Delft3D can show which kind of sediment is deposited where and in which quantities (Figure 11).

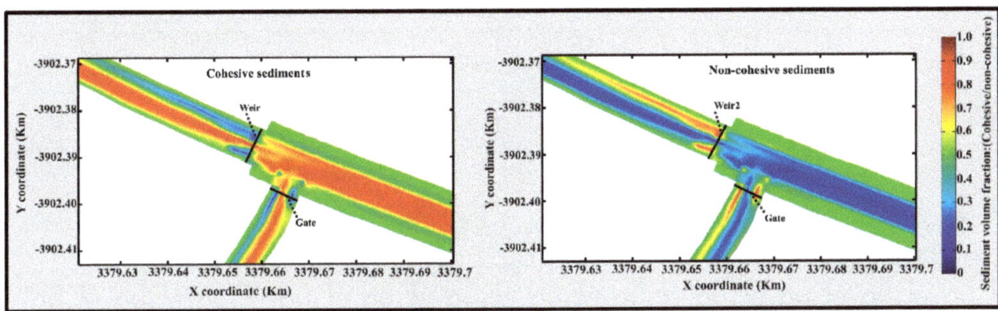

Figure 11. The difference in sediment distribution between the cohesive and non-cohesive sediments at the diversion.

The deposition and distribution of both kinds of sediments are different where the large amounts of cohesive sediments pass through the minor canal (Figure 11). On the other hand, less non-cohesive sediments enter the minor canal (Figure 11) since it is rapidly deposited in the upstream part. (For more details, see the PowerPoint which contains movies showing the difference in distributions between the cohesive and non-cohesive sediment in the system, also movies showing the difference in the distribution of both sediments within the cross-section in the minor canal. The link for the supplementary data is: https://drive.google.com/drive/folders/1Wlw9SQSqGgRBLxyIoQ5FOjqdYAmxXFVV?usp=sharing).

3.2. Scenario 2: Effect of Weirs 1 and 2

3.2.1. Effect of the Upstream Weir (Weir 1)

To see the effect of weir 1, we compare the sedimentation while reducing or raising the crest height of the weir. Raising the weir to 0.6 m increases the deposition slightly because of the obstruction of the water flow and creation of a backwater curve which leads to an increase in the water level and water depth. Combined with a constant discharge this leads to reduced velocity, reduced sediment transport capacity and hence more sediment deposition. This effect is noticeable only upstream of the weir and in the upstream part of the major canal, with a negligible effect on the downstream part of the major canal (Figure 12).

Lowering or removing weir 1 leads to reduce deposition because water moves freely without structures disturbing its movement so the sediment transport capacity is sufficient to move sediments. The effect is noticeable upstream of the weir and in the upstream part of the major canal, with a negligible effect in the downstream part of the major canal. The effect of the lowering or increasing of the weir height has little effect on the minor canal (Figure 12).

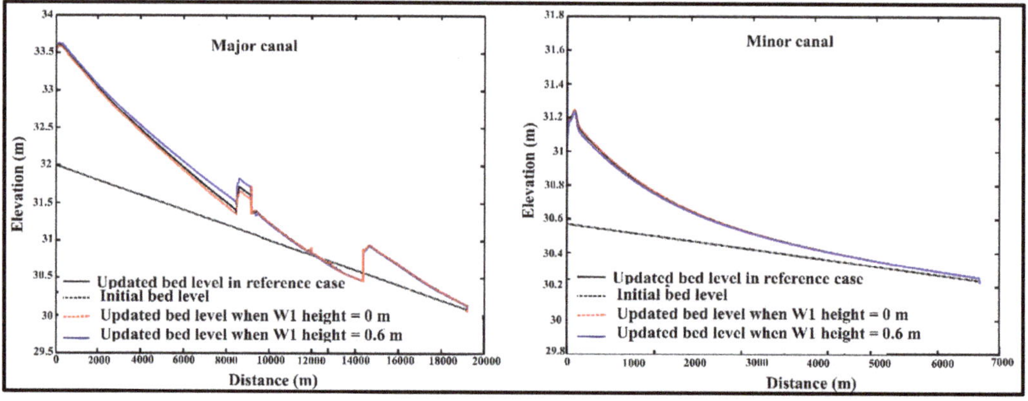

Figure 12. The effect of the upstream weir on sedimentation in the major canal (left panel) and the minor canal (right panel).

Changing the upstream weir settings reduces the sediment deposition significantly in the major canal while it has a negligible impact on the minor canal bed morphology (Figure 12).

3.2.2. Effect of the Downstream Weir (Weir 2)

To evaluate the effect of weir 2, the weir has been raised and lowered in a similar way as weir 1 and compared the results with the reference case, the results shown in Figure 13 were too close. For this reason, changing the crest height of weir 2 has a little impact on sediment transport in the major and minor canals (Figure 13). Lowering and raising the downstream weir does not reduce the negative impacts of sedimentation, where the reduction in the deposition in both canals is very small.

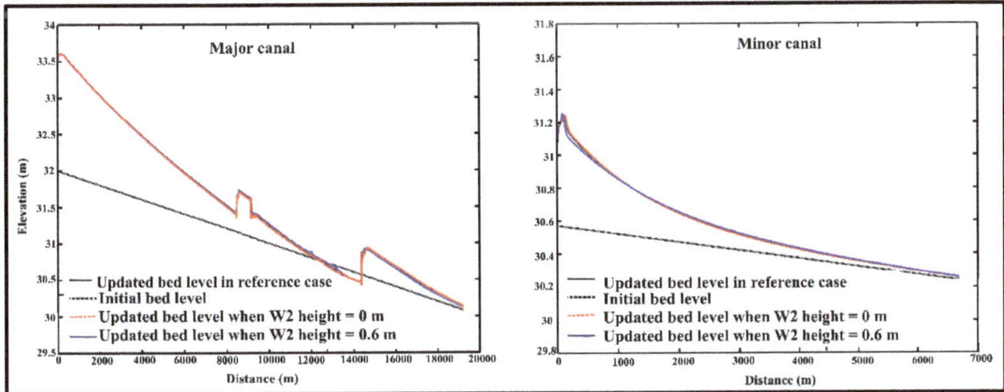

Figure 13. The effect of the downstream weir on sedimentation in the major canal (left panel) and the minor canal (right panel).

3.2.3. Effect of Both Weirs

In this scenario, we lower and raise both weirs simultaneously to see if there is a bigger impact on sediment transport. By comparing the results with results of the reference case, similar results were got as shown in Figures 12 and 13.

3.3. *Effect of Gate Settings*

3.3.1. Constant Gate Opening

To see the effect of changing gate settings on sedimentation in the major and minor canal, the model was run with different gate openings to 0.2 m; 0.4 m; 0.6 m and 0.8 m and compared the modelling results with the reference case. Lowering the gate has a small impact on the major canal but substantially reduces sediment deposition in the minor canal. In case of gate openings equal to 0.2 m and 0.4 m sediment deposition almost fully blocks the canal reducing the flow into the minor canal to close to zero. The deposition in the first kilometers of the minor canal occurs due to the effect of weir 2. Due to the disturbance in the flow caused by weir 2 the water entering the minor canal is well mixed and loaded with sediment. The backwater curve due to weir 2 causes sediment deposition (Figures 5–14). Lowering the gate reduces the deposition in the minor canal but at the same time, only a small amount of water can pass through the half-blocked canal which will not be sufficient to meet crop water requirements. Raising the gate can flush the sediment away.

Figure 14 presents the effect of different fixed gate openings on the sediment deposition patterns in the major and minor canals. Reducing the gate height has a negligible impact on the major canal but a significant impact on sedimentation in the minor canal. However, reducing the gate also means less water entering the minor canal which may lead to insufficient water delivery to crops. Even though the gate setting of 0.8 m reduces the sediments deposition less than the other gate settings compared to the reference case, the larger opening ensures sufficient water to meet crop water requirements. The large sediment deposition is located at the upstream part of the minor canal and the subsequent narrowing of the canal is visible in the field and on Google Earth imagery (Figure 15) providing further evidence that modelling results mimic the actual situation.

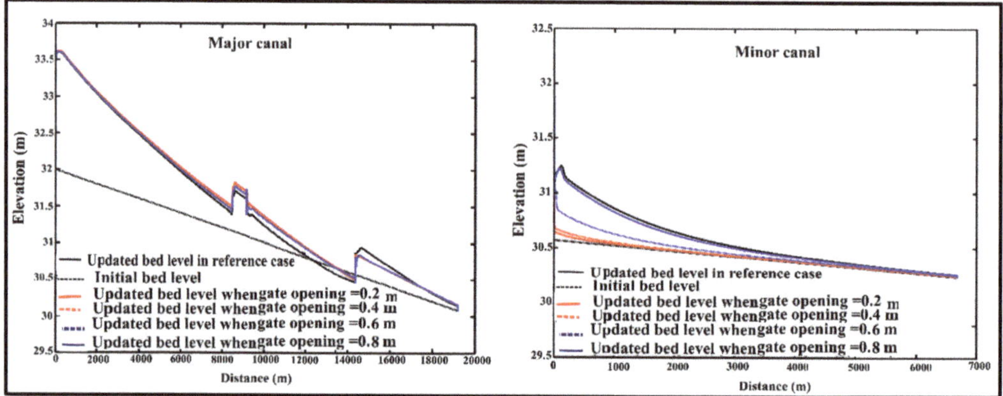

Figure 14. The effect of different fixed gate openings in time on sedimentation in the major canal (left panel) and the minor canal (right panel).

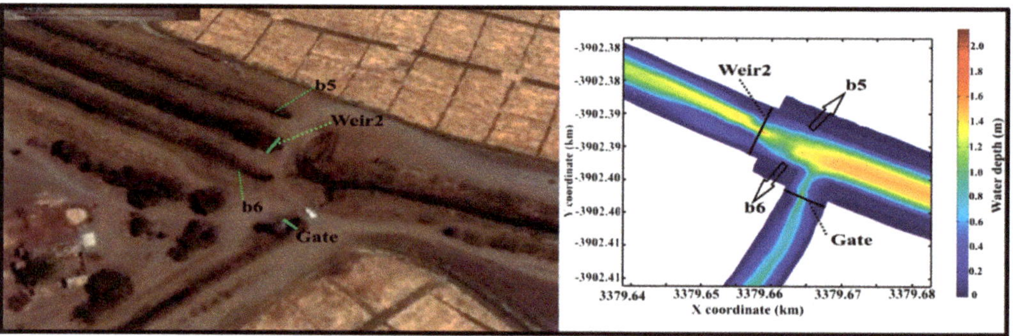

Figure 15. Comparing the Delft3D model results with actual field conditions as captured by Google Earth.

3.3.2. Variable Gate Openings Following Different Operation Plans

To test the impact of changing gate operation on the sedimentation in the canals, we formulate two different operation plans with different openings and time intervals based on the crop water requirements which change with crop growth stage. We prepare the first operation plan as shown in Figure 16 based on the data from Reference [1]. However, we simplified it by reducing the number of closing and opening the gate while keeping the same water distribution.

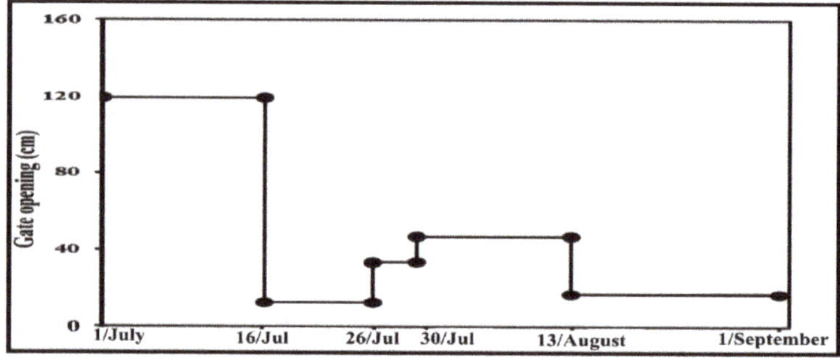

Figure 16. Operation plan with varying openings and time intervals based on crop water requirements.

The second operation plan is prepared, by fully closing and opening the gate at varying time intervals taking into account crop water requirements (Figure 17).

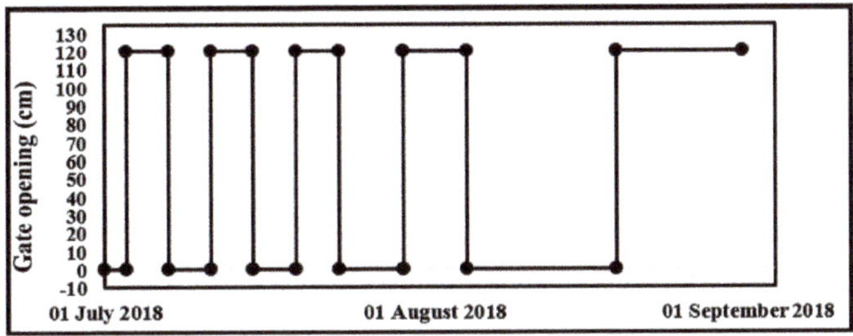

Figure 17. Operation plan with fully opening or closing the gate but with varying time intervals taking into account changing crop water requirements.

The first operation has a limited impact on sedimentation in the major and the minor canals as compared to the reference case. On other hand compared to the reference case operation plan 2 leads to a reduction in sediment deposition by half in the minor canal but limited impact on the major canal (Figure 18) while still meeting crop water requirements in a satisfactory manner. During the closure of the gate in the second operation plan the sediment accumulates near the gate and entrance to the minor canal. This is flushed away after fully opening the gate.

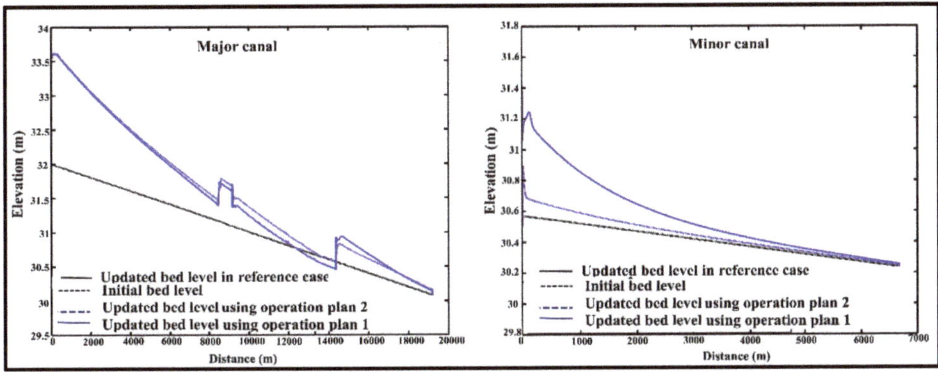

Figure 18. Effect of gate operation plan (in which the gate is either fully closed or opened at different time intervals) on sedimentation of the major canal (left panel) and left canal (right panel).

4. Discussion

Many factors affect the flow and the sediment movement in the irrigation canals. The offtakes diverting to the branch canals and field outlets catering for different water requirement, the changes in the canal geometry (contraction or widening) and other parameters all affect hydrodynamic and morphologic parameters that determine canal performance and capacity to transport sediment. In this paper, we illustrated how the location and the settings of weirs and gates do affect hydrodynamic and morphologic parameters.

Comparing scenarios to reduce sediment deposition in major and minor canals.

Table 4 summarizes the results of the scenarios related to the impacts of the weir and gate settings on the amount of cohesive and non-cohesive sediment in the major and minor canal. The last column provides a qualitative assessment of whether the weirs and gate settings in the scenario can meet the crop water requirements (CWR), based on the quantity of water that can be delivered to the outlet.

Table 4. The impacts of operation in the sediment deposition as compared to the reference scenario.

Scenarios	Description		Major Canal		Minor Canal		Meeting
			Cohesive	Non-Cohesive	Cohesive	Non-Cohesive	CWR
Scenario 2 (Weir effects)	Upstream Weir	w1 = 0	−0.5%	−0.5%	No change	No change	3
		w1 = 0.6	+1%	+1%	+2%	+2%	2
	Downstream Weir	w2 = 0	No change	No change	+3%	+3%	2
		w2 = 0.6	No change	No change	−1.2%	−1.2%	3
	Both Weirs	w1 = w2 = 0	−2%	−2%	+3%	+4%	2
		w1 = w2 = 0.6	+1%	+1%	−1.1%	−1.1%	2
Scenario 3 (Gate effects)	Fixed gate height	g = 0.2	+4%	+4%	Block	Block	0
		g = 0.4	+4%	+4%	Block	Block	0
		g = 0.6	+3%	+3%	Partially block	Partially block	1
		g = 0.8	+1%	+1%	−19%	−19%	2
	Operation	Plan1	−0.1%	−0.1%	No change	No change	3
		Plan2	+3%	+3%	−54%	−55%	2

where (−) denotes a reduction and (+) an increase in sediments deposition as compared to the reference scenario; CRW = crop water requirement is assessed qualitatively in which 0 = no water, 1 = insufficient water for crops, 2 = more or less sufficient water to satisfy CWR, 3 more than sufficient water to satisfy CWR.

The upstream weir (w1 in Figure 1) has some impact on the deposition in the upstream of the major canal while in the downstream part and in the minor canal the effect is negligible. Raising the weir height disturbs the water flow and creates a backwater curve which leads to an increased water level and with constant discharge reduced velocity, ultimately resulting in reduced velocity, reduced sediment transport capacity and deposition of sediments. The downstream weir (w2) has less impact on both canals. Simultaneously lowering or raising both weirs resemble the results of the individual weir settings.

The results of scenario 3 with fixed gate settings reveal a relatively small impact on sedimentation in the major canal but a potentially large impact on the minor canal. Lowering the gate less than 0.6 m leads to substantial sediment deposition at the entrance and upstream of the minor canal. The sediment deposition in the first kilometer of the minor canal occurs because (1) due to the flow disturbance caused by downstream weir, water entering the minor canal carries the eroded sediment and (2) due to the small gate opening, less water flows in the minor canal leading to lower sediment carrying capacity and hence deposition further downstream. Over time this leads to a complete or partial blockage of the minor canal which will adversely affect the capacity to deliver sufficient water to meet crop water requirements. It should be noted however that the Delft3D model may not be able to accurately model the local sedimentation near the gate and subsequent canal blockage. We used the 3D hydrostatic mode with a limited resolution which cannot resolve the full 3D details of the local flow near structures. Modelling the flow and sediment dynamics at an even higher resolution would be desirable but is out of scope for operational reasons (mostly due to significantly increased simulation times).

Compared to the reference case the two operation plans (both based on crop water requirements but one with variable gate settings, the other with variable time intervals) have a limited impact on the major canal. However, the second operation plan reduces the sediment deposition in the minor canal by more than 50%. In this operation plan during the closure of the gate, sediment is deposited upstream the gate; the subsequent full opening of the gate flushes the sediment away. This could be incorporated as a convenient maintenance practice.

Table 4 shows that the best operation scenarios are (1) fixed gate opening at 0.8 m where crop water requirement can be met in a satisfactory manner while reducing sediment deposition in the minor canal (2) operation plan 2 with either fully closing or opening the gate at variable intervals. Sediment accumulated during gate closure can be flushed away by fully opening the gate.

Reference [17] found in one of her operation scenarios that reduced inflows during the high sedimentation period by 51% led to sediment reduction of 48%. In this paper in our first operation scenario, the same timings and gate settings were used as used by Reference [1] but kept the flow and (high) sediment concentration constant. The effect on sedimentation in the canals is small. Hence we conclude that in Osman's scenario, the reduction of sediment-laden flow was the dominating factor in reducing sediment accumulation in the canal. Our second operation scenario shows the beneficial role of the intermittently opening and closing gate, even with high flow and sediment load. In this scenario, we assume constant sedimentation load temporally and spatially. In practice, sediment concentration in canals varies: some canals have very little or no sediments while others are suffering from high concentrations. Further, in some months sediment loads in the river are more severe than in others. Adjusting the timing of gate operation by closing the gate during periods of high sediment loads in the river to avoid sediments entering the minor canals can further reduce sediment problems.

The use of Delft3D for simulating sediment deposition in irrigation systems has significant advantages—(1) the 2D and 3D mode show where in the canal, longitudinal and in cross-sections, deposition and erosion takes place; (2) the RTC feature allows for including weirs and gates that can be adjusted during the simulation, to mimic gate operations; (3) the model handles non-steady flow well. This is important in irrigation systems where structures (gates and weirs) in the canal disturb the water flow; (4) the model can handle both cohesive and non-cohesive sediment and their interaction. The latter is important where irrigation systems use natural rivers which typically carry a mix of sediments. The Delft3D model helps to understand the mechanism of sediment transport, to predict the location, quantity and type of sediment accumulation under different operational scenarios. This information is essential for the design of operation and maintenance plans that will be effective in reducing sediment problems in irrigation systems.

As any other numerical model, Delft3D has limitations—(1) being developed for rivers, Delft3D does not simulate well the effects of sidewall roughness which makes the model inappropriate for narrow rectangular canals; (2) Delft3D and other 2D/3D hydrostatic models cannot predict local scour because vertical accelerations of the flow are ignored, turbulence modelling is limited and the sediment transport formulations are based on smooth flow conditions. For local scour detailed 3D non-hydrostatic models are needed with non-equilibrium sediment transport pickup and deposition processes [36]; (3) Delft3D does not take into consideration the effect of consolidation of (cohesive) sediments and makes no distinction between newly deposited fluffy material and old consolidated materials [37].

Finally there are two model implementation issues that need attention—(1) due to the much higher resolution than typical 1D models, simulation time can be extremely long, especially for large irrigation networks, despite useful tools such as Domain Decomposition, Flexible Mesh and Morphologic Factor, for example, in this study the cup time was 3.5 days and 2days for 3D and 2D modelling respectively; (2) the model implements the Q-h relationship as boundary condition in the downstream (i.e., water level as a function of the outflowing discharge Q). When the canal becomes dry and the water depth H drops to zero, this boundary does not reopen when the canal starts carrying water again. This situation frequently occurs in irrigation canals that are intermittently wet and dry depending on the water allocation plan. Most of these limitations are not insurmountable to solve since the model Delft3D is continuously developed further.

5. Conclusions

Efficient and well-executed canal operation plans can substantially improve hydraulic performance and reduce sediment problems which may lead to lower maintenance costs and as the result may increase crop production. This requires the proper operation of gates and finding the right balance between providing sufficient water for crop production and reducing sedimentation by the reduced sediment-laden flow. Our scenarios in the Gezira scheme in Sudan show how adjusting gate settings and varying timing of opening can be effective in reducing sedimentation in the secondary, distributary and field canals while meeting crop requirements in a satisfactory manner.

The Delft3D model, originally designed for rivers, was validated using measured field data from a previous study. The model was able to represent the actual condition (as shown in Figures 4–15).

The biggest advantages of the model (as compared to previous sediment studies) proved its ability to model both cohesive and non-cohesive sediments and its 2D mode. The latter allowed viewing flow parameters and sediments pattern within the cross-section, near offtakes, near gates and weirs and in the longitudinal direction. Determining the exact position of the sediment accumulation will help to reduce the maintenance costs and efforts and will also help the stakeholders to decide on the best operation to meet the crop water requirements while simultaneously minimizing sediment problems. Using a 3D model for cohesive and non-cohesive sediment, this study provides a substantial step forward in modelling the effect of structures on sediment behavior in irrigation canals and the use of gate operation to reduce sediment problems. Further studies are needed, in particular on the use of 3D models for large canal networks and with a better resolution around control and regulation structures. Running time and model stability are challenges here. Also, studies about the effect of gate operation with variable sediment concentrations will refine our scenarios.

Supplementary Materials: The PowerPoint and other helpful movies are available online at: https://drive.google.com/drive/folders/1Wlw9SQSqGgRBLxyIoQ5FOjqdYAmxXFVV?usp=sharing.

Author Contributions: S.A.T. carried out this research, conceived ideas and wrote the paper. B.J. took part in the implementation, analysis, guiding the ideas. F.X.S. participated in guiding the ideas. C.d.F. supervised and took part in the implementation, analysis, guiding the ideas and in the revising process.

Funding: This research received no external funding.

Acknowledgments: The Author would like to thank the Iraqi Ministry of Higher Education and Scientific Research and the Ministry of Water resources as well for funding my scholarship. Also, I would like to thank Deltares in Delft, the Netherlands for their technical help in providing the new version of the Delft3D and all modelling courses and workshops.

Conflicts of Interest: The authors declare no conflict of interest.

References

1. Osman, I. Impact of Improved Operation and Maintenance on Cohesive Sediment Transport in Gezira Scheme, Sudan: UNESCO-IHE. Ph.D. Thesis, CRC Press, Boca Raton, FL, USA, 2015.
2. Nawazbhutta, M.; Shahid, B.A.; Van Der Velde, E.J. Using a hydraulic model to prioritize secondary canal maintenance inputs: Results from Punjab, Pakistan. *Irrig. Drain. Syst.* **1996**, *10*, 377–392. [CrossRef]
3. Belaud, G.; Baume, J.P. Maintaining equity in surface irrigation network affected by silt deposition. *J. Irrig. Drain. Eng.* **2002**, *128*, 316–325. [CrossRef]
4. Jian, H.U. Study on mathematical modeling for non-uniform sediment transport in an irrigation district along the lower reach of the Yellow River. *J. Chin. Inst. Water Resour. Hydropower Res.* **2008**, *1*, 005.
5. Jinchi, H.; Zhaohui, W.; Zhang, Q. A study on sediment transport in an irrigation district. In Proceedings of the 15th International Congress on Irrigation and Drainage, Hague, The Netherlands, 4–11 September 1993; pp. 1373–1384.
6. Mendez, N.J. Sediment Transport in Irrigation Canals: UNESCO-IHE. Ph.D. Thesis, CRC Press, Boca Raton, FL, USA, 1998; p. 308.
7. Paudel, K.P. Role of Sediment in the Design and Management of Irrigation Canals: Sunsari Morang Irrigation Scheme, Nepal: UNESCO-IHE. Ph.D. Thesis, CRC Press, Boca Raton, FL, USA, 2010.
8. Depeweg, H.; Paudel, K. Sediment transport problems in Nepal evaluated by the SETRIC model. *Irrig. Drain.* **2003**, *52*, 247–260. [CrossRef]
9. Munir, S. Role of Sediment. Transport in Operation and Maintenance of Supply and Demand Based Irrigation Canals: Application to Machai Maira Branch Canals: UNESCO-IHE. Ph.D. Thesis, CRC Press, Boca Raton, FL, USA, 2011.
10. Celik, I.; Rodi, W. Modeling suspended sediment transport in nonequilibrium situations. *J. Hydraul. Eng.* **1988**, *114*, 1157–1191. [CrossRef]
11. Liu, W.C.; Hsu, M.H.; Kuo, A.Y. Modelling of hydrodynamics and cohesive sediment transport in Tanshui River estuarine system, Taiwan. *Mar. Pollut. Bull.* **2002**, *44*, 1076–1088. [CrossRef]

12. Lopes, J.; Dias, J.; Dekeyser, I. Numerical modelling of cohesive sediments transport in the Ria de Aveiro lagoon, Portugal. *J. Hydrol.* **2006**, *319*, 176–198. [CrossRef]
13. Guan, W.B.; Wolanski, E.; Dong, L.-X. Cohesive sediment transport in the Jiaojiang River estuary, China. *Estuar. Coast. Shelf Sci.* **1998**, *46*, 861–871. [CrossRef]
14. Van Rijn, L.C.; Van Rossum, H.; Termes, P. Field verification of 2-D and 3-D suspended-sediment models. *J. Hydraul. Eng.* **1990**, *116*, 1270–1288. [CrossRef]
15. Wu, Y.; Falconer, R.; Uncles, R. Modelling of water flows and cohesive sediment fluxes in the Humber Estuary, UK. *Mar. Pollut. Bull.* **1999**, *37*, 182–189. [CrossRef]
16. Theol, A.S.; Jagers, B.; Suryadi, F.; de Fraiture, C. The use of Delft3D for Irrigation Systems Simulations. *Irrig. Drain.* **2019**, *68*, 318–331. [CrossRef]
17. Osman, S.I.; Schultz, B.; Osman, A.; Suryadi, F. Effects of different operation scenarios on sedimentation in irrigation canals of the Gezira Scheme, Sudan. *Irrig. Drain.* **2017**, *66*, 82–89. [CrossRef]
18. Theol, S.; Jagers, B.; Suryadi, F.; De Fraiture, C. The use of 2D/3D models to show the differences between cohesive and non-cohesive sediments in irrigation systems. *Submitt. Am. J. Irrig. Drain. Eng. (ASCE)* **2019**. forthcoming.
19. Lesser, G.R.; Roelvink, J.V.; Van Kester, J.; Stelling, G. Development and validation of a three-dimensional morphological model. *Coast. Eng.* **2004**, *51*, 883–915. [CrossRef]
20. Morianou, G.G.; Kourgialas, N.N.; Karatzas, G.P.; Nikolaidis, N.P. Hydraulic and sediment transport simulation of Koiliaris River using the MIKE 21C model. *Procedia Eng.* **2016**, *162*, 463–470. [CrossRef]
21. Lesser, G.R. An Approach to Medium-Term Coastal Morphological Modelling: UNESCO-IHE. Ph.D. Thesis, CRC Press, Boca Raton, FL, USA, 2009; p. 255.
22. Villaret, C.; Hervouet, J.-M.; Kopmann, R.; Merkel, U.; Davies, A.G. Morphodynamic modeling using the Telemac finite-element system. *Comput. Geosci.* **2013**, *53*, 105–113. [CrossRef]
23. Kemp, L. The Evolution of Sandbars along the Colorado River Downstream of the Glen Canyon Dam. Master's Thesis, Delft University of Technology, Delft, The Netherlands, 2010.
24. Gebrehiwot, K.A.; Haile, A.M.; De Fraiture, C.; Chukalla, A.D.; Tesfa-alem, G.E. Optimizing flood and sediment management of spate irrigation in Aba'ala Plains. *Water Resour. Manag.* **2015**, *29*, 833–847. [CrossRef]
25. Flokstra, C.; Jagers, H.R.A.; Wiersma, F.E.; Mosselman, E.; Jongeling, T.H.G. *Numerical Modelling of Vanes and Screens*; Development of Vanes and Screens in Delft3D-MOR; Delft University of Technology: Delft, The Netherlands, 2003.
26. De Jong, J. Modelling the Influence of Vegetation on the Morphodynamics of the River Allier. Master's Thesis, Technology University of Delft, Delft, The Netherlands, 2005.
27. Van der Wegen, M.; Jaffe, B.; Roelvink, J. Process-based, morphodynamic hindcast of decadal deposition patterns in San Pablo Bay, California, 1856–1887. *J. Geophys. Res. Earth Surf.* **2011**, *116*. [CrossRef]
28. Partheniades, E. Erosion and deposition of cohesive soils. *J. Hydraul. Div.* **1965**, *91*, 105–139.
29. Deltares. *Delft3D-Flow User Manual*; Deltares: Delft, The Netherlands, 2016; Available online: https://oss.deltares.nl/documents/183920/185723/Delft3D-FLOW_User_Manual.pdf (accessed on 28 May 2016).
30. Van Rijn, L.C. *Principles of Sediment Transport in Rivers, Estuaries and Coastal Seas*; Aqua Publications: Amsterdam, The Netherlands, 1993; Volume 1006.
31. Gismalla, Y.A. Sedimentation problems in the Blue Nile reservoirs and Gezira scheme: A review. *Gezira J. Eng. Appl. Sci.* **2009**, *4*, 1–12.
32. Osman, I.S.; Schultz, B.; Osman, A.; Suryadi, F. Simulation of Fine Sediment Transport in Irrigation Canals of the Gezira Scheme with the Numerical Model FSEDT. *J. Irrig. Drain. Eng.* **2016**, *142*. [CrossRef]
33. Li, L. A Fundamental Study of the Morphological Acceleration Factor. Available online: http://resolver.tudelft.nl/uuid:2780f537-402b-427a-9147-b8652279a83e (accessed on 23 August 2010).
34. Roelvink, D.; Boutmy, A.; Stam, J.-M. A simple method to predict long-term morphological changes. *Coast. Eng. Proc.* **1998**, *1*. [CrossRef]
35. Krone, R.B. *Flume Studies of the Transport of Sediment in Estuarial Shoaling Processes*; University of California: Oakland, CA, USA, 1962.

36. Thanh, N.V.; Chung, D.H.; Nghien, T.D. Prediction of the local scour at the bridge square pier using a 3D numerical model. *Open J. Appl. Sci.* **2014**, *4*, 34. [CrossRef]
37. Zhou, Z.; van der Wegen, M.; Jagers, B.; Coco, G. Modelling the role of self-weight consolidation on the morphodynamics of accretional mudflats. *Environ. Model. Softw.* **2016**, *76*, 167–181. [CrossRef]

Article

Numerical Study of Remote Sensed Dredging Impacts on the Suspended Sediment Transport in China's Largest Freshwater Lake

Jianzhong Lu *, Haijun Li, Xiaoling Chen and Dong Liang

State Key Laboratory of Information Engineering in Surveying, Mapping and Remote Sensing, Wuhan University, Wuhan 430079, China; lihaijun@whu.edu.cn (H.L.); xiaoling_chen@whu.edu.cn (X.C.); liangdong@whu.edu.cn (D.L.)

* Correspondence: lujzhong@whu.edu.cn; Tel.: +86-27-6877-8755

Received: 30 October 2019; Accepted: 19 November 2019; Published: 21 November 2019

Abstract: As the largest freshwater lake in China, Poyang Lake plays an important role in the ecosystem of the Yangtze River watershed. The high suspended sediment concentration (SSC) has been an increasingly significant problem under the influence of extensive sand dredging. In this study, a hydrodynamic model integrated with the two-dimensional sediment transport model was built for Poyang Lake, considering sand dredging activities detected from satellite images. The sediment transport model was set with point sources of sand dredging, and fully calibrated and validated by observed hydrological data and remote sensing results. Simulations under different dredging intensities were implemented to investigate the impacts of the spatiotemporal variation of the SSC. The results indicated that areas significantly affected by sand dredging were located in the north of the lake and along the waterway, with a total affected area of about 730 km^2, and this was one of the main factors causing high turbidity in the northern part of the lake. The SSC in the northern area increased, showing a spatial pattern in which the SSC varied from high to low from south to north along the main channel, which indicated close agreement with the results captured by remote sensing. In summary, this study quantified the influence of human induced activities on sediment transport for the lake aquatic ecosystem, which could help us to better understand the water quality and manage water resources.

Keywords: hydrodynamic model; remote sensing; sand dredging; suspended sediment concentration (SSC); spatiotemporal analysis; Poyang Lake

---

1. Introduction

As one of the most important factors of water environment variations, suspended sediment carries a large amount of pollutants, including nutrients and heavy metals, affecting the water turbidity, bottom elevation evolution and hydrodynamic process in the long term. The numerical model has been widely used to study the suspended sediment transport process, calibrating and validating the physical parameters, and analyzing the relationships between sediment transport, erosion, and sedimentation [1–7]. Human activities like sand dredging in the navigation channel and harbor, building artificial islands by dredger, and lifting sediment into water in the coastal zones have great impacts on the suspended sediment concentration (SSC) in an aquatic ecosystem [8]. These processes and influences can be described with the help of the numerical model [9–11]. At the same time, remote sensing technology, along with in-situ observations, has been widely applied to understand the spatial distribution of suspended sediment and monitor the dredging effects on the SSC at the spatial scale [12–16]. Considering that the numerical model is able to simulate the water flow and sediment transport in any spatial and temporal resolution, and reveal the physical mechanism when remote sensing is applied to monitor water and sediment at a low cost and at a large-scale, it is of great significance to combine the numerical model and remote sensing to study water flow and sediment transport scientifically [17–21].

Poyang Lake, the largest freshwater lake in China, plays an important role in the local ecosystem and has received widespread attention. Nevertheless, the inundated area of the lake has shrunk sharply, and drought has occurred quite frequently in recent years [22]. There are many reasons for this, including economic development, human activities, and climate change. The SSC has been increased significantly and the water quality has severely deteriorated under the influence of extensive sand dredging, which is having a serious impact on the natural environment, and on living and production around the lake [7,23–26]. The distribution pattern and spatial-temporal variations of SSC in the surface waters was explored using satellite images [27–30], and the potential relationship between the SSC and number of dredging vessels could be explained the dredging impacts on the sediment budget assessments in Poyang Lake [23,31]. The Poyang Lake has suffered intensive sand dredging, while the water has been disturbed, leading to extremely high turbidity. However, the impact mechanism of sand dredging on the suspended sediment concentration in Poyang Lake has not yet been well investigated.

In this study, we aimed to understand the temporal-spatial impacts of sand dredging activities on suspended sediment transport in the Poyang Lake, and a hydrodynamic integrated suspended sediment transport model was employed. The models were used to simulate the sand dredging impact on the sediment transport process continuously, using historical meteorological and hydrological data along with bottom topological data. The simulated hydrodynamic results were validated by the observed data at hydrology gauging stations. Based on the hydrodynamic model, a two-dimensional sediment transport model of Poyang Lake was developed considering sand dredging activities detected from satellite images. Finally, contrastive scenario simulations were implemented to analyze the spatiotemporal variation characteristics of SSC under different intensities of sand dredging. The impacts of sand dredging on the spatiotemporal distribution pattern of SSC were investigated to help us understand water quality variation under intensive human induced activities.

2. Study Area and Data

The Poyang Lake (115°50′ E–116°50′ E, 28°00′ N–29°50′ N) is an important hydrological subsystem in the middle and lower Yangtze River, located in the north of central Jiangxi Province of China. Separated by Songmenshan Island in the middle of lake, Poyang Lake is geographically divided into two parts, including the narrow north lake and broad south lake (Figure 1). The lake has a storage capacity of 27.6 billion m^3 and an average water depth of 8.4 m [32]. As a unique inland freshwater lake, there is a high variability in the water level, and the inundation area fluctuates from less than 1000 km^2 in the dry season to over 3000 km^2 in the wet season [33].

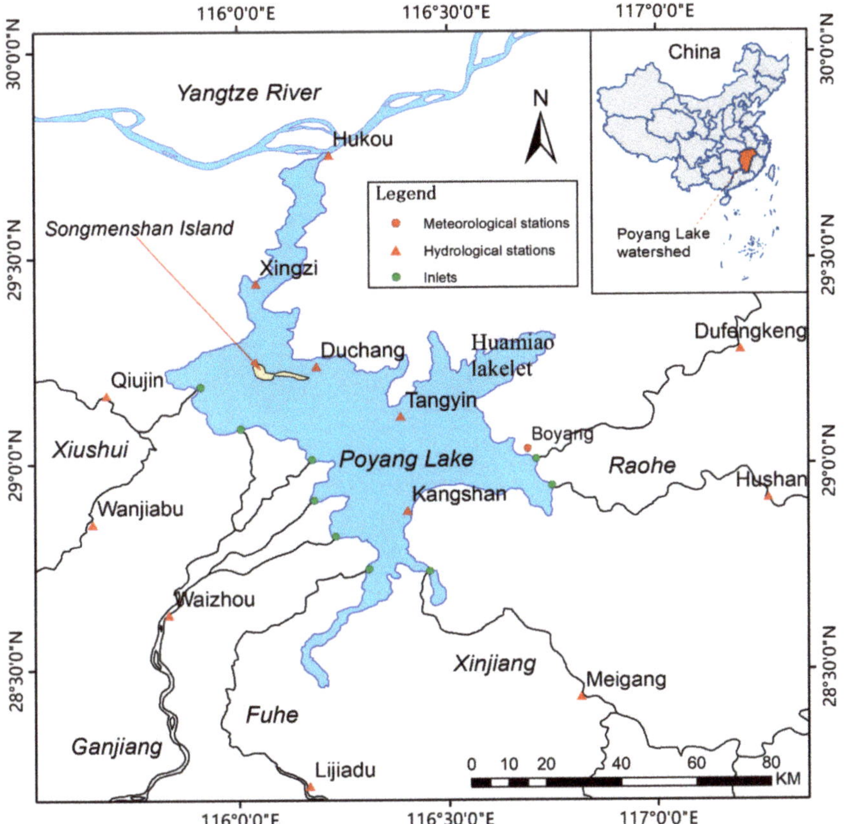

Figure 1. The location of Poyang Lake, the inflow tributaries, and the hydrological and meteorological stations. The daily river discharge and water level data during 2011 to 2015 at different hydrology gauging stations were collected from the Yangtze River Water Resources Commission and Hydrological Bureau of Jiangxi Province. The water level and water flux records were collected at Hukou station, which is located at the junction of Poyang Lake and the Yangtze River. Meteorological data at Boyang meteorological station, representing the weather that occurred over the Poyang Lake, was collected from the China Meteorological Data Sharing Service System (http://cdc.nmic.cn/). Field observations were conducted from July 15 to 23 in 2011 to measure in-situ data, including the turbidity, water depth, and suspended sediment concentration (SSC) of 50 water samples (Figure 2). Additionally, a satellite image Landsat ETM+ (Enhanced Thematic Mapper plus) was used to extract information on the sand dredging activities, in order to simulate the suspended sediment transport considering the effect of dredging activities during July 2011.

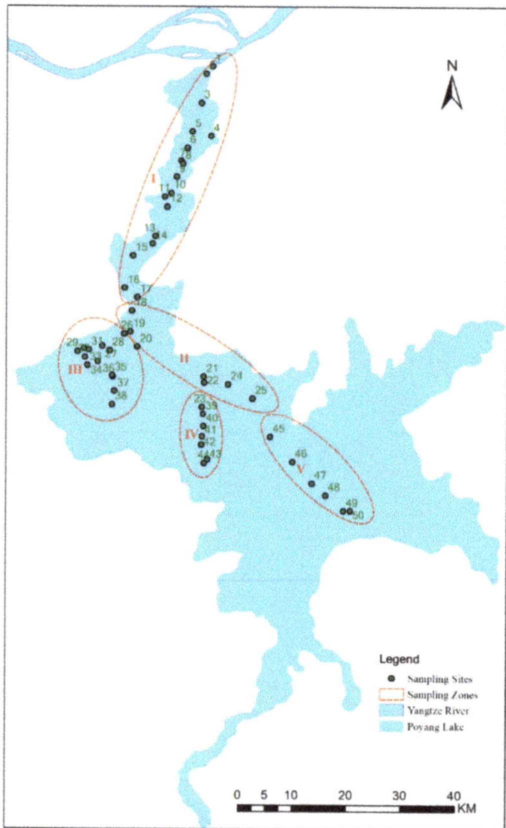

Figure 2. Spatial distribution of 50 samples in Poyang Lake during 15–23 July 2011.

As the largest freshwater lake in China, Poyang Lake has abundant water resources, with five major tributaries flowing into the lake including Xiushui River, Ganjiang River, Fuhe River, Xinjiang River, and Raohe River. The hydrological data were collected daily from the discharging rivers, including at the Qiujin, Wangjiabu, Waizhou, Lijiadu, Meigang, Hushan, and Dufengkeng hydrology gauging stations. The discharges of these rivers drain into the lake by nine main estuaries, passing through the narrow channel in the north lake and flowing into the Yangtze River. The Yangtze River has a great impact on regulating the flow from the lake, and the discharges may reverse at times from July to September, with a frequency of 708 days in 45 years [34].

Water quality has been reported to be declining in recent years in Poyang Lake. Suspended sediment is one of the major factors affecting the water quality. The mean SSC in the south lake was close to 20 mg/L, and slightly higher than 20 mg/L, in 2000, with a highest value greater than 60 mg/L in 2006; it was much higher in the north lake than the south lake [24]. In the most recent two decades, sand dredging has been rapidly increasing in the Poyang Lake since it was banned in the Yangtze River in 2001. Since then, sand dredging activities have been continuously carried out, though there was a ban on dredging in 2008, which has led to a significant increase in the SSC in the lake and serious negative impacts on the Poyang Lake ecosystem [33].

3. Methods

3.1. Remote Sensing Image Processes

Due to the heterogeneity of the high turbidity in Poyang Lake, it is difficult to efficiently detect dredging vessels by remote sensing with moderate or low resolution. In order to solve this problem, a vessel detection algorithm based on 30 m resolution images acquired by Landsat ETM+ (Enhanced Thematic Mapper plus) in our previous study [26] was used here to monitor the sand dredging vessels, in order to simulate the suspended sediment transport considering the effect of dredging activities during July 2011. The optimal algorithm is sufficient, with an average detection rate of 87.6% for a large lake (>3000 km^2 in the wet season), to monitor sand dredging activities using medium resolution optical remote sensing images. The vessel detection algorithm provided an effective solution for monitoring sand dredging dynamics, as well as useful information for managing sand dredging in freshwater environments and assessing the potential impacts on aquatic ecosystems.

The temporal-spatial resolution of Moderate Resolution Imaging Spectroradiometer (MODIS) images allows monitoring of hydrology changes, the water field, and water quality at a large scale. The MODIS 250 m resolution daily reflectance products, MOD09GQ and MYD09GQ, were used to extract the water body and calculate the total inundated area for Poyang Lake from 2011 to 2015, to validate the hydrodynamic simulated results. Under the influence of cloud cover and sensor-induced noise, available data were limited to 196 images during 2011 to 2015. To extract the water body precisely, a simple normalized difference vegetation index (NDVI) with the threshold method [35], aided by visual interpretation, was used. Additionally, the MODIS surface reflectance products were collected to derive the SSC distribution, to verify the model results during the simulated time using an empirical method proposed by Cui et al. [24], in which an exponential model of the MODIS Aqua red band was applied to estimate SSC from MODIS Aqua images in this study.

3.2. Hydrodynamic-Sediment Transport Model Setup for Poyang Lake

In this study, the Delft3D numerical modeling system developed by WL Delft Hydraulics in the Netherlands [36], which has been successfully applied in aquatic environment system simulations like hydrodynamic processes, sediment transport, and water quality, was used to set up the hydrodynamic model of Poyang Lake. This model system includes several modules, including flow, sediment, morphology, particle tracing, and orthogonal curvilinear coordinates in the horizontal direction. The Delft3D system has been developed for the modeling of unsteady water flow, temperature, salinity, and cohesive/non-cohesive sediment transport in shallows seas, estuarine and coastal areas, and rivers and lakes [37]. The Delft3D-Flow module performs hydrodynamic calculations by solving continuity and horizontal momentum equations for given initial and boundary conditions in two or three dimensions [38]. The Delft3D-Online is integrated with the Delft3D-Flow module, which enables the calculation of the fine sediment transport model by the two-dimensional advection–dispersion equation. The two dimension sediment transport can be described as Equations (1)–(3):

$$h\frac{\partial c}{\partial t} + h\frac{\partial uc}{\partial x} + h\frac{\partial vc}{\partial x} = \frac{\partial}{\partial x}\left(D_x h\frac{\partial c}{\partial t}\right) + \frac{\partial}{\partial y}\left(D_y h\frac{\partial c}{\partial y} + E - D\right), \tag{1}$$

$$E = \begin{cases} M\left(\frac{\tau_b}{\tau_e} - 1\right), & \tau_b > \tau_e \\ 0, & \tau_b \le \tau_e \end{cases}, \tag{2}$$

$$D = \begin{cases} w_s c\left(1 - \frac{\tau_b}{\tau_d}\right), & \tau_b < \tau_d \\ 0, & \tau_b \ge \tau_d \end{cases}, \tag{3}$$

where Dx, and Dy are the diffusion coefficient in the x and y direction; h is the water depth; c is the sediment transport volume; E is the erosion rate; D is the deposition flux of suspended matter; M is the erosion rate; τ_b is the bed stress; w_s is the settling velocity of suspended matter; τ_e is the critical shear stress for erosion; and τ_d is the critical shear stress for deposition.

The largest area of water body was chosen to delineate the land–water boundary among the derived results with MODIS images from 2011 to 2015. The study area was divided into about 19,000 orthogonal curvilinear grids with the resolution ranging from 300 to 500 m, and the bottom elevation was interpolated over the cells using a grid cell averaging method. The measured daily discharges of the five major incoming tributary rivers divided into nine estuaries were set as the upstream boundary conditions. Considering that Ganjiang is divided into four branches, the discharges of four estuaries were allocated by the daily measured total discharge of Waizhou station by different ratios [39]. The downstream boundary condition was set as the observed daily average water level at Hukou station. The observed water level of Duchang station, located in the middle of Poyang Lake, was set as the initial water level of the whole lake, and the initial velocity was set to zero. The time step was set to 5 min, considering the Courant–Friedrichs–Lewy (CFL) condition and computing costs. Detailed parameter configurations can be found in our previous study [38]. Based on the flow model, the suspended sediment transport model can be applied. The observed daily sediment load of the corresponding hydrological stations of five major incoming tributary rivers were added into the upstream boundary from July 1 to 31, 2011, as well as the sediment load of four branches of Ganjiang, which were allocated by ratios that were the same as the flow model above. Except for the water level, the sediment runoff could be set as 50 mg/L and the initial suspended sediment concentration could be set as zero, referring to the previous study [40].

The sand dredging activities have been so extensive that their effects on the suspended sediment transport cannot be ignored in Poyang Lake. The Landsat ETM+ satellite images could be used to detect the location of dredging areas and dredging vessels [26]. Several sand dredging areas were recorded during field observations from July 15 to 23, 2011. Using a Landsat image on July 4, 2011, three main sand dredging areas and dredging vessels were detected (Figure 3). Area A is located in the north of Songmenshan Island, with five dredging vessels, area B is located in southeast of Songmenshan Island, with four dredging vessels, and area C is located in the northwest of Tangyin, with eight dredging vessels. Considering that the SSC in the dredging center is 150 mg/L when the dredging power reaches 5500 m^3/h [41], and the fact that the dredging power is about 10,000 t/h in Poyang Lake and the measured SSC is 277.4 mg/L around the dredging area, the SSC in the dredging center could be set as 300 mg/L in the simulation for a single dredging vessel. As the size of a grid cell was larger than the resolution (30 m) of the Landsat ETM+ images, there were often several dredging vessels in the same grid cell. To simplify the model properly, a single point source of suspended sediment was configured to simulate the total effect of dredging vessels in the same grid cell and dredging area, with corresponding SSCs of 1500, 1200, and 2500 mg/L in dredging areas A, B, and C, which was consistent with the different intensities of the dredging areas in 2011 [26]. This assumed that those dredging vessels worked at the same location and had the same working hours from 8:00 to 18:00 every day during July 1 to 31, 2011. To analyze the effect of sand dredging on the SSC, another scenario was also simulated under the same conditions without dredging.

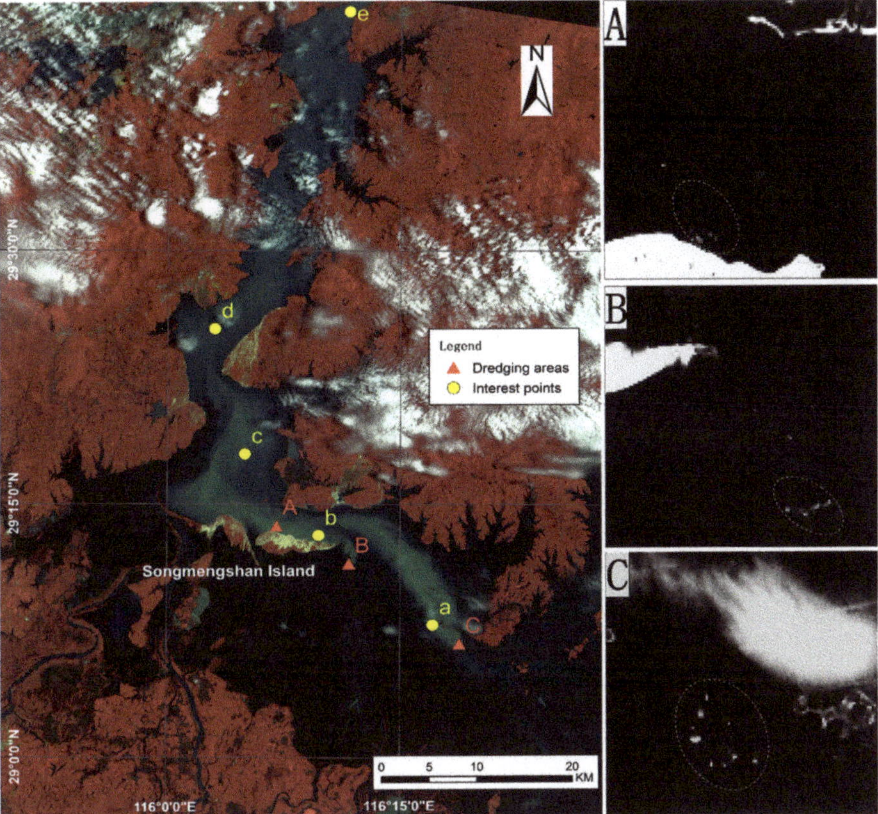

Figure 3. Three main sand dredging areas (A, B, and C in the subfigure) and five points of interest (POIs a, b, c, d, and e) in Poyang Lake on Landsat ETM+ images composed by Band 4, 3, and 2, acquired on July 4, 2011, and zoomed-in for the main dredging areas A, B, and C on Band 7, which is sensitive to dredging vessels.

4. Results and Discussion

4.1. Hydrodynamic Model Validation

The observed water levels at the Xingzi, Duchang, and Kangshan hydrology gauging stations were used to validate the simulated water level by the Poyang Lake hydrodynamic model. The simulated water levels showed close agreement with the observations at the three gauging stations (Figure 4). The coefficients of determination (R^2) of the results at Xingzi, Duchang, and Kangshan were 0.996, 0.989, and 0.885, respectively, and the root mean square error (RMSE) varied from 0.26 (Xingzi) to 0.63 m (Kangshan), while the respective error (RE) was 1.80%, 2.12%, and 3.35%, indicating that the Delft3D model has the capability to catch the high-dynamic changes in water levels of Poyang Lake. The results also showed lower accuracy during the annual dry period, the reason for which may be as follows: (1) the shorelines changed from a lake into a river channel during the dry period on account of water recession, and the hydrodynamic characteristics changed a lot while the model parameter conditions remained the same as during the previous wet period, meaning the uniform model condition may result in more error during the dry period, or (2) sand dredging activities have changed the bottom bed seriously since 2001, and the bathymetric data used was measured in 2000. Consequently, the outdated bottom elevation data induced errors.

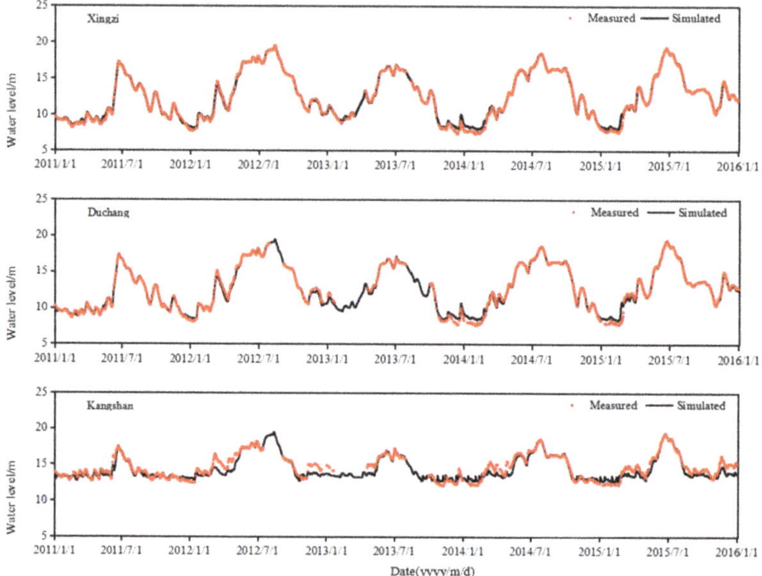

Figure 4. Comparison between model results and observations of the water level at the Xingzi, Duchang, and Kangshan gauging stations.

To explore the water exchange process between Poyang Lake and the Yangtze River, simulated discharges at the Hukou station from 2011 to 2015 were calculated (Figure 5). The simulated results were in reasonable agreement with the measured data at Hukou, with an RMSE of 1557.83 m^3/s, R^2 of 0.869, and RE of 18.70%. Further, the blocking effect and flow backward phenomenon of the Yangtze River was caught during the wet periods.

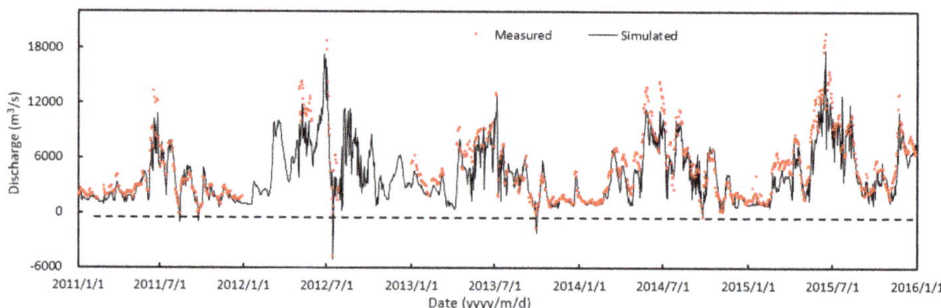

Figure 5. Comparison between model results and observations of outlet discharge at Hukou.

Based on the daily measured water level at Duchang station, located in the middle of Poyang Lake, and combined with the simulated results, the maximum and minimum inundated area, corresponding time of model and MODIS, the water level on the day, and the ratio of maximum and minimum area can be calculated. The maximum inundated area occurs in the wet period, mainly in June or July, while the minimum area occurs in the dry period from December to February. The simulated ratios of the maximum and minimum area are close to those derived from MODIS images, with the largest of 4.25, which reflects the high-dynamic changes of the water field of the lake. The result of a large inundation area change rate was similar to the findings by Feng et al. [21]. Figure 6 shows the comparison of the maximum and minimum inundated area of the model results and those derived from

MODIS images. In general, the maximum and minimum inundated areas show similar morphology and water distribution of Poyang Lake—the shape of the lake is almost full in the wet period and a river with some lakelets scattered around occurs in the dry period—indicating the annual high-dynamic changes of the lake. In the south of Songmenshan Island, close to the estuary of Xiu River and Ganjiang River, a water body exists in the simulated results but not in the MODIS images, which may be due to the simple NDVI threshold used to extract the water field in the satellite images. The spectral characteristics in those mud areas near the estuary differ from a natural water body like the main lake, causing errors in the extraction with NDVI. In addition, there was a good consistency between the model simulated water area and the MODIS extracted results (Figure 7), with an RMSE of 283.47 km^2, R^2 of 0.829, and RE of 15.90%, suggesting that the calibrated model could be further employed in suspended sediment transport simulation.

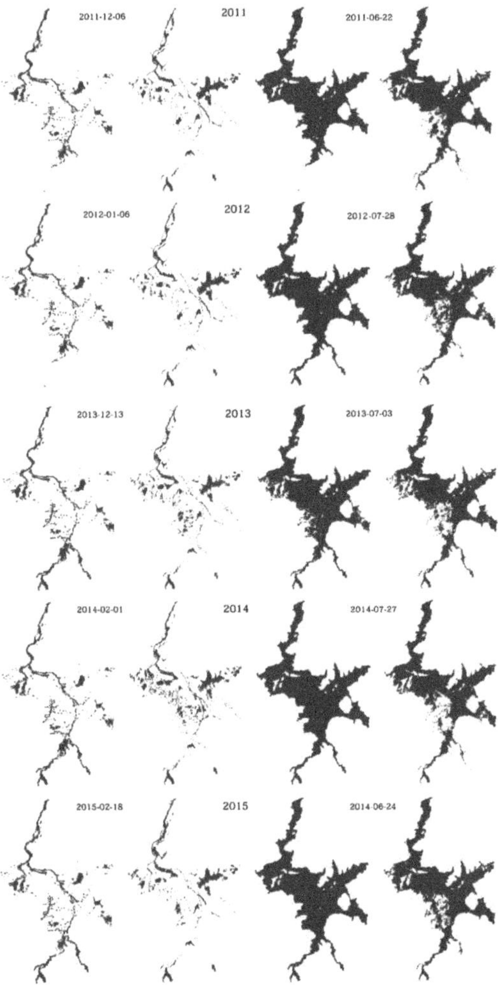

Figure 6. Comparison of the maximum and minimum inundated areas of the model results (left of the date) and those derived from Moderate Resolution Imaging Spectroradiometer (MODIS) images (right of the date).

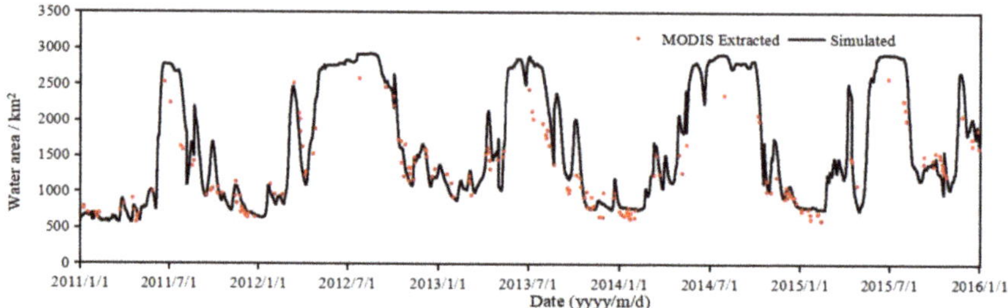

Figure 7. Comparison between MODIS extracted water areas and model simulated results.

4.2. Simulation of Sand Dredging Effects on the Suspended Sediment Concentration

The SSC, simulated with/without sand dredging, was also validated by 50 samples of in-situ observation during the cruises of July 15 to 23 July, 2011 (Figure 8). When sand dredging activities were ignored in the simulation, there were large deviations between the simulated results and the in-situ observed data. For some samples, like Site 18, 20, 24, and 25, the SSC values observed were higher than 100 mg/L but much lower in the model without dredging, and increased a lot to a level close to the measured values in the dredging model. After integrating the sand dredging into the model, the simulated results were improved, with an R^2 of 0.831 and RMSE of 15.5 mg/L, which were validated against the observed data. The results show the simulation of the SSC in Poyang Lake could be made more effective by considering sand dredging activities.

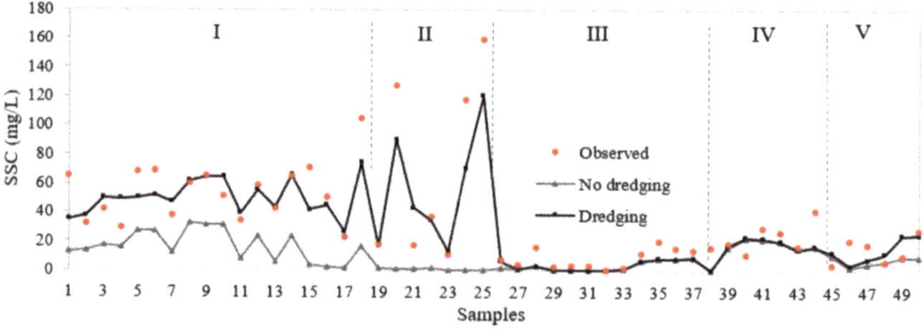

Figure 8. Comparison between the SSC, simulated with or without dredging, and in-situ observations during July 15 to 23, 2011.

Three main dredging areas showed different effects on the samples in different zones of the lake (Figure 2). Sites 1 to 17 were located in Zone I, namely the long and narrow channel in the north of the lake, where upstream suspended sediment transport affected the SSC here synthetically. Sites 18 to 25 were located in Zone II, close to the dredging areas A and B, where the SSC increased greatly under the influence of dredging. However, the SSC showed a stable trend in Zones III, IV, and V, which were located in the upstream of a dredging area or in the south main lake with the obstruction of Songmenshan Island.

Using the SSC results derived from cloud free MODIS satellite images on July 4 and 20, 2011, the SSC simulated by the suspended sediment transport model with or without sand dredging could be compared (Figure 9). In the model without dredging, the SSC of the lake was low in general, with a maximum of about 80 mg/L in the north. The spatial pattern varied from the results derived from MODIS. After sand dredging was considered, the SSC in the northern area increased, with a spatial pattern of the SSC which varied from high to low from south to north along the main channel, indicating a close agreement with the MODIS derived results. Around the center of the dredging activities, the SSC showed abnormally high values—much higher than those retrieved from remote sensing. The potential factors affecting this may include the following: (1) the grid size was much larger than the resolution of the Landsat images (30 m) used to detect sand dredging vessels, which meant that there were several vessels in the same grid to be simulated as a single point source of suspended sediment. Hence, the aggregation of multiple dredging vessels caused a high SSC for the center grid cell in the dredging areas, and using fine grid cells to represent multiple vessels in a cell may eliminate the abnormally high SSC in the center of dredging, or (2) the corresponding relationship and calculation of dredging power and point of SSC referenced could be closer to the Poyang Lake regionally, and improved to give more accurate simulations.

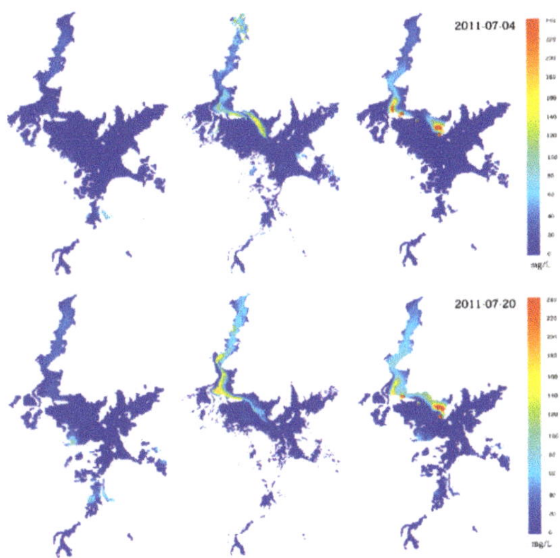

Figure 9. Comparison of the SSC spatial distribution, simulated with (left) or without (right) dredging, and MODIS derived (center) on July 4 and 20, 2011.

4.3. Spatial and Temporal Impacts of Dredging on the Suspended Sediment in Poyang Lake

To investigate the affected area and variation of the spatial pattern quantitatively, the maximum increment of the SSC was defined as the difference between the largest SSC in stable status with dredging and the SSC in normal status with no dredging. The spatial distribution of the maximum increment of the SSC under different dredging intensities is presented as Figure 10. Generally, the maximum increment of the SSC caused by sand dredging varied from large to small from south to north, upstream near the dredging area to downstream along the channel. Additionally, it increased relatively with dredging intensity. Specifically, the maximum of the SSC near the center of the dredging areas reached about 70 mg/L when the dredging intensity was 300 mg/L (single dredging vessel). The maximum increment of the SSC in the north of the lake increased dramatically, especially in the center of the waterway channel, with a maximum of 300 mg/L under a dredging intensity of 1500 mg/L (five dredging vessels). Although fluctuation existed in the water area of different concentrations, the total area affected under these different dredging intensities was stable to an extent across an area of 730 km^2.

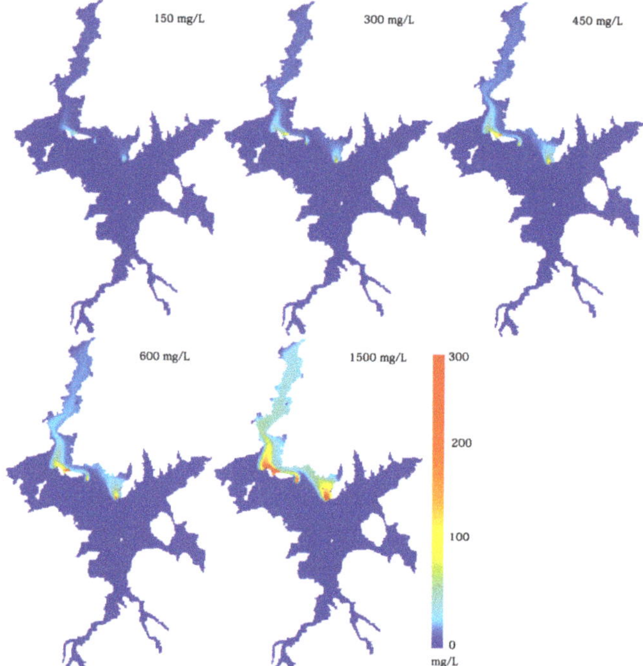

Figure 10. The spatial distribution the maximum increment of the SSC under different dredging intensities.

Considering that the suspended sediment is transported with the water flow from dredging centers to downstream in the waterway channel in the north of the lake, the sediment diffused slightly from the dredging center, causing a slight increase of the total affected area. This suggested that sand dredging in these three dredging areas showed nearly no effects in the south of the main lake and river inlets, but huge effects in the north of the lake, including the area west of Tangyin station, north of Songmenshan Island, the waterway channel, and the outlet to the Yangtze River, which was one of the main reasons for high turbidity water in the northern lake.

From south to north along the main channel, five points of interest (POIs, a, b, c, d, and e) were chosen for further exploration of the effects of dredging on the SSC. Figure 11 shows the maximum increment of the SSC and the recovery time to normal status for the five POIs under different dredging intensities. In general, there were similar SSCs for all the points when the dredging intensity was 150 mg/L, but it increased gradually along with the dredging intensity, showing a spatial pattern of the maximum increment of the SSC varying from large to small from south to north along the channel. However, the recovery time to normal status of the points showed a relatively opposite trend of the maximum increment of the SSC, which varied from short to long from south to north along the channel. This suggested a positive correlation between the maximum increment of the SSC and the dredging intensity, as well as a relationship between the recovery time and the dredging intensity. The closer to the dredging area the POI was, the larger the maximum increment at the POI, and the recovery time was longer downstream than upstream.

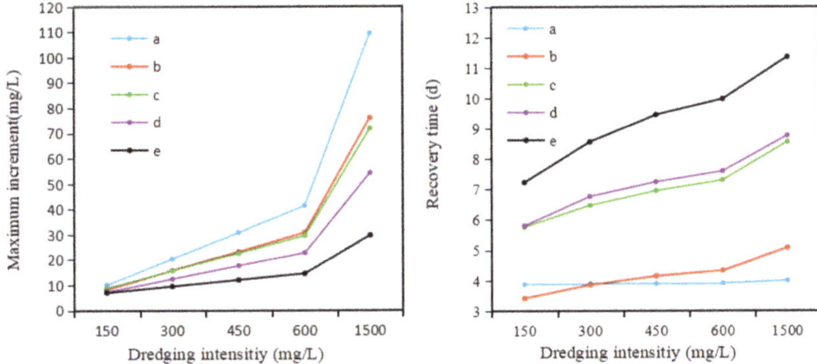

Figure 11. The maximum increment of the SSC and the recovery time to normal status for the five POIs under different dredging intensities.

Specifically, there was a similar level of SSC and trend line at the POIs b and c, but the recovery times were varied. The reasons for this may be the following: (1) the POI b was close to dredging area A and was heavily influenced, while the POI c was further away from the center of the dredging area but was influenced by dredging area A combined with area B, resulting in a similar maximum increment variation as POI b; (2) from the perspective of the velocity field simulated in the north of the lake (Figure 12), the depth averaged velocity in the narrow channel (in POI b) was much larger than in the open water area (in POI c), indicating a larger diffusion rate in this area, meaning water exchange occurred more quickly with the upstream discharge than in open water for the SSC under the influence of sand dredging activities. Therefore, the high turbidity induced by sand dredging would quickly recover to normal status in this narrow channel in the case that sand dredging stops.

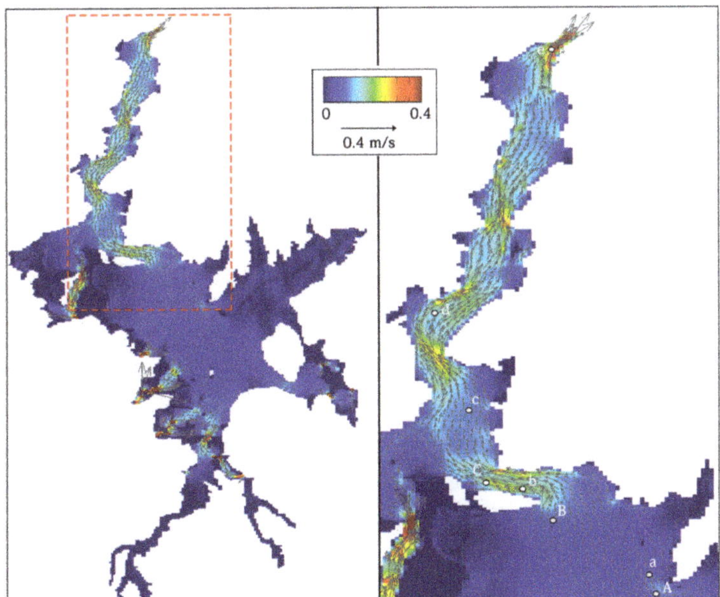

Figure 12. The simulated velocity field in the north of Poyang Lake.

The recovery time to normal status for Huamiao lakelet, together with the simulated velocity field, is shown in Figure 13. The recovery time to normal status exceeded 18 days and was longer in the northern lakelet. According to the velocity field that was simulated, the water flow rate was much slower here, and the velocity was lower than 0.1 m/s for most of the grid cells, decreased gradually from open water to the lakeshore, forming a closed annular flow field separated in the bay. This means a low exchange rate for water flow, and also a low diffusion rate for the SSC, resulting from sand dredging activities.

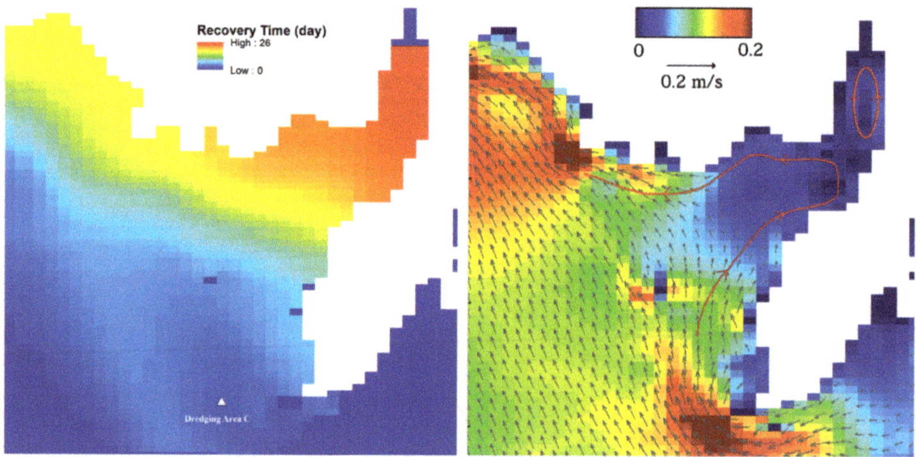

Figure 13. The recovery time to normal status and corresponding velocity field in the Huamiao lakelet, southeastern Poyang Lake.

As with the recovery time for the SSC to normal status after sand dredging stopped, the water age could be used to describe the characteristics for the water residence and transport process. Li et al. indicated a lower exchange rate and a longer time for water residence with a greater water age, and vice versa [42]. Qi et al. simulated the spatial and temporal distribution pattern of the water age of Poyang Lake based on a hydrodynamic model named EFDC in 2011 [43]. It showed that it is difficult for the suspended sediment to diffuse in an area with a higher water age. Compared with the distribution of the water age in the Huamiao lakelet, the time to recover to normal status for the SSC after sand dredging stopped varied from short to long from the open water to the lakeshore. Additionally, the water age of the northern lake presented an increasing trend, which was in good agreement with the spatial pattern of the recovery time for the SSC in the north of the lake.

5. Conclusions

The main purpose of this study was to reveal the characteristics of hydrodynamic variations and their impacts on the SSC, caused by sand dredging activities at the spatial and temporal scale, in Poyang Lake, the largest freshwater lake with high turbidity in China. With the help of remote sensing technology and numerical simulation, a sediment transport method integrated with sand dredging was built for the Poyang Lake, considering the sand dredging activities. The hydrodynamic process was well validated against observed data and remote sensing results, showing the dramatic changes of the inundation area in the dry and wet seasons annually and inter-annually. Both the numerical model and remote sensing have the capability to observe the high-dynamic changes in the hydrologic regime of the lake.

Integrated with the validated hydrodynamic model, the two-dimensional sediment transport model was employed with consideration of sand dredging activities detected from Landsat images. Based on the observed data and remote sensing results, the sand dredging areas were determined to set up scenario simulations. This is a feasible way to monitor dredging activities with a moderate resolution. The simulation of the SSC, which indicated close agreement with the MODIS derived results, could be more effective by considering the dredging activities in Poyang Lake. Further explorations of spatiotemporal variations of the SSC were conducted with the help of corresponding scenario simulations designed under different dredging intensities. The three main dredging areas showed nearly no effects on the SSC in the southern main lake, but notable effects in downstream areas in the north of the lake, with a total affected area of 730 km^2.

The dredging activities were one of the main factors causing high turbidity in the north of Poyang Lake. The proposed suspended sediment transport model integrated with sand dredging was proven to effectively simulate the influence of dredging activities quantitatively, which improved the model accuracy for human induced high turbidity. However, because of difficulty quantifying sediment release from the pump during sand dredging, the dredging intensities were approximately estimated by the number of dredging vessels, and then integrated with the sediment transport model with scenario simulations of different intensities in this study. In fact, this study demonstrated that the proposed method enables us to evaluate the impacts of dredging activities on the suspended sediment pattern at the spatial and temporal scale. In the case that the monitoring instrument is equipped on the dredging pump, the real time sediment release into the water could be observed as a point source of sediment integrated into the model. This study provided an analysis tool to understand water quality under an intensive impact of human activities for water resource management.

Author Contributions: Conceptualization, J.L. and X.C.; methodology and validation, H.L.; write, J.L. and H.L.; review and editing, D.L.; Project administration and funding acquisition, J.L.

Funding: This work was funded by the Frontier Project of Applied Foundation of Wuhan (2019020701011502), the Natural Science Foundation of Hubei Province (2019CFB736), the Fundamental Research Funds for the Central Universities (2042018kf0220), the open foundation of Jiangxi Engineering Research Center of Water Engineering Safety and Resources Efficient Utilization (OF201601), and the LIESMARS Special Research Funding.

Acknowledgments: We would also thank the Delft Hydro-Morphodynamics for providing open source code of Delft3D-Flow.

Conflicts of Interest: The authors declare no conflict of interest.

References

1. Vos, R.J.; Brummelhuis, P.G.J.T.; Gerritsen, H. Integrated data-modelling approach for suspended sediment transport on a regional scale. *Coast. Eng.* **2000**, *41*, 177–200. [CrossRef]
2. Miller, R.L.; Castillo, C.E.D.; Chilmakuri, C.; Mccorquodale, J.A. Using MULTI-temporal MODIS 250 m Data to Calibrate and Validate a Sediment Transport Model for Environmental Monitoring of Coastal Waters. In Proceedings of the IEEE International Workshop on the Analysis of Multi-Temporal Remote Sensing Images, Biloxi, MS, USA, 16–18 May 2005; Volume 62, pp. 200–204. [CrossRef]
3. Kunte, P.D.; Zhao, C.; Osawa, T.; Sugimori, Y. Sediment distribution study in the Gulf of Kachchh, India, from 3D hydrodynamic model simulation and satellite data. *J. Mar. Syst.* **2005**, *55*, 139–153. [CrossRef]
4. Elias, E.P.L.; Cleveringa, J.; Buijsman, M.C.; Roelvink, J.A.; Sive, M.J.F. Field and model data analysis of sand transport patterns in Texel Tidal inlet (The Netherlands). *Coast. Eng.* **2006**, *53*, 505–529. [CrossRef]
5. Fettweis, M.; Nechad, B.; Van den Eynde, D. An estimate of the suspended particulate matter (SPM) transport in the southern North Sea using SeaWiFS images, in situ measurements and numerical model results. *Cont. Shelf Res.* **2007**, *27*, 1568–1583. [CrossRef]
6. Chen, X.; Lu, J.; Cui, T.; Jiang, W.; Tian, L.; Chen, L.; Zhao, W. Coupling remote sensing retrieval with numerical simulation for SPM study—Taking Bohai Sea in China as a case. *Int. J. Appl. Earth Obs. Geoinform.* **2010**, *12*, 203–211. [CrossRef]
7. Liang, D.; Lu, D.; Chen, X.; Zhang, L. Numerical simulation of hydrological and hydrodynamic responses to channel erosion in China's largest freshwater lake. *Appl. Ecol. Environ. Res.* **2019**, *17*, 6865–6886. [CrossRef]
8. Hossain, S.; Eyre, B.D.; McKee, L.J. Impacts of dredging on dry season suspended sediment concentration in the Brisbane River Estuary, Queensland, Australia. *Estuar. Coast. Shelf S.* **2004**, *61*, 539–545. [CrossRef]

9. Thibodeaux, L.; Duckworth, K. The Effectiveness of Environmental Dredging: A Study of Three Sites. *Remediation* **2001**, *11*, 5–33. [CrossRef]
10. Alam, S.; Matin, M.A. Application of 2D morphological model to assess the response of Karnatuli River due to capital dredging. *J. Water Resour. Ocean. Sci.* **2013**, *2*, 40–48. [CrossRef]
11. Jeroen, S. The Influence of Dredging Activities on the Morphological Development of the Columbia River Mouth. MSc Thesis, Department of Hydraulic Engineering Section of Coastal Engineering, Delft University of Technology, TU Delft, Delft, The Netherlands, 2012.
12. Cheng, X.; Wan, Y.; Cao, B. Identification of sand dredges in Yangtze River based on ASAR remote sensing data. In Proceedings of the 17th International Conference on Geoinformatics, IEEE, Fairfax, VA, USA, 12–14 August 2009; pp. 1–5. [CrossRef]
13. Evans, R.D.; Murray, K.L.; Field, S.N.; Moore, J.A.; Shedrawi, G.; Huntley, B.G.; Fearms, P.; Broomhall, M.; Mckinna, L.I.; Marrable, D. Digitise this! A quick and easy remote sensing method to monitor the daily extent of dredge plumes. *PLoS ONE* **2012**, *7*, 16733–16737. [CrossRef]
14. He, M.X.; Hu, L.; Hu, C. Harbour dredging and fish mortality in an aquaculture zone: Assessment of changes in suspended particulate matter using multi-sensor remote-sensing data. *Int. J. Remote Sens.* **2014**, *35*, 4383–4398. [CrossRef]
15. Islam, M.A.; Wang, L.; Smith, C.; Reddy, R.; Lewis, A.; Smith, A. Evaluation of satellite remote sensing for operational monitoring of sediment plumes produced by dredging at Hay Point, Queensland, Australia. *J. Appl. Remote Sens.* **2007**, *1*, 011506. [CrossRef]
16. Sipelgas, L.; Raudsepp, U. Monitoring of HARBOR dredging Using Remote Sensing and Optical In Situ Data. In Proceedings of the IEEE International Geoscience and Remote Sensing Symposium, Cape Town, South Africa, 12–17 August 2009; pp. II-476–II-478. [CrossRef]
17. Lu, J.; Chen, X.; Zhang, P.; Huang, J. Evaluation of spatiotemporal differences in suspended sediment concentration derived from remote sensing and numerical simulation for coastal waters. *J. Coast. Conserv.* **2017**, *21*, 197–207. [CrossRef]
18. Jordan, Y.C.; Ghulam, A.; Hartling, S. Traits of surface water pollution under climate and land use changes: A remote sensing and hydrological modeling approach. *Earth-Sci. Rev.* **2014**, *128*, 181–195. [CrossRef]
19. Milzow, C.; Kgotlhang, L.; Kinzelbach, W.; Meier, P.; Bauer-Gottein, P. The role of remote sensing in hydrological modelling of the Okavango Delta, Botswana. *J. Environ. Manag.* **2009**, *90*, 2252–2260. [CrossRef]
20. Oey, L.-Y.; Ezer, T.; Hu, C.; Muller-Karger, F.E. Baroclinic tidal flows and inundation processes in Cook Inlet, Alaska: Numerical modeling and satellite observations. *Ocean. Dynam.* **2007**, *57*, 205–221. [CrossRef]
21. Van Dongeren, A.; Plant, N.; Cohen, A.; Roelvink, D.; Haller, M.C.; Catalan, P. Beach Wizard: Nearshore bathymetry estimation through assimilation of model computations and remote observations. *Coast. Eng.* **2008**, *55*, 1016–1027. [CrossRef]
22. Feng, L.; Hu, C.; Chen, X.; Cai, X.; Tian, L.; Gan, W. Assessment of inundation changes of Poyang Lake using MODIS observations between 2000 and 2010. *Remote Sens. Environ.* **2012**, *121*, 80–92. [CrossRef]
23. Wu, G.; Cui, L. Remote sense-based analysis of sand dredging impact on water clarity in Poyang Lake. *Acta Ecol. Sin.* **2008**, *28*, 6113–6120. (in Chinese).
24. Cui, L.; Qiu, Y.; Fei, T.; Wu, G. Using remotely sensed suspended sediment concentration variation to improve management of Poyang Lake, China. *Lake Reserv. Manag.* **2013**, *29*, 47–60. [CrossRef]
25. Lai, X.; Shankman, D.; Huber, C.; Yeou, H.; Huang, Q.; Jiang, J. Sand mining and increasing Poyang Lake's discharge ability: A reassessment of causes for lake decline in China. *J. Hydrol.* **2014**, *519*, 1698–1706. [CrossRef]
26. Li, J.; Tian, L.; Chen, X.; Li, X.; Lu, J.; Feng, L. Remote-sensing monitoring for spatio-temporal dynamics of sand dredging activities at Poyang Lake in China. *Int. J. Remote Sens.* **2014**, *35*, 6004–6022. [CrossRef]
27. Chen, X.; Wu, Z.; Tian, L.; Chen, L. Inversion model for dynamic monitoring of suspended sediment: A case study on Poyang Lake. *Sci. Technol. Rev.* **2007**, *25*, 19–22. [CrossRef]
28. Wu, G.; Cui, L.; Duan, H.; Fei, T.; Liu, Y. An approach for developing Landsat-5 TM-based retrieval models of suspended particulate matter concentration with the assistance of MODIS. *ISPRS J. Photogramm.* **2013**, *85*, 84–92. [CrossRef]
29. Wu, G.; Cui, L.; He, J.; Duan, H.; Fei, T.; Liu, Y. Comparison of MODIS-based models for retrieving suspended particulate matter concentrations in Poyang Lake, China. *Int. J. Appl. Earth Obs.* **2013**, *24*, 63–72. [CrossRef]

30. Zhang, P.; Wai, O.W.H.; Chen, X.; Lu, J.; Tian, L. Improving Sediment Transport Prediction by Assimilating Satellite Images in a Tidal Bay Model of Hong Kong. *Water* **2014**, *6*, 642–660. [CrossRef]
31. Leeuw, J.; Shankman, D.; Wu, G.; de Boer, W.F.; Burnham, J.; He, Q.; Yesou, H.; Xiao, J. Strategic assessment of the magnitude and impacts of sand mining in Poyang Lake, China. *Reg. Environ. Chang.* **2009**, *10*, 95–102. [CrossRef]
32. Zhang, P.; Lu, J.; Feng, L.; Chen, X.; Zhang, L.; Xiao, X.; Liu, H. Hydrodynamic and Inundataion modeling of China's Largest Freshwater Lake aided by remote sensing data. *Remote Sens.* **2015**, *7*, 4858–4879. [CrossRef]
33. Feng, L.; Hu, C.; Chen, X.; Tian, L.; Chen, L. Human induced turbidity changes in Poyang Lake between 2000 and 2010: Observations from MODIS. *J. Geophys. Res.-Oceans* **2012**, *117*. [CrossRef]
34. Gao, J.; Jia, J.; Kettner, A.J.; Xing, F.; Wang, Y.; Xu, X.; Yang, Y.; Zou, X.; Gao, S.; Qi, S.; et al. Changes in water and sediment exchange between the Yangtze River and Poyang Lake under natural and anthropogenic conditions, China. *Sci. Total Environ.* **2014**, *481*, 542–553. [CrossRef]
35. Huete, A.; Didan, K.; Miura, T.; Rodriguez, E.P.; Gao, X.; Ferreira, L.G. Overview of the radiometric and biophysical performance of the MODIS vegetation indices. *Remote Sens. Environ.* **2002**, *83*, 195–213. [CrossRef]
36. Delft Hydraulics. *Delft3d-Flow User Manual: Simulation of Multi-Dimensional Hydrodynamic Flows and Transport Phenomena, Including Sediments*; Deltares: Delft, The Netherlands, 2014; pp. 1–683.
37. Lesser, G.; Roelvink, J.; van Kester, J.; Stelling, G. Development and validation of a three-dimensional morphological model. *Coast. Eng.* **2004**, *51*, 883–915. [CrossRef]
38. Zhang, P.; Chen, X.; Lu, J.; Zhang, W. Assimilation of remote sensing observations into a sediment transport model of China's largest freshwater lake: Spatial and temporal effects. *Environ. Sci. Pollut. Res.* **2015**, *22*, 18779–18792. [CrossRef] [PubMed]
39. Tang, L.; Xiao, Y.; Zhou, H.; Luo, J.; Tang, H. Research on the effect factors of flow diversion ratio in Ganjiang River. *Hydro-Sci. Eng.* **2011**, *27*, 64–68. [CrossRef]
40. Zhang, P.; Chen, X.; Lu, J.; Zhang, W.; Xiao, X. Suspended sediment transport modeling of Poyang Lake in the wet season based on remote sensing data. *Geomat. Inform. Sci. Wuhan Univ.* **2017**, *42*, 369–376. [CrossRef]
41. Pennekamp, J.G.S.; Epskamp, R.J.C.; Rosenbrand, W.F.; Mullie, A.; Wessel, G.L.; Arts, T.; Deibel, I.K. Turbidity caused by dredging: Viewed in perspective. *Terra Aqua* **1996**, *64*, 10–17.
42. Li, Y.; Acharya, K.; Chen, D.; Stone, M. Modeling water ages and thermal structure of Lake Mead under changing water levels. *Lake Reserv. Manag.* **2010**, *26*, 258–272. [CrossRef]
43. Qi, H.; Lu, J.; Chen, X.; Sauvage, S.; Sanchez-Perez, J.-M. Water age prediction and its potential impacts on water quality using a hydrodynamic model for Poyang Lake, China. *Environ. Sci. Pollut. Res.* **2016**, *23*, 13327–13341. [CrossRef]

Article

Validation of a Novel, Shear Reynolds Number Based Bed Load Transport Calculation Method for Mixed Sediments against Field Measurements

Gergely T. Török [1,2,*], János Józsa [1,2] and Sándor Baranya [2]

1 MTA-BME Water Management Research Group; Hungarian Academy of Science, Budapest University of Technology and Economics, Műegyetem rakpart 3, H-1111 Budapest, Hungary; jozsa.janos@epito.bme.hu
2 Department of Hydraulic and Water Resources Engineering, Faculty of Civil Engineering, Budapest University of Technology and Economics, Műegyetem rakpart 3, H-1111 Budapest, Hungary; baranya.sandor@epito.bme.hu
* Correspondence: torok.gergely@epito.bme.hu; Tel.: +36-1-463-2248

Received: 30 July 2019; Accepted: 27 September 2019; Published: 30 September 2019

Abstract: In this study, the field measurement-based validation of a novel sediment transport calculation method is presented. River sections with complex bed topography and inhomogeneous bed material composition highlight the need for an improved sediment transport calculation method. The complexity of the morphodynamic features (spatially and temporally varied bed material) can result in the simultaneous appearance of the gravel and finer sand dominated sediment transport (e.g., parallel bed armoring and siltation) at different regions within a shorter river reach. For the improvement purpose of sediment transport calculation in such complex river beds, a novel sediment transport method was elaborated. The base concept of it was the combined use of two already existing empirical sediment transport models. The method was already validated against laboratory measurements. The major goal of this study was the verification of the novel method with a real river case study. The combining of the two sediment transport models was based on the implementation of a recently presented classification method of the locally dominant sediment transport nature (gravel or sand transport dominates). The results were compared with measured bed change maps. The verification clearly referred to the meaningful improvement in the sediment transport calculation by the novel manner in the case of spatially varying bed content.

Keywords: bed load transport; shear Reynolds number; bed-armoring; bed-change; Danube; gravel–sand mixture; 3D CFD modeling

---

1. Introduction

Sediment transport modeling is a recently still developing topic of morphodynamic investigations. Although researchers elaborate an increasingly accurate description of the sediment motion, there is still no one generally accurately applicable sediment transport model. The selection of the applied appropriate sediment transport method for a given case with even unique morphodynamic features (such as grain size, homogeneity of the bed content, bed slope, bed shear stress, hydraulic flow regime, the measure of armoring, etc.) must be preceded by a careful preliminary examination. There are, however, a large amount of empirically derived bed load transport formulas [1–10]. A comprehensive collection of the most widely applied formulas can be found in the Sedimentation Engineering Handbook [11]. The collection contains the most relevant sediment transport models, such as the ones from Meyer-Peter and Müller [1], from Einstein [2], Ashida and Michiue [4], Parker, Klingeman and McLean [5], surface-based relation of Parker [6], two-fraction relation of Wilcock and Kenworthy [7], surface-based relation of Wilcock and Crowe [8], relation of Wu et al. [9] and of Powell et al. [10].

The summary provides a short description of the hydraulic and sediment conditions of the experiments for which the given bed load formulas are developed (such as grain size, homogeneity of the bed material, bed slope, range of the bed shear stress and the measure of armoring). Although the applicability limits of the formulas are usually not defined in an exact way, they can be indirectly concluded based on the conditions of the benchmark experiments.

Török et al. [12] elaborated on a novel calculation method, which does not represent a new sediment transport model. The method says that by the combined and parallel application of the present models the applicability range can be increased. The laboratory measurements based validation evinced that the novel method can result in significantly more accurate sediment transport calculation [12]. However, the novel method was not yet verified with field measurements based comparative investigation.

2. Case Study

A problematic reach of the upper Hungarian Danube reach (river km (rkm) 1796–rkm 1794, Figure 1) has undergone major morphological changes during the last decades. Many studies presented in the literature [13–19] that because of installations of many river regulation measures (e.g., groin fields, ripraps and hydropower plant at rkm 1819) in the last decades, intensive gravel formations [19], important bed level incision [20] and bed armoring processes [16,19] could be detectable, mainly in the main channel [21]. In contrast, the bed content is much finer in the groin fields, causing siltation and erosion of the finer sediments during flood waves [19,22].

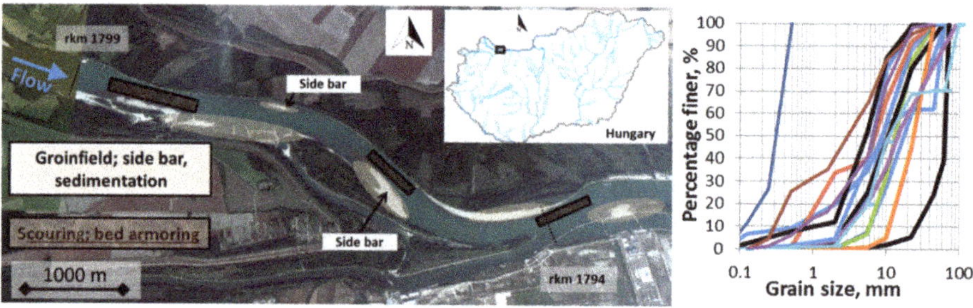

Figure 1. The sketch of the investigated Danube study reach (**left**) and grain size distributions taken from the investigated reach (**right**). The characteristic water discharges are mean flow Q_m = 2000 m^3/s, range of bankfull discharge Q_{bf} = 4300–4500 m^3/s, 2, 10 and 100-year flood event Q_2 = 5950 m^3/s, Q_{10} = 7950 m^3/s and Q_{100} = 10,400 m^3/s [23].

The river reach can be characterized by the following parameters: the channel bed width at mean water-stage ranges between 150 m and 350 m [19] with the average water surface gradient of 0.0002–0.00025. As Figure 1 shows, the river section is regulated by conventional structures, such as groins and the banks are protected against erosion by ripraps. Also, sidearms, islands, gravel bars, confluence zone can be observed, which refer to the complex topography of the river section. Bed material samples were taken from the main stream, groin fields and gravel bars. Some of the grain size distributions can be seen in Figure 1 right. The figures refer to very diverse spatial bed contents (0.32 mm < d_{50} < 70.5 mm, where d_{50} is the median grain size). Such a wide dispersion of the bed content is a unique feature of the Danube River (~rkm 1600–rkm 1800); at the Lower Austrian Danube, (~rkm 1885, 90 km upstream), the Danube flows through a gravel bed, where d_{50} is 21.1 mm without any finer fractions [24]. In turn, the middle Hungarian Danube (~200 km downstream) has a typical sandy bed with d_{50} < 0.05 mm [25]. The complexity of the topography and bed content suggest spatially and temporally varied sediment transport nature [22]. That is in some places the gravel, elsewhere the sand transport dominates [26]. This kind of individual complexity was presented e.g., by the field measurements of Török and Baranya [19], or in [27,28].

Based on field measurements, the researchers could make approximate estimates regarding the ongoing local- and reach-scale morphodynamic processes. For instance, the essential bed changes in the last decades caused important water management related problems and also difficulties in navigation, which could be detected with the field measurements [19,20,27–29]. However, the studies emphasize that additional studies are required for even more reliable morphodynamic investigation. Computational modeling offers an alternative for analyzing and predicting the recent and expected processes [30,31]. That is, the numerical modeling based investigation could be an important complementary manner with meaningful added value.

For this reason, the reliable calculation of the morphological changes is a major interest to researchers and the application of a 3D computational fluid dynamics (CFD) sediment transport model became justified. However, the choice of the applied sediment transport model was not obvious. Many formulas can be found in the literature (e.g., [1–4,32]). Most of them are developed focusing on a given morphodynamic process (e.g., Wilcock and Crowe model: bed armoring [8]; van Rijn model: sand erosion and deposition [33] etc.). However, in the case of the examined river section, spatially and temporally varied sediment transport nature occurs. That is, none of the existing sediment transport formula are expected to operate reliably for both the sand and coarse bed material, within a given river reach.

3. Materials and Methods

3.1. Introducing the Combining Sediment Transport Calculation Method

A novel combined approach [12] of the van Rijn and the Wilcock and Crowe bed load sediment transport formulas were applied because of the complex and spatially varied bed material and dominant sediment transport nature. From now, the Wilcock and Crowe formula will be indicated with *W&C*, while the van Rijn will be with *vR*. A brief summary of these models can be read in the Appendix A. The combined manner was already presented and validated against laboratory measurements [12]. In that study, laboratory experiments were used, which were carried out in an 11 m long, 1 m wide straight flume with non-uniform bed material. A groin was installed 5 m far from the outlet, which caused scouring, bed armoring and sand aggradation processes in the channel [34].

Török and Baranya [22,26] pointed out a novel decision criterion, which is a suitable method for indicating whether the sand or rather the sand transport dominates locally. The herein presented combining method was based on this statement. Namely, if the shear Reynolds number (Re^*) is below 300, the sand transport is prevalent. Otherwise, if the Re^* occurs above 400 the gravel transport dominates. Based on these, the combined calculation method said that the local bed load sediment transport rate was calculated as the following ($q_{bi,W\&C}$ is the sediment transport rate calculated by *W&C* and $q_{bi,vR}$ is the rate by *vR*):

$$q_{bi} = \begin{cases} q_{bi,W\&C} \; if \; Re^* > 400 \\ q_{bi,vR} \; if \; Re^* \leq 300 \\ else \\ f \cdot q_{bi,W\&C} + (1-f) \cdot q_{bi,vR} \end{cases}, \tag{1}$$

where

$$f = \frac{1}{100}(Re^* - 300). \tag{2}$$

These models do not contain any variable to calibrate, the constants (e.g., the constant von Karman, or the lift coefficient) and base equations (e.g., the original Shields curve [35] was used in the van Rijn model) were defined as they are published and recommended in the original papers [8,33].

In addition to the bed load transport estimation, the suspended sediment transport was calculated in each computational grid according to the suspended *vR* formula [36]. Thus, the bed load was calculated according to Equations (1) and (2), while the suspended load was estimated by the van Rijn equation.

3.2. Applied 3D Flow Model

The numerical model used in this study [37,38] solved the 3D Reynolds-averaged Navier–Stokes (RANS) equations with the k-ε turbulence closure (see e.g., [39]) by using a finite-volume method and the (semi-implicit method for pressure linked equations) SIMPLE algorithm [40,41] on a 3D non-orthogonal grid. At the boundaries, where the fluid flow cannot be considered as a free turbulence zone, the wall law was applied for the velocity profile calculation [42]. The momentum equations were in the complete form, describing the hydrodynamic effects in all directions. The roughening impact of the vegetation in the flood plain area was described as an energy loss term in the Navier–Stokes equations [43], and could be specified for each cell. Using this option, the effect of the vegetation was taken into account as a drag-effect.

In order to eliminate the boundary effect, the computational grid was longer in both upstream (rkm 1801) and downstream (rkm 1793.5) direction than the investigated ~ rkm 1795 and rkm 1799 river reach. The applied grid can be seen in Figure 2.

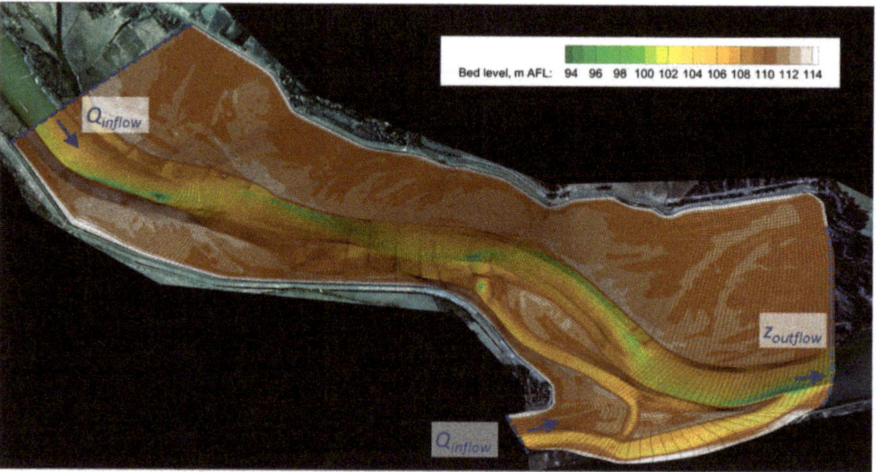

Figure 2. The computational grid of the investigated Danube study reach.

The study side was discretized with 355 cells in the streamwise direction and 150 cells in the lateral direction, respectively, resulting in the streamwise direction in an average resolution of 18 m and transversely 5 m in the main channel, while 13 m in the floodplain area. Vertically 11 layers were defined, at every ten percent of the depth, resulting in a maximum of 585,750 computational cells. The resolution of the grid was set based on previous 3D model studies [37,44–48] and Section 3.2.2 demonstrates its applicability.

The bed material was discretized by five fractions, which were: $d_1 = 0.3125$ mm, $d_2 = 1.25$ mm, $d_3 = 5.7$ mm, $d_4 = 16.2$ mm and $d_5 = 56.57$ mm. The ripraps and groins were characterized by $d = 300$ mm. According to field measurement based considerations, the active layer thickness was set to 0.5 m.

3.2.1. Parameterization

For instance, Parker mentioned [11] that the bed material of most river reaches is less complex and the grain sizes happen in a narrow range. In turn, in rare cases—e.g., the herein studied river section—the occurring grain sizes cover a significantly wider range (silt–gravel), resulting in a very complex spatial distribution. The bed material cannot be supposed as spatially uniform because of this (as many studies do at most river sections [44,46]), which makes the allocation of the bed material a less obvious method. According to Baranya [49], a relation can be stated between the calculated local bed shear stress value by 3D flow model at the mean-water stage and the local d_{50}. Thus, applying the fitted function, a transitional and continuous d_{50} map can be estimated based on the calculated bed shear stress distributions. This method was used based on 33 bed material samples [19,26]. The standard deviation of the calculated d_{50} to the measured d_{50} was 3.2 mm.

As the boundary conditions for the RANS equations, at the inflow boundaries the water discharge, at the outflow boundary, the water level was set. The discharge time series (Figure 3) and the water levels of the Danube were defined based on the measured time series. Additionally, the flow discharge time series of the Mosoni-Danube was set based on a 1D numerical Danube model [50].

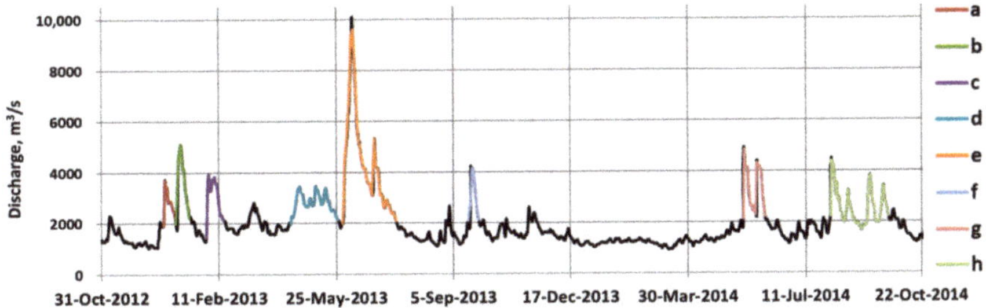

Figure 3. Discharge time series at rkm 1801 for the period October 2012–October 2014. The different colored periods marked with letters a–h display the flood waves (including the 100-year historical flood wave from 2013, e) which exceed the bed-forming flow discharge ($Q > 2100$ m^3/s).

An essential part of the model setup is the correct set of the inflow sediment rate. For this purpose, the flow discharge dependence of the suspended load [51] and the bed load [19] functions were used.

The riverbed topography of the main river channel was available from 2012 and 2014. The initial bed geometry was set according to the map from 2012. The calculated bed change map could be prepared for this two-year-long period, which included the historical flood wave from 2013 (Figure 3). This bed change map was used as a benchmark for the validation purpose (Figure 4).

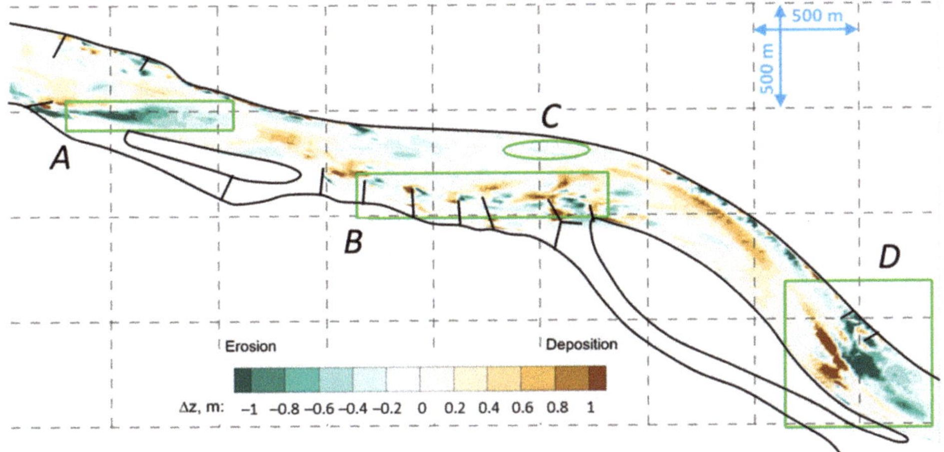

Figure 4. Measured bed changes for the period October 2012–October 2014.

Four regions in the river reach were highlighted by green rectangles and an ellipse, marked by A, B, C and D. In these places, the following bedforms and morphodynamic processes were detected by field measurements [16,19]. At region A and D, the blue spots refer to a pronounced scouring downstream of the groins. In region B a groin field could be found. Here, local bed changes took place, both scouring processes (blue spots) and sediment depositions (brown spots). The ellipse (C) and the brown spot in region D show the places where gravel bars were located. As these phenomena basically determine the reach-scale morphodynamic processes, a key question is whether the novel sediment transport calculation manner introduced a more reliable estimation of them.

The numerical simulation of the 2 years, 722 days long period demands very large computational capacity. According to the preliminary estimation calculations, the simulation of one model variant for such a long time period would take around half a year. Therefore, to reduce the duration of the simulation, only the periods that exceeded the bed-forming flow discharge ($Q > 2100$ m^3/s) [19,24] were simulated. It meant that the inflow boundary condition was set to a continuous flow discharge series, including the highlighted periods in Figure 3a–h, respectively and omitting the intermediate periods. In this range, 66% of the annual bed load amount passed [23]. In turn, it is emphasized that the ignored 34% of the annual bed load yield was still significant. That is, the numerical model neglects the simulation of bed changes that take place during the lower water regime. However, many studies (including papers regarding the investigated river reach) display that the major bed changes such as scouring, bar formations and flushing of the groin fields are expected rather during floods [16,19,24,52]. That is, the mean and lower water stages are less important in the view of bed changes. Based on the above, the bed changes caused by the eight flood waves (Figure 3a–h, a total of 211 days) approach well the real two years changes. Hence, the calculated bed changes were comparable to the measured.

3.2.2. 3D Flow Model Validation

The herein applied 3D CFD model was already adapted and validated for the investigated Hungarian reach of River Danube, which was published in previous research works, e.g., [45,46,53,54]. Those studies have already demonstrated the reliable application of the 3D flow model. Regardless of these, the flow model validation was elaborated for the peak of the historical flood wave in 2013. The cross-sectional acoustic Doppler current profiler (ADCP) flow measurements by the North-Transdanubian Water Directorate regarding the peak stage of the flood wave were used as benchmark flow values. Figure 5 shows the measured (left) and the calculated (right) cross-sectional velocity distributions, from exactly the same points. However, as the ADCP instrument is not able to measure at the direct water - and bed surfaces, the measured cross-sections are smaller slices, respectively. Furthermore, regarding the comparison it is important to mention that the moving boat ADCP measurements record the momentary flow conditions, which include the effect of turbulence, resulting in velocity pulsation in 0.1 m/s order of magnitude [25,53,55]. In turn, the RANS model calculates the time-averaged velocity values.

Comparing the measured and calculated cross-sections, the following remarks can be stated. The velocity values were in the same ranges (0–3 m/s). The highest velocities (yellow and red spots) were calculated at the same place as the cross-sections than in the real case, which underlined the reliable estimation of the main stream. The locations of the lower velocities (blue and light green spots) were calculated as trustworthy also. That is the calculated flow pattern could be realized reliable at the groin fields (e.g., at the right-bank sides of cross-section III, IV and V, Figure 5) and gravel bars also (e.g., at the right-bank sides of cross-section VII). Finally, the root-mean-square deviation (RMSD) between the measured and calculated velocities could be calculated for all the seven cross-sections. These values can be seen in Table 1.

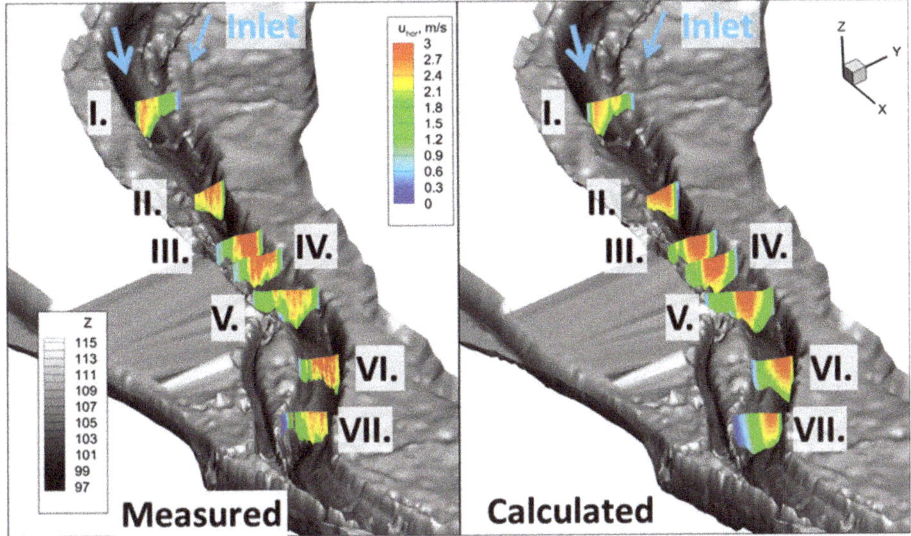

Figure 5. Measured and calculated cross-sectional horizontal velocity distributions.

Table 1. The root-mean-square deviation (RMSD) between the measured and calculated velocities, the maximum cross-sectional horizontal velocity and the percentage deviation of them for the seven cross-sections.

Cross-Section	I.	II.	III.	IV.	V.	VI.	VII.
RMSD, m/s	0.35	0.19	0.17	0.19	0.47	0.42	0.27
Max v_{hor}, m/s	2.73	2.84	3.01	3.14	3.13	2.88	2.62
Average dif., %	12.7	6.7	5.7	6.2	14.9	14.6	10.4

Considering the maximum cross-sectional horizontal velocities (Table 1, third row), the RMSD values displayed that the deviation of the flow calculation was below ~12% of the maximum values. The only two sections, where the inaccuracy happens ~15% were cross-sections V and VI. In these cross-sections, the higher deviation suggested that the numerical model estimated a narrower main stream as the measurements display. Around these cross-sections, the varying geometry increasec the transverse flow structure. However, a known limitation of the time-averaged RANS modeling was just this phenomenon, which is the underestimation of the secondary flow strength [56–58]. Next to this section, the flow model could be considered as validated for higher flood waves too, taking into account the available measurement data. For a more comprehensive validation, fixed boat ADCP measurements would have been expedient, which could be used to obtain time-averaged velocity values and also to estimate local bed shear stresses. The model validation for the 100-year flood wave should be evaluated in this light.

4. Results

4.1. Comparison of the Calculation Methods

In order to confirm the better operation of the novel Re^* dependent combined method, the bed change calculation by the van Rijn, Wilcock and Crowe and the combined method were compared.

The simulations were performed only for the d and e flood waves because of the significant computational time (see Figure 3d,e). The initial model setup (the flow field, water levels and the bed material) was given for each run from the results of the model runs for the first three flood waves by the Re^* dependent combined method. The flood wave d was a relatively low (peak is around 3460 m^3/s), but durable (~2 months long) flood wave, while the e was the historical one with a peak higher than 10,000 m^3/s. Thus, the comparative analysis presented the operational characteristics of the sediment transport models for both the durable lower and also for the extreme water regimes. As a benchmark, the measured bed change map indicated the extent of the possible bed changes. However, the measured and calculated maps could not be compared directly, because the measured belonged to the whole two-year-long period (Figure 3).

The calculated bed change maps for the flood wave d (Figure 3) are presented in Figures 6–8.

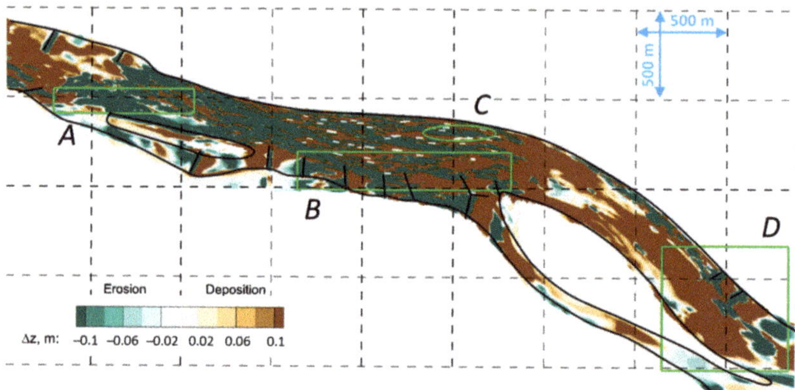

Figure 6. Calculated bed changes by the van Rijn (vR) formula for a 2.5 month-long period (Figure 3).

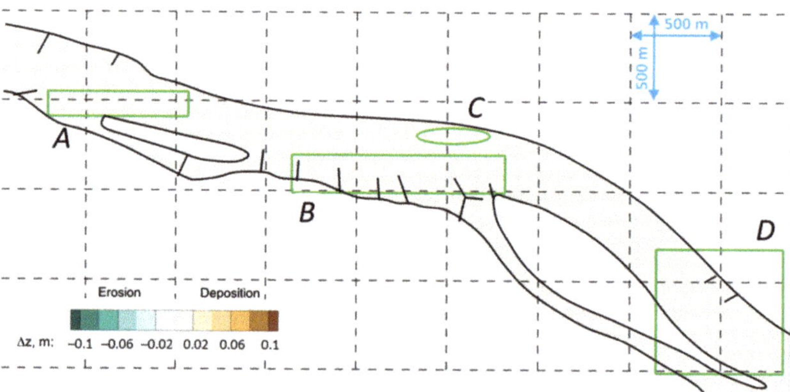

Figure 7. Calculated bed changes by the Wilcock and Crowe (*W&C*) formula for a 2.5 month-long period (Figure 3).

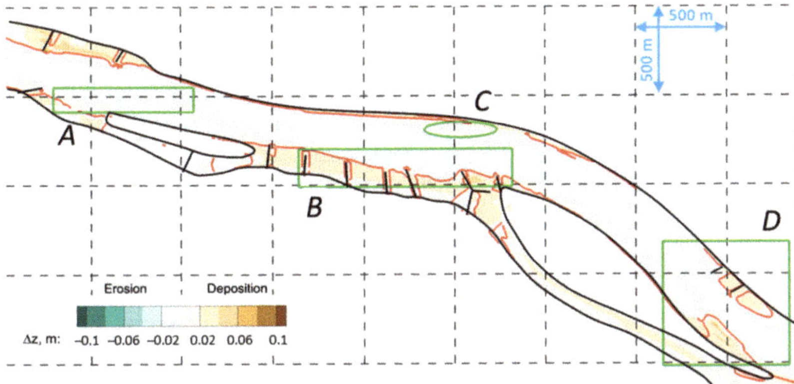

Figure 8. Calculated bed changes by the combined method for a 2.5 month-long period (Figure 3).

As the results show, the *vR* model estimated unrealistic changes both spatially and in magnitude. The unrealistically huge bed changes suggest that the *vR* model did not seem to be an appropriate model for the given Danube reach, particularly not for bed change calculation in the main channel.

The *W&C* sediment transport model estimated a more stable bed surface than the *vR* (Figure 7). In this particular case, the bed surface seemed so resistant that the mean flow field was too weak to cause any significant bed changes. The motion of the very fine, basically suspended inlet load was calculated by the *W&C* model as bed load. Therefore, that part of the inlet sediments settled progressively along the channel. However, because of the quite low suspended load [51], the bed level rise caused by sedimentation was negligible (<0.005 m).

Figure 8 shows the bed changes calculated by the combined method. The red lines illustrate the borderline which separated the areas where the *vR* or the *W&C* model was activated in the initial moment of the model run. Accordingly, it can be seen that the *vR* formula was invoked at the near-bank areas, at the groin fields and also at a smaller part of the Vének lower gravel bar. At these regions, more significant (~0.05 m) sedimentation was estimated. That is, at these less hydraulically rough parts of the river bed, the deposition of both the finer bed load and suspended load were expected, which areas can be detected by the Re^* [22]. In turn, in the navigation channel, no considerable bed change happened. According to the suspended form of the *vR* formula, the finer suspended load passes over the calculation domain, while the bed surface remains still, calculated by the *W&C* formula. This assumption is consistent with the conclusions of the field measurements [19]; the main channel seems to be armored enough to be resistant at the mean water regime.

The other flood wave, for which the comparative analysis of the three calculation methods (*vR*: Figure 9, *W&C*: Figure 10 and combined: Figure 11) was established was the historical flood wave from 2013 (Figure 3). Regarding this hydrological case, the *vR* model estimated also an unrealistic bed change map (Figure 9.). This was mainly true for the main channel. There, the motion of the coarser grains was probably overestimated, resulting in huge erosions and depositions. In turn, at the near-bank regions, at gravel bars and at the groin fields, the changes seemed to be partly in the expectable order of magnitude. However, it is clearly visible that the *vR* formula is not an appropriate choice for the morphological change calculation of such a complex river reach.

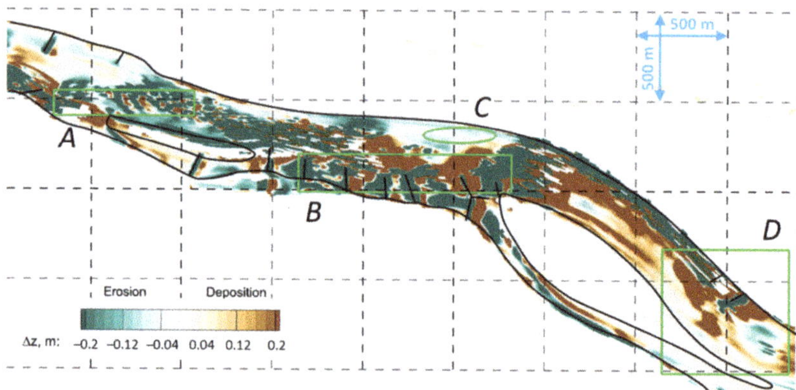

Figure 9. Calculated bed changes by the *vR* formula for the historical flood wave (Figure 3).

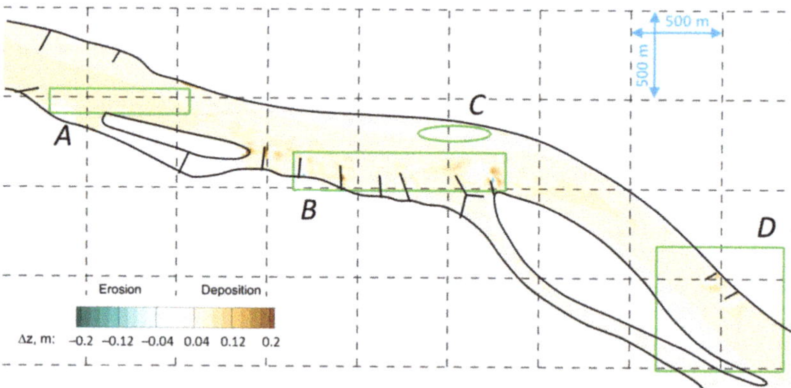

Figure 10. Calculated bed changes by the *W&C* formula for the historical flood wave (Figure 3).

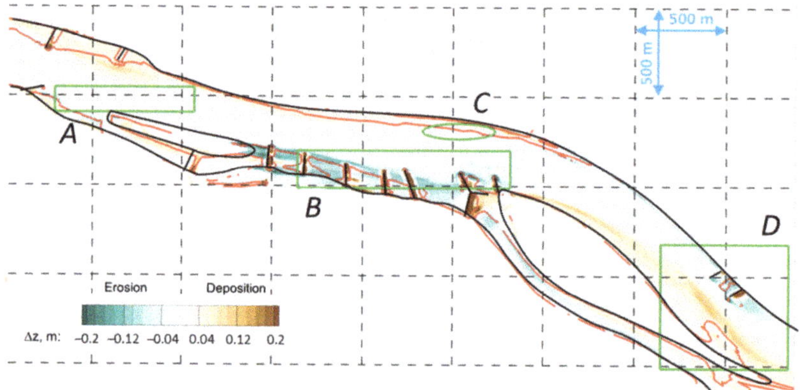

Figure 11. Calculated bed changes by the combined method for the historical flood wave (Figure 3).

The *W&C* model calculated more realistic bed changes (Figure 10), especially in the main channel. However, the measurements at the lower gravel bar and also at the whole main channel showed significant ($\Delta z > \pm 0.2$ m) changes. Based on these, the *W&C* formula likely overestimated the stability of the channel. A remarkable bed level increase can be pointed out, which moderated towards the downstream direction. These bed level changes could be explained by the settling of the inlet finer load, which could not be taken into account as suspended load. The publication of Török et al. [59] pointed out that in the case of mixed bed content, the Shields diagram predicts lower critical bed shear stress for the bed load of the finer, sand particles, than the reference shear stress of the *W&C* model. Accordingly, the *W&C* model estimates respectively higher stability for the sand particles, than the Shields diagram and thus the *vR* model. This leads to such kind of deposition pattern in the main stream which is unrealistic compared to the measured bed changes map [19].

The combined method predicts (Figure 11) more significant bed changes, compared to the one resulted by the *W&C* model. The red borderline suggested that as at the flood wave d, the combined method calculated the sediment transport by the *vR* formula in the near-bank regions. During the flood wave e, the remarkable erosion at the groin field B meant that the groin field got flushed and the earlier deposited finer sands got eroded. Also, notable changes took place at the vicinity of the gravel bar at region D. Here, the widening of the downstream sides of the gravel bar can be seen, in accordance with the measured bed change map in Figure 4. Around the groin pair at the left bank, on the opposite side of the gravel bar, the blue spots refer to erosion. This process was probably the result of a similar, flushing process like the one which happened at the upstream groin field. In turn, in the main and navigation channel, no considerable bed change happened. According to the suspended form of the *vR* formula, the finer suspended load passed over the calculation domain, while the bed surface remained still, calculated by the *W&C* formula. The conclusions of the field measurements [19] referred also to the resistant main channel.

The results indicate that the interactions between the sediment transport at different channel sections (groin fields, gravel bars and main channel) could not be estimated by the *vR* or *W&C* formulas. However, the expedient combination of them gives an opportunity to deal with the interaction-mechanism between the local- and reach-scale processes.

Analyzing the calculated bed changes in the marked boxes, the following assumptions can be stated. In this part, the results of the *vR* model were skipped because of the unrealistic bed change calculations. At region A, the real bed level deepening (Figure 4) could not be reproduced by any method. In region B, only the combining method was able to predict significant depositions and erosions. Accordingly, the depositions probably took place during the lower water regime, while the bed level incision occurred during the flood waves. Thus, the measured bed level changes during the two years were formed most likely indeed during the whole two-year-long period. In region C, all two model results suggested that the gravel bar was in a stable state. Besides region A, the bed level increase in the main channel between region C and D (Figure 4) could not be pointed out by any sediment transport formula. Finally, in region D the model results showed that the widening of the gravel bar and also the erosion at the vicinity of the groin pair occurred rather during the higher flood wave than during the slighter flood waves, or mean water regime.

4.2. Measured Data-Based Verification of the Combined Method

A quantitative assessment of the tested sediment transport formulas was performed based on the results of the comparative analysis. First, in accordance with the conclusions of the field measurement based investigation, it was assumed that the measured erosion and deposition took place mainly during higher water regimes in region D, when the flow discharge was higher than the bed forming discharge [19,23]. Thus, the measured and calculated data could be considered as indirectly comparable in this region. Therefore, from the measured bed level change maps, the total volume of the erosion and deposition could be calculated. Furthermore, counting the number of days of the higher water levels, the average daily rate of the volume of both the erosion and deposition could be estimated. Likewise, based on the calculated bed change maps and knowing the duration of the historical flood wave, the daily average volume changes of the deposition and erosion could be estimated.

Table 2 shows data about these volumes. The table presents the ratio of the calculated volume to the measured, regarding the deposition and erosion separately, for each sediment transport formula. A value of 1 would indicate a perfect match to the measured volume change. Table 2 shows that the *vR* model was the one which overestimates most both the deposition and erosion volumes. The *W&C* model calculated the deposition rate more accurately, but still indicated more than the measured. This was partly explained by the lack of the suspended sediment calculation. In turn, the *W&C* model estimated negligible low erosion. In total, the more reliable results were provided by the combined method. With this, the erosions at the near-bank parts were calculated more accurately by the *vR* model, resulting in sediment feed for region D. Thus, because of the capturing of the coming sediments, the widening of the deposition could be better represented. As the Re^* dependent criterion activated the *vR* formula at the groin pair, the bed level incision at its vicinity was also estimated better.

Table 2. The average daily volume changes for calculated (ΔV_c) and measured (ΔV_m) volume ratio values $\Delta V_c/\Delta V_m$ for region *D*. A value of 1 would indicate the perfect match.

		Sediment Transport Model		
		van Rijn	**Wilcock and Crowe**	**Re^* Based Combined**
The rate of the calculated to measured volumes	Deposition	48.7	4.9	3.5
	Erosion	7.2	0	0.7

Even though the measured and calculated bed changes could not be compared directly, the nature, the magnitudes and the locations of the remarkable bed changes suggested the greater aptitude of the combined method.

As it was discussed in Section 3.2.1, the calculated bed changes were elaborated only for the higher flow discharges (>2100 m^3/s, Figure 3), so the results could not be compared with the measured changes directly. Therefore, to achieve a notionally common scale, the bed change values were normalized. That is both the measured and calculated bed changes got divided by the highest bed level decrease or increase the value of the main channel:

$$\Delta z_{norm} = \left\{ \begin{array}{c} if\ \Delta z > 0 \rightarrow \frac{\Delta z}{\Delta z_{max}} \\ else \rightarrow \frac{\Delta z}{|\Delta z_{min}|} \end{array} \right\}. \tag{3}$$

Thus, the occurring values develop between −1 and 1, where 1 indicates the maximum deposition height along the river reach, while −1 belongs to the biggest erosion ($\Delta z_{max,\ meas} = 1.5\ m$, $|\Delta z_{min,\ meas}| = 1.8\ m$; $\Delta z_{max,\ calc} = 0.5\ m$, $|\Delta z_{min,\ meas}| = 0.25\ m$). The measured bed changes of both the erosions and depositions were consequently higher than the calculated. That is, the numerical model underestimated the magnitudes of the bed changes. This could be partly explained by the ignoring of the third part of the annual bed load in the numerical model estimation.

By the comparison of the measured (Figure 12) and calculated normalized bed changes (Figure 13) the following remarks can be stated. The modeled bed changes did not represent the remarkable erosion at region A. It is noted that this difference also meant a sediment supply loss for the downstream in the model calculation. Since the model did not manifest any bed level decrease, the bed material was probably finer here than it was set in the model. At region B, the measured scours appeared in the calculated results (blue spots). However, not as concentric scours, but rather as lengthwise formations. The numerical model represented depositions close to the measured magnitude at region B also. However, their location iwa not accurate; the brown spots occur between the groins, instead of in the front of the groins, like in Figure 12.

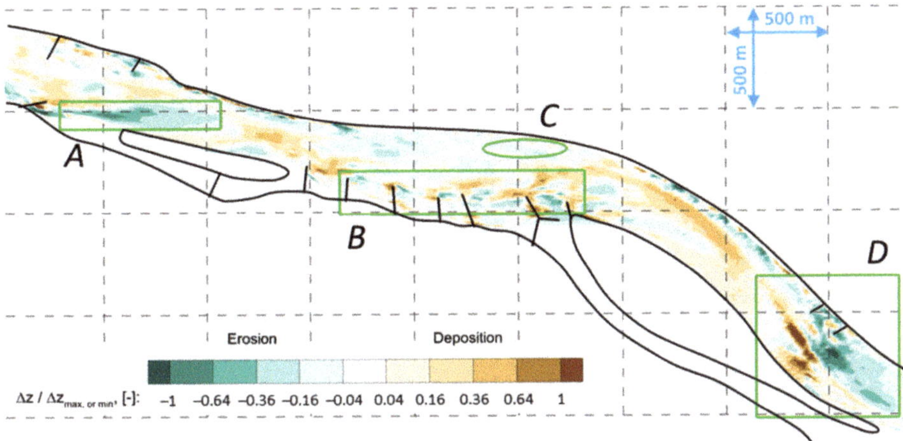

Figure 12. Normalized bed changes of the measured values (Figure 4) regarding the whole period October 2012–October 2014.

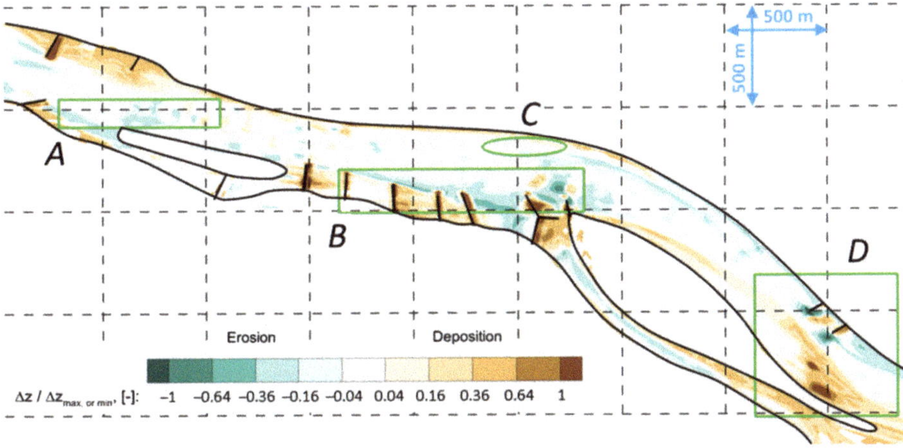

Figure 13. Normalized bed changes of the calculated values. The calculation was elaborated for the eight flood waves in the period October 2012–October 2014 (Figure 3).

At the gravel bar in region C, negligible bed changes were measured. Likewise, the numerical model predicted a stable bed surface. Finally, the combined method pointed out the growing of the lower part of the Vének lower gravel bar (region D, Figure 13). The measured bed changes indicated two, separable depositions. These double depositions were also represented by the model results. Like the measured changes, the model also calculated scouring in front of the left bank groin pair. However, the extension of it was considerably lower than the scale of the measured deepening. In turn, the lengthwise deposition between the groins (between regions C and D) was also indicated in both measured and calculated maps. However, the calculated deposition occurred in a significantly lower range and formed at the right bank side and not in the main stream. Also, an important match was that neither the measured nor the calculated values suggested any essential large-scale bed changes in the main channel.

A significant difference between the measured and calculated bed changes happened in region A. Here, the effect of the potential error in the initial bed material was further examined. An investigation was performed, which was based on the assumption that the initial bed material was set inaccurately around region A. The bed material samples around $d_{50} \approx 0.01$ m. Considering the grain-size distributions [19], a still realistic, but considerably finer bed material was presupposed. Therefore, the model was set up by 30% lower d_{50}, which was $d_{50} \approx 0.007$ m. With this only one difference, the model was run for the historical flood wave. Figure 14 presents the bed changes at region A, and at the downstream of it.

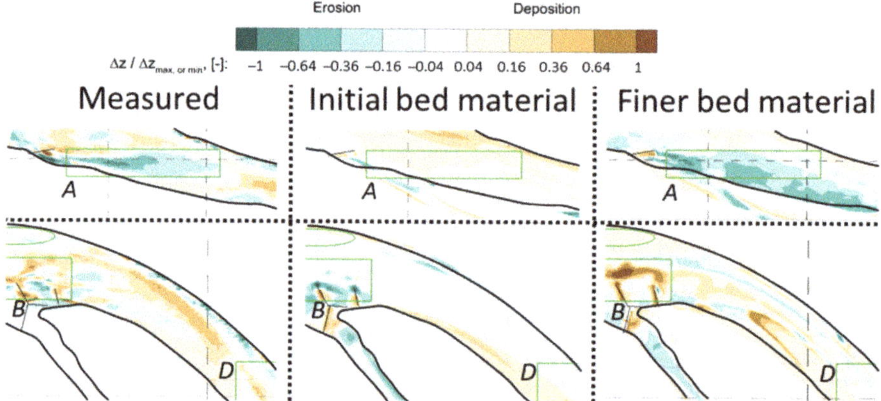

Figure 14. Calculated bed changes by the combining method for the historical flood wave. The model in the middle figure was set up with the initial, while in the right figure with finer bed material.

The right side of Figure 14 shows the bed changes in the case of finer bed material. It can be seen that the 30% decrease of the d_{50} resulted in major erosion at region A. Considering the measured changes in the left figure it is obvious that the decreasing of the d_{50} led to a better match to the real bed changes. The lower row of Figure 14 represents the bed changes at the downstream. Here, important deposition formations could be measured (left Figure 14) in front of the right bank groin pair (region B) and also in the main stream, between the two gravel bars (between region B and D). These changes could not be represented by the original model setup (middle Figure 14). In turn, in the case of finer bed material (right Figure 14), the model predicted important depositions in these regions. The extension of the deposition in front of the groin pair (region B) was very similar to the measured. Also, the lengthwise extension of the deposition downstream (between regions B and D), in the main stream were also reproduced. However, the location of it was not correct. It seems that the model underestimated the crosswise sediment transport, which is a known limitation of the Reynolds averaged description of the flow field [56–58,60]. Concluding, the herein presented investigation suggested that the bed material at region A was finer than the predicted d_{50} allocation for the original model setup. With this assumption, the deposition nature at the downstream has become also detectable.

5. Discussion

There are existing proven models in the literature that work reliably for given morphodynamic cases (e.g., the van Rijn model for hydraulic smooth regimes, which mainly occurs in clear sand bed; or the Wilcock and Crowe model for hydraulic rougher regimes, which develops in coarser bed surfaces). However, there are river reaches, within which hydraulic smoother and rougher zones can form, resulting in a completely different type of sediment transport, such as bed armoring and silt aggradation parallelly. For such cases, Török et al. developed a novel sediment transport calculation method and validated it against laboratory measurements.

In this study, its validation against field measurements was introduced. The results showed that the combined application of the Wilcock and Crowe and the van Rijn models could significantly increase the precision of numerical bed change calculations. The results showed that the combined method calculated both erosion and deposition the most reliably. It estimated the volume of the scouring by order of magnitude more precisely than the other two models. Additionally, the results also evinced that the location, extension and shape of the bed formations (e.g., scours, deposition, bar evolvement) could be calculated more reliably by the novel combined approach. These results were consistent with the laboratory measurements based on validation of the calculation method.

The sediment transport interaction between the sand and gravel dominated parts of a river reach played an essential role in the bed changes. For instance, the flushing of the silt dominated groin fields can feed the downstream gravel bars, leading to the growth of the bars. The application of the combined approach really makes sense, within river reaches, where such well-separated morphodynamic situations prevail (e.g., sand aggradation in groin fields; bed armoring in the main stream). In such cases, the local-scale morphodynamic processes are calculated by the proven models, while the interaction between them leads to a more accurate reach-scale calculation. The comparison of the numerical simulations to the measured data presented that such reach-scale interactions can be taken into account by the novel combining method.

6. Conclusions

The validation against both laboratory and field measurements suggests that not necessarily the development of completely new models can lead to the evolution of the sediment transport calculation. This study highlights that the optimally combined use of sediment transport models can be a promising alternative in the sediment transport calculation. In this context, it is important to state that regardless of whether a sediment transport model is applied in combination or simply, the determination of its applicability limits by any morphodynamic variables in a well-defined way is vital. In this study, a shear Reynolds number based description was applied which allowed for the authors an easily described combination criteria. However, the essence of the combining theory does not insist on any variable for the applicability limit description. However, the herein presented example highlights the need for such a kind of description method and also the potential of the parallel use of any sediment transport models.

Author Contributions: Conceptualization, methodology, software, validation, writing—original draft preparation, visualization, G.T.T., resources, data curation, G.T.T. and S.B.; formal analysis, writing—review and editing, S.B.; supervision, project administration, funding acquisition, S.B. and J.J.

Funding: This research received no external funding.

Acknowledgments: This research was supported by MTA TKI (Office for Research Groupsof the Hungarian Academy of Sciences). Support of grant BME FIKP-VÍZ by EMMI (Ministry of Human Capacities) is also kindly acknowledged. The authors acknowledge the funding of the OTKA (Hungarian Scientific Research Fund) FK 128429 grant. The research was partly supported by the ÚNKP-19-4 New National Excellence Program of the Ministry for Innovation and Technology. The last author acknowledges the funding from the János Bolyai fellowship of the Hungarian Academy of Sciences.

Conflicts of Interest: The authors declare no conflict of interest.

Appendix A

Appendix A.1 Wilcock and Crowe Bed Load Transport Formula

The well-known surface-based bed load formula of *W&C* [8] was developed for mixed gravel/sand sediments. The gravel ranged in size from 2.0 to 64 mm, the sand from 0.5 to 2.0 mm. They conducted laboratory experiments in a tilting laboratory flume, in which both the water and the sediments were recirculated. Five different sediment mixtures were investigated, varying the sand content of the sediment mixtures from 6.2% up to 34.3% [60]. The mixture grain-size distribution used in the experiments of Török et al. [34] and of Wilcock et al. [60] shows a good match. In case of the experiments of Wilcock et al. [60], d_m grain size of the initial bed materials was in the range of 4.1–10.5 mm, whereas d_m was 8.7 mm the measurements of Török et al. [34]. The *W&C* formula is therefore expected to be capable of describing the gravel/sand bed load transport processes in our experiments. In the case of such non-uniform bed material, the interaction between (hiding and exposure) the particles of different sizes plays an essential role in the stability of the sediments and can result in bed armoring. The *W&C* formula is reported to be capable of predicting the transient conditions of bed armoring or scouring processes. The preliminary numerical simulations of Török et al. [46] also supported this statement.

The transport formula uses the collapse similarity hypothesis, which estimates the fractional sediment transport rate (w_i*) as the function of the ratio of the shear stress (τ) and the reference shear stress (τ_{ri}). The reference shear stress τ_{ri} is the value of τ at which w_i* is equal to a small but already perceptible value w_i* = 0.002. That is the reference shear stress has a similar physical meaning to the critical shear stress (τ_c), which indicates the initiation of motion of particles. The dimensionless transport rate is obtained according to the following equation:

$$W_i^* = \begin{cases} 0.002\theta^{7.5} & if\ \theta < 1.35 \\ 14\left(1 - \frac{0.894}{\theta^{0.5}}\right)^{4.5} & if\ \theta \geq 1.35 \end{cases}, \tag{A1}$$

where

$$\theta = \tau / \tau_{ri}. \tag{A2}$$

The volumetric transport rate per unit width q_{bi} can be expressed from the following formula:

$$W_i^* = \frac{(s-1)gq_{bi}}{F_i u_*^3}, \tag{A3}$$

where F_s is the proportion of sand in surface size distribution.

For the estimating of τ_{ri}, the *W&C* model suggests that the stability of the sediment particles depends on the sand content. Thus τ_{ri} can be calculated by the following steps:

$$\tau_{rm}^* = 0.021 + 0.015e^{-20F_S}, \tag{A4}$$

where τ^*_{rm} is the dimensionless reference shear stress for the mean size of the bed surface. The reference shear stress for the mean size can be obtained from the particle Froude number equation:

$$\tau_{rm}^* = \frac{\tau_{rm}}{g(\rho_S - \rho_W)d}. \tag{A5}$$

Then, the last step is the estimation of the reference shear stress for the particle sizes separately:

$$\frac{\tau_{ri}}{\tau_{rm}} = \left(\frac{D_i}{D_{sm}}\right)^b, \tag{A6}$$

where

$$b = \frac{0.67}{1 + e^{\left(1.5 - \frac{D_i}{D_{sm}}\right)}}. \tag{A7}$$

Appendix A.2 Van Rijn Bed Load Transport Formula

The development of the *vR* formula is based on laboratory and field measurements. Van Rijn presented a mathematical method for the computation of bed load transport rate. The formula is basically the numerical solution of the equations of motions for a solitary particle. The mathematical model was calibrated based only on the experiments of Fernandez Luque [61,62], where the motions of uniform particles (d = 0.18 mm) under steady flow (u_* = 0.04 m/s) were monitored.

The lifting term of the equation of motion for a particle is derived as [63]:

$$F_L(shear) = \alpha_L \rho \nu^{0.5} D^2 v_r \left(\frac{\partial u}{\partial z}\right)^{0.5}, \tag{A8}$$

where F_L is the lift force, α_L is the lift coefficient, ν is the kinematic viscosity coefficient, ρ is the density of the fluid, v_r is the relative particle velocity and $\partial u/\partial z$ is the velocity gradient. The mathematical method uses the turbulent wall law Equation (A9) for calculation of the vertical flow velocity distribution:

$$U(z) = \frac{u_*}{\kappa} ln\left(\frac{z}{z_0}\right), \tag{A9}$$

where $U(z)$ is the horizontal velocity as the function of the z, z is the level above the bed, u_* is the bed shear velocity, κ is constant von Karman ($\kappa = 0.41$), and z_0 is the zero-velocity level above the bed.

The volumetric bed load transport rate is defined as:

$$\frac{q_b}{[(s-1)g]^{0.5} D_{50}^{1.5}} = 0.053 \frac{T^{2.1}}{D_*^{0.3}}, \tag{A10}$$

where q_b is the bed load transport per unit width, D_{50} is the particle size, s is the specific density, g is the acceleration of gravity, T is the transport stage parameter, can be represented as:

$$T = \frac{(u'_*)^2 - (u_{*,cr})^2}{(u_{*,cr})^2}, \tag{A11}$$

where $u_{*,cr}$ is the critical bed shear velocity according to Shields [35], $u'_* = \left(g^{0.5}/C'\right)\overline{u}$ is the bed shear velocity related to grains, where C' is the Chézy-coefficient related to grains and $\overline{u}$ is the mean flow velocity.

In the Equation (A10), D_* is the particle parameter defined as:

$$D_* = D_{50}\left[\frac{(s-1)g}{\nu^2}\right]^{1/3} \tag{A12}$$

Finally, the vR formula was verified based on 56 field measurements and 524 flume data. The bed load formula is recommended to use for particles in the range of 200–2000 µm (i.e., below 2 mm). As such the formula is not expected to reliably calculate the motion of gravels and the interaction between particles of different sizes and to describe such a process, like bed armoring. However, the model can presumably describe the transport of finer grains.

References

1. Meyer-Peter, E.; Müller, R. Formulas for Bed-Load Transport. In Proceedings of the IAHSR 2nd Meeting, Stockholm, Sweden, 7–9 June 1948.
2. Einstein, H.A. *The Bed-Load Function for Sediment Transportation in Open Channel Flows*; United States Department of Agriculture, Economic Research Service: Washington, DC, USA, 1950.
3. Egiazaroff, I.V. Calculation of Nonuniform Sediment Concentrations. *J. Hydraul. Div.* **1965**, *91*, 225–247.
4. Ashida, K.; Michiue, M. Study on hydraulic resistance and bedload transport rate in alluvial streams. *Trans. Jpn. Soc. Civ. Eng.* **1972**, *206*, 59–69. [CrossRef]
5. Parker, G.; Klingeman, P.; McLean, D. Bedload and size distribution in natural paved gravel bed streams. *J. Hydraul. Eng.* **1982**, *108*, 544–571.
6. Parker, G. Surface-based bedload transport relation for gravel rivers. *J. Hydraul. Res.* **1990**, *28*, 417–436. [CrossRef]

7. Wilcock, P.R.; Kenworthy, S.T. A two-fraction model for the transport of sand/gravel mixtures. *Water Resour. Res.* **2002**, *38*, 12-1–12-12. [CrossRef]
8. Wilcock, P.R.; Crowe, J.C. Surface-based transport model for mixed-size sediment. *J. Hydraul. Eng. ASCE* **2003**, *129*, 120–128. [CrossRef]
9. Wu, W.; Wang, S.S.Y.; Jia, Y. Nonuniform sediment transport in alluvial rivers. *J. Hydraul. Res.* **2000**, *38*, 427–434. [CrossRef]
10. Powell, D.M.; Reid, I.; Laronne, J.B. Evolution of bed load grain size distribution with increasing flow strength and the effect of flow duration on the caliber of bed load sediment yield in ephemeral gravel bed rivers. *Water Resour. Res.* **2001**, *37*, 1463–1474. [CrossRef]
11. Parker, G. Transport of Gravel and Sediment Mixtures. In *Sedimentation Engineering*; Garcia, M., Ed.; American Society of Civil Engineers: Reston, VA, USA, 2008; pp. 165–251. ISBN 978-0-7844-0814-8.
12. Török, G.T.; Baranya, S.; Rüther, N. 3D CFD Modeling of Local Scouring, Bed Armoring and Sediment Deposition. *Water* **2017**, *9*, 56. [CrossRef]
13. Rákóczi, L.; Sass, J. A Felső-Duna és a Szigetközi mellékágak mederalakulása a dunacsúni duzzasztómű üzembe helyezése után (Changes of the channel of the Hungarian Upper Danube and of the side river arms of the Szigetköz upon putting the Dunacsúny I. river barrage into operati. *Vízügyi Közlemények* **1995**, *77*, 46–75.
14. Hankó, Z. Gondolatok a Duna Szap és Szob közötti szakaszának fejlesztéséről (Considerations related to the development of the Danube reach between Szap and Szob). *Vízügyi Közlemények* **2000**, *82*, 285–299.
15. Holubová, K.; Capeková, Z.; Szolgay, J. Impact of hydropower schemes at bedload regime and channel morphology of the Danube River. In Proceedings of the River Flow 2004: Second International Conference on Fluvial Hydraulics, Napoli, Italy, 23–25 June 2004; CRC Press: Napoli, Italy, 2004; pp. 135–142.
16. Baranya, S.; Józsa, J.; Török, G.T.; Ficsor, J.; Mohácsiné Simon, G.; Habersack, H.; Haimann, M.; Riegler, A.; Liedermann, M.; Hengl, M. A Duna hordalékvizsgálatai a SEDDON osztrák-magyar együttműködési projekt keretében (Introduction of the joint Austro-Hungarian sediment research under the SEDDON ERFE-project). *Hidrológiai Közlöny* **2015**, *95*, 41–46.
17. Hankó, Z.; Starosolszky, Ö.; Bakonyi, P. Megvalósíthatósági tanulmány a Duna környezetének és hajózhatóságának fejlesztésére (Danube Environmental and navigation Project, Feasibility Study). *Vízügyi Közlemények* **1996**, *78*, 291–315.
18. Goda, L. A Duna gázlói Pozsony-Mohács között (Shallows of the River Danube between Pozsony, Bratislava and Mohács. *Vízügyi Közlemények* **1995**, *77*, 71–102.
19. Török, G.T.; Baranya, S. Morphological Investigation of a Critical Reach of the Upper Hungarian Danube. *Period. Polytech. Civ. Eng.* **2017**, *61*, 752–761. [CrossRef]
20. Holubová, K.; Comaj, M.; Lukac, M.; Mravcová, K.; Capeková, Z.; Antalová, M. *Final Report in DuRe Flood Project—Danube Floodplain Rehabilitation to Improve Flood Protection and Enhance the Ecological Values of the River in the Stretch between Sap and Szob*; Danube Transnational Programme: Bratislava, Slovakia, 2015.
21. Varga-Lehofer, D.T. A Felső-Magyarországi Duna Morfológiai Változásainak Elemzése (Investigation of the Morphological Changes of the Hungarian Upper Danube). Bachelor's Thesis, Budapest Univerity of Technology and Economics, Budapest, Hungary, 2014.
22. Török, G.T.; Józsa, J.; Baranya, S. A Shear Reynolds Number-Based Classification Method of the Nonuniform Bed Load Transport. *Water* **2019**, *11*, 73. [CrossRef]
23. Török, G.T. Methodological Improvement of Morphodynamic Investigation Tools for Rivers with Non-Uniform Bed Material. Ph.D. Thesis, Budapest Univerity of Technology and Economics, Budapest, Hungary, 2018.
24. Liedermann, M.; Gmeiner, P.; Kreisler, A.; Tritthart, M.; Habersack, H. Insights into bedload transport processes of a large regulated gravel-bed river. *Earth Surf. Process. Landf.* **2018**, *43*, 514–523. [CrossRef]
25. Baranya, S. Three-Dimensional Analysis of River Hydrodynamics and Morphology. Ph.D. Thesis, Budapest University of Technology and Economics, Budapest, Hungary, 2009.
26. Török, G.T.; Baranya, S. A shear Reynolds number based sediment transport classification method for complex river beds. In Proceedings of the 8th International Symphosium on Environmental Hydraulics, Notre Dame, IN, USA, 4–7 June 2018.

27. Rákóczi, L. *A Duna Szap-Gönyű Közötti Szakaszának Hajózási Viszonyait Javító Beavatkozások Vizsgálata (Investigation of Interventions for Navigation Improvment at the Danube Channel between Szap and Szob)*; Technical Report; Budapest University of Technology and Economics: Budapest, Hungary, 2004.
28. Baranya, S.; Goda, L.; Józsa, J.; Rákóczi, L. Complex hydro- and sediment dynamics survey of two critical reaches on the Hungarian part of river Danube. In Proceedings of the IOP Conference Series: Earth and Environmental Science, Bled, Slovenia, 2–4 June 2008; Volume 4.
29. Habersack, H.; Haimann, M.; Baranya, S.; Józsa, J.; Riegler, A.; Sindelar, C.; Liedermann, M.; Ficsor, J.; Simon, G.M.; Hengl, M. Gemeinsame österreichisch-ungarische Sedimentforschung im Rahmen des EFRE-Projektes SEDDON. *Österreichische Wasser Und Abfallwirtschaft* **2014**, *66*, 340–347. [CrossRef]
30. Wu, K.; Yeh, K.-C.; Lai, Y.G. A Combined Field and Numerical Modeling Study to Assess the Longitudinal Channel Slope Evolution in a Mixed Alluvial and Soft Bedrock Stream. *Water* **2019**, *11*, 735. [CrossRef]
31. Sattar, A.M.A.; Bonakdari, H.; Gharabaghi, B.; Radecki-Pawlik, A. Hydraulic Modeling and Evaluation Equations for the Incipient Motion of Sandbags for Levee Breach Closure Operations. *Water* **2019**, *11*, 279. [CrossRef]
32. Bogárdi, J. *A Hordalékmozgás Elmélete*; Vilmos, I., Ed.; Akadémiai Kiadó: Budapest, Hungary, 1955.
33. Van Rijn, L.C. Sediment Transport, Part I: Bed Load Transport. *J. Hydraul. Eng.* **1984**, *110*, 1431–1456. [CrossRef]
34. Török, G.T.; Baranya, S.; Rüther, N.; Spiller, S. Laboratory analysis of armor layer development in a local scour around a groin. In Proceedings of the International Conference on Fluvial Hydraulics, RIVER FLOW 2014, Lausanne, Switzerland, 7–10 July 2014; Taylor and Francis Group: Lausanne, Switzerland, 2014; pp. 1455–1462.
35. Shields, A. Application of Similarity Principles and Turbulence Research to Bed-Load Movement. *Mitt. Preuss. Versuchsanst. Wasserbau Schiffbau* **1936**, *26*, 47.
36. van Rijn, L.C. Mathematical modelling of morphological processes in the case of suspended sediment transport. Ph.D. Thesis, Civil Engineering and Geoscience, TU Delft, Delft, The Netherlands, 1987.
37. Reidar, B.; Olsen, N. A Three-Dimensional Numerical Model for Simulation of Sediment Movements in Water Intakes with Moving Option; Trondheim, Norway. 2018. Available online: http://folk.ntnu.no/nilsol/ssiim/manual5.pdf (accessed on 30 September 2019).
38. Olsen, N.R.B. Numerical Modelling and Hydraulics; Online Manuscript. 2012. Available online: http://folk.ntnu.no/nilsol/tvm4155/flures6.pdf (accessed on 30 September 2019).
39. Pope, S.B. *Turbulent Flows*; Cambridge University Press: Cambridge, UK, 2000; ISBN 9780521598866.
40. Patankar, S.V. *Numerical Heat Transfer and Fluid Flow*; Minkowycz, M.J., Sparrow, E.m., Eds.; McGraw-Hill Book Company: New York, NY, USA, 1980; ISBN 0-07-048740-5.
41. Glock, K.; Tritthart, M.; Habersack, H.; Hauer, C. Comparison of Hydrodynamics Simulated by 1D, 2D and 3D Models Focusing on Bed Shear Stresses. *Water* **2019**, *11*, 226. [CrossRef]
42. Schlichting, H. *Boundary-Layer Theory*; McGraw-Hill: New York, NY, USA, 1979.
43. Fischer-Antze, T.; Stoesser, T.; Bates, P.; Olsen, N.R.B. 3D numerical modelling of open-channel flow with submerged vegetation 3D numerical modelling of open-channel flow with submerged vegetation Modélisation numérique 3D d'un écoulement en canal avec vegetation submergée. *J. Hydraul. Res.* **2013**, *39*, 303–310. [CrossRef]
44. Fischer-Antze, T.; Reidar, B.; Olsen, N.; Gutknecht, D. Three-dimensional CFD modeling of morphological bed changes in the Danube River. *Water Resour. Res.* **2008**, *44*. [CrossRef]
45. Baranya, S.; Józsa, J. Numerical and laboratory investigation of the hydrodynamic complexity of a river confluence. *Period. Polytech. Civ. Eng.* **2007**, *51*, 3–8. [CrossRef]
46. Török, G.T.; Baranya, S.; Rüther, N. Three-dimensional numerical modeling of non-uniform sediment transport and bed armoring process. In Proceedings of the 18th Congress of the Asia & Pacific Division of the International Association for Hydro-Environment Engineering and Research 2012, Jeju, Korea, 19–23 August 2012.
47. Bihs, H.; Olsen, N.R.B. Numerical Modeling of Abutment Scour with the Focus on the Incipient Motion on Sloping Beds. *J. Hydraul. Eng.* **2011**, *137*, 1287–1292. [CrossRef]
48. Rüther, N.; Olsen, N.R.B. Modelling free-forming meander evolution in a laboratory channel using three-dimensional computational fluid dynamics. *Geomorphology* **2007**, *89*, 308–319. [CrossRef]

49. Baranya, S. River Bed Material Mapping to Support Habitat Assesment in Large Rivers. In Proceedings of the 12th International Symposium on Ecohydraulics, Tokyo, Japan, 19–24 August 2018.
50. Krámer, T.; Szilágyi, J.; Józsa, J. Mértékadó árvízszintek: Országos felülvizsgálat után (High Water Levels in Hungary: after the reconsideration). *Mérnök Újság* **2015**, *1–2*, 22–25.
51. Ficsor, J. Lebegtetett Hordalék Vizsgálata a Felsö-Magyarorszaági Duna-Szakaszon (Study of Suspended Sediment Transport on the Upper Hungarian Reach of the Danube River). Master's Thesis, Budapest University of Technology and Economics, Budapest, Hungary, 2016.
52. Luo, P.; Mu, D.; Xue, H.; Ngo-duc, T.; Dang-dinh, K.; Takara, K.; Nover, D.; Schladow, G. Flood inundation assessment for the Hanoi Central Area, Vietnam under historical and extreme rainfall conditions. *Sci. Rep.* **2018**, *8*, 12623. [CrossRef]
53. Baranya, S.; Józsa, J. Flow analysis in River Danube by field measurement and 3D CFD turbulence modelling. *Period. Polytech. Civ. Eng.* **2006**, *50*, 57–68.
54. Török, G.T. Vegyes szemcseösszetételű folyómedrek numerikus vizsgálata (Numerical investigation of non-uniform river bed). *Hidrológiai Tájékoztató* **2013**, 22–24. Available online: http://www.hidrologia.hu/mht/letoltes/hidrologiai_tajekoztato_2013.pdf (accessed on 30 September 2019).
55. Guerrero, M.; Lamberti, A. Flow Field and Morphology Mapping Using ADCP and Multibeam Techniques: Flow Field and Morphology Mapping Using ADCP and Multibeam Techniques: Survey in the Po River. *J. Hydraul. Eng.* **2011**, *137*, 1576–1587. [CrossRef]
56. Koken, M.; Constantinescu, G. An investigation of the flow and scour mechanisms around isolated spur dikes in a shallow open channel: 1. Conditions corresponding to the initiation of the erosion and deposition process. *Water Resour. Res.* **2008**, *44*, 1–19. [CrossRef]
57. Catalano, P.; Wang, M.; Iaccarino, G.; Moin, P. Numerical simulation of the flow around a circular cylinder at high Reynolds numbers. *Int. J. Heat Fluid Flow* **2003**, *24*, 463–469. [CrossRef]
58. Roulund, R.; Sumer, B.M.; Fredsøe, J.; Michelsen, J. Numerical and experimental investigation of flow and scour around a circular pile. *J. Fluid Mech.* **2005**, *534*, 351–401. [CrossRef]
59. Török, G.T.; Baranya, S.; Rüther, N. Validation of a combined sediment transport modelling approach for the morphodynamic simulation of the upper Hungarian Danube River. In Proceedings of the 19th EGU General Assembly, Vienna, Austria, 8–13 April 2018; Volume 19, p. 15749.
60. Baranya, S.; Olsen, N.R.B.; Stoesser, T.; Sturm, T. Three-dimensional rans modeling of flow around circular piers using nested grids. *Eng. Appl. Comput. Fluid Mech.* **2012**, *6*, 648–662. [CrossRef]
61. Wilcock, P.R.; Kenworthy, S.T.; Crowe, J.C. Experimental Study of the Transport of Mixed Sand and Gravel. *Water Resour. Res.* **2001**, *37*, 3349–3358. [CrossRef]
62. Fernandez Luque, R. Erosion and Transport of Bed-load Sediment. Bachelor's Thesis, Delft Technical University, Delft, The Netherland, 1974.
63. Fernandez Luque, R.; van Beek, R. Erosion and Transport of Bed-load Sediment. *J. Hydraul. Res.* **1976**, *14*, 127–144. [CrossRef]

Article

Secondary Flow Effects on Deposition of Cohesive Sediment in a Meandering Reach of Yangtze River

Cuicui Qin [1,*], Xuejun Shao [1] and Yi Xiao [2]

1 State Key Laboratory of Hydroscience and Engineering, Tsinghua University, Beijing 100084, China
2 National Inland Waterway Regulation Engineering Research Center, Chongqing Jiaotong University, Chongqing 400074, China
* Correspondence: qincuicui2009@163.com; Tel.: +86-188-1137-0675

Received: 27 May 2019; Accepted: 9 July 2019; Published: 12 July 2019

Abstract: Few researches focus on secondary flow effects on bed deformation caused by cohesive sediment deposition in meandering channels of field mega scale. A 2D depth-averaged model is improved by incorporating three submodels to consider different effects of secondary flow and a module for cohesive sediment transport. These models are applied to a meandering reach of Yangtze River to investigate secondary flow effects on cohesive sediment deposition, and a preferable submodel is selected based on the flow simulation results. Sediment simulation results indicate that the improved model predictions are in better agreement with the measurements in planar distribution of deposition, as the increased sediment deposits caused by secondary current on the convex bank have been well predicted. Secondary flow effects on the predicted amount of deposition become more obvious during the period when the sediment load is low and velocity redistribution induced by the bed topography is evident. Such effects vary with the settling velocity and critical shear stress for deposition of cohesive sediment. The bed topography effects can be reflected by the secondary flow submodels and play an important role in velocity and sediment deposition predictions.

Keywords: secondary flow; cohesive; deposition; 2D depth-averaged model; meandering; Yangtze River

---

1. Introduction

Helical flow or secondary flow caused by centrifuge force in meandering rivers plays an important role in flow and sediment transport. It redistributes the main flow and sediment transport, mixes dissolved and suspended matter, causes additional friction losses, and additional bed shear stress, which are responsible for the transverse bed load sediment transport [1–3]. Moreover, the secondary flow may affect lateral evolution of river channels [4–6]. Extensive researches have been conducted about secondary flow effects on flow and sediment transport, especially bed load in a singular bend [7] or meandering channels of laboratory scale [8] and rivers of field scale [9,10]. However, few researches focus on suspended load transport. In China, sediment transport in most rivers is dominated by suspended load, such as Yangtze River and Yellow River. On the Yangtze River, the medium diameter of sediment from upstream is ~0.01 mm [11], which has taken on cohesive properties to some extent [11,12]. More importantly, these cohesive suspended sediments have been extensively deposited in several reaches which have blocked the waterway in Yangtze River [13]. As most of these reaches are meanders with a central bar located in the channel, to what extent the secondary flow has affected the cohesive suspended sediment deposition should be investigated.

Cohesive sediment deposition is controlled by bed shear stress [14], which is determined by the flow field. In order to investigate the secondary flow effects on cohesive sediment transport, its effects on flow field should be considered first. Secondary flow redistributes velocities, which means the high velocity core shifts from the inner bank to the outer bank of the bend [15,16]. Saturation of secondary flow takes place in sharp bends [17]. Due to the inertia, the development of secondary flow lags behind the curvature called the phase lag effect [18]. All these findings mainly rely on laboratory experiments or small rivers with a width to depth radio less than 30 [19] probably resulting in an exaggeration of secondary flow. When it comes to natural meandering or anabranching rivers, especially large or mega rivers, secondary flow may be absent or limited in a localized portion of the channel width [20–24]. However, those researches are only based on field surveys and mainly focused on influences of bifurcation or confluence of mega rivers with low curvatures and significant bed roughness [23] at the scale of individual hydrological events. On contrary, Nicholas [25] emphasized the role of secondary flow played in generating high sinuosity meanders via simulating a large meandering channel evolution on centennial scale. Maybe it depends on planimetric configurations, such as channel curvature, corresponding flow deflection [26] and temporal scales. Therefore, whether secondary flow exists and has the same effects on the flow field in a meandering mega river as that in laboratory experiments should be further investigated. Besides, the long-term hydrograph should be taken into account.

As to its effects on bed morphology, secondary flow induced by channel curvature produces a point bar and pool morphology by causing transversal transport of sediment, which in turn drives lateral flow (induced by topography) known as topographic steering [9] which plays an even more significant role in meandering dynamics than that curvature-induced secondary flow [3]. The direction of sediment transport is derived from that of depth-averaged velocity due to the secondary flow effect, which has been accounted for in 2D depth-averaged models and proved to contribute to the formation of local topography [4–6,27], especially bar dynamics [28,29], and even to channel lateral evolution [4,6,30]. Although Kasvi et al. [31] has pointed that the exclusion of secondary flow has a minor impact on the point bar dynamics, temporal scale effects remain to be investigated as the authors argued for only one flood event has been considered in their research and the inundation time may affect the effects of secondary flow [32]. Those researches have enriched our understandings of mutual interactions of secondary flow and bed morphology. However, they mainly focused on bed load sediment transport, whereas the world largest rivers are mostly fine-grained system [21] and are dominated by silt and clay, such as Yangtze River [11,12]. Fine-grained suspended material ratio controls the bar dynamics and morphodynamics in mega rivers [23,33]. As is known, such fine-grained sediment is common in estuarine and coastal areas. However, how they work under the impacts of secondary flow in mega rivers is still up in the air and the temporal effects of secondary flow should be investigated.

Numerical method provides a convenient tool for understanding river evolution in terms of hydrodynamics and morphodynamics in addition to the laboratory experiments and field surveys. The 2D depth-averaged model is preferable because it keeps as much detailed information as possible on the one hand and remains practical for investigation of long-term and large-scale fluvial processes on the other hand. The main shortcoming of the 2D depth-averaged model is that the vertical structure of flow has been lost due to the depth-integration of the flow momentum and suspended sediment transport equations, and thus the secondary flow effects on the flow field and suspended sediment transport are neglected. These effects can be retrieved by incorporating closure correction submodels into the 2D depth-averaged model. In order to account for these effects on the flow field, various correction submodels have been proposed by many researchers [34–38]. The differences among these models are whether or not they consider (1) the feedback effects between main flow and secondary flow and (2) the phase lag effect of the secondary flow caused by inertia. Models neglecting the former one are classified as linear models, in contrast to nonlinear models which consider such effects [1,38]. The nonlinear models [1,39] based on the linear ones are more suitable for flow simulation of sharp bends [1,2]. The phase lag effect, which is obviously pronounced in meandering channels [40], has been thought to be important in sharp bends especially with pronounced curvature variations [2], and proven to influence bar dynamics considerably [29]. Although the performances of those above mentioned models have been extensively tested by laboratory scale bends, their applicability to field meandering rivers, especially mega rivers, needs to be further investigated. Besides, which model is preferable in flow simulation of meandering channels of field mega scale remains to be answered.

To consider the secondary flow effects on the suspended sediment transport, closure submodels should be coupled to the sediment module of the 2D depth-averaged models in a similar way to the flow module [41]. However, as to the cohesive sediment transport, it is mainly related to the bed shear stress determined by the flow field. Besides, according to field survey of two reaches of Yangtze River by Li et al. [11], cohesive sediment transport is controlled by the depth-averaged velocity. Therefore, only the secondary flow effects on flow field are considered to further analyze their effects on bed morphology here. In addition, the turbulence models should be considered in the 2D depth-averaged model, especially when there are recirculating flows [34]. Based on the previous research work [34,36], the depth-averaged parabolic eddy viscosity model can be applied.

This paper aims to investigate the secondary flow effects on cohesive sediment deposition in meandering reach of field mega scale during an annual hydrography. The following questions will be addressed; (1) whether secondary flow effects on the flow field can be reflected by typical secondary flow correction models in such mega meandering rivers as laboratory meandering channels, (2) which model should be given priority to flow simulation in meandering channels of such scale, and (3) what the temporal influence of secondary flow is on bed morphology variations associated with cohesive sediment deposition. The contents of this paper are as follows; three secondary flow submodels referring to the aforementioned different effects have been selected from the literature—Lien et al. [37], Bernard [35], and Blankaert and de Vriend [1] models—to reveal secondary flow impacts and distinguish their performances on flow simulation in meandering channels of this mega scale first, and the preferable model is selected. Then, the corresponding model is applied to investigate secondary flow effects on bed morphology variations related with cohesive sediment deposition during an annual hydrograph. Finally, the correction terms representing secondary flow effects have been analyzed to justify their functionalities and performances of these models in meandering channels of such scale. Besides, the roles of cohesive sediment played in secondary flow effects have been investigated as well. The main contributions of this paper are three-fold: (1) the L model has been found to outperform the other models in flow simulation of the field mega scale meandering reach; (2) the bed topography effects have been identified to be reflected by the secondary flow submodels, and the transverse bed topography plays a more important role than the longitudinal one and results in the great improvements of velocity and sediment deposition predictions of the L model in this reach; and (3) secondary flow effects on cohesive sediment deposition become obvious during the last period of an annual hydrography when the sediment concentration is low and the transverse bed topography has been formed. Such effects on the predicted amount of deposition vary with the cohesive sediment properties.

2. Methods

A 2D depth-averaged model (Section 2.1, referred to as the N model hereafter) has been improved by considering secondary flow effects and cohesive sediment transport. Secondary flow module (Section 2.2) incorporates three different submodels to reflect its different effects, together with the sediment module (Section 2.3) are described briefly. All the equations are solved in orthogonal curvilinear coordinates.

2.1. Flow Equations

The unsteady 2D depth-averaged flow governing equations are expressed as follows [42]

$$\frac{\partial Z}{\partial t} + \frac{1}{J}\left[\frac{\partial (h_2 HU)}{\partial \xi} + \frac{\partial (h_2 HV)}{\partial \eta}\right] = 0 \tag{1}$$

$$\frac{\partial (HU)}{\partial t} + \frac{1}{J}\left[\frac{\partial (h_2 HUU)}{\partial \xi} + \frac{\partial (h_1 HVU)}{\partial \eta}\right] - \frac{HVV}{J}\frac{\partial h_2}{\partial \xi} + \frac{HUV}{J}\frac{\partial h_1}{\partial \eta} + \frac{gH}{h_1}\frac{\partial Z}{\partial \xi} + \frac{HUg\left|H\vec{u}\right|}{(CH)^2} = \frac{v_e H}{h_1}\frac{\partial E}{\partial \xi} - \frac{v_e H}{h_2}\frac{\partial F}{\partial \eta} - S_\xi \tag{2}$$

$$\frac{\partial(HV)}{\partial t}+\frac{1}{J}\left[\frac{\partial(h_2HUV)}{\partial\xi}+\frac{\partial(h_1HVV)}{\partial\eta}\right]+\frac{HUV}{J}\frac{\partial h_2}{\partial\xi}-\frac{HUU}{J}\frac{\partial h_1}{\partial\eta}+\frac{gH}{h_2}\frac{\partial Z}{\partial\eta}+\frac{HUg\left|H\vec{u}\right|}{(CH)^2}=\frac{\upsilon_eH}{h_1}\frac{\partial E}{\partial\eta}-\frac{\upsilon_eH}{h_2}\frac{\partial F}{\partial\xi}-S_\eta \quad (3)$$

$$E=\frac{1}{J}\left[\frac{\partial(h_2U)}{\partial\xi}+\frac{\partial(h_1V)}{\partial\eta}\right]F=\frac{1}{J}\left[\frac{\partial(h_2V)}{\partial\xi}-\frac{\partial(h_1U)}{\partial\eta}\right] \quad (4)$$

where ξ and η = longitudinal and transverse direction in orthogonal curvilinear coordinates, respectively; h_1 and h_2 = metric coefficients in ξ and η directions, respectively; $J = h_1h_2$; g = acceleration gravity, m/s^2; $\vec{u} = (U, V)$ depth-averaged resultant velocity vector and (U, V) = depth-averaged velocity in ξ and η directions, separately; H = water depth; Z = water surface elevation; C = Chezy factor; ν_e = eddy viscosity; and S_ξ and S_η = correction terms related to the vertical nonuniform distribution of velocity.

2.2. Secondary Flow Equations

In order to consider different effects of secondary flow on flow, three secondary flow models are selected from literature to calculate the dispersion terms (S_ξ, S_η) in Equations (2) and (3). Among them, the Lien et al. [37] (L) model has been widely applied, which ignores the secondary flow phase lag effect and is suitable for fully developed flows. As secondary flow lags behind the driving curvature due to inertia [2], it will take a certain distance for secondary flow to fully develop, especially in meandering channels. There are several models using a depth-averaged transport equation to consider these phase lag effect, such as the Delft-3D [43] model, Hosoda et al. [44] model, and Bernard [35] model. The Delft-3D model has two correction coefficients to calibrate and Hosoda model is complex to use. In addition, both of them focus on flow simulation in channels with a single bend. In contrast, Bernard (B) model is simple, practicable and has been validated by several meandering channels. Moreover, the sidewall boundary conditions considered by B model is more reasonable, that is, the production of secondary flow approaches zero on the sidewalls [16]. Therefore, the B model is selected as another representative model. Because the above mentioned two models are linear models which are theoretically only applicable to mildly curved bends, a simple nonlinear (NL) model [1] is selected as a typical model to reflect the saturation effect of secondary flow [17] in sharply curved bends. All of the three models can reflect the velocity redistribution phenomenon caused by secondary flow at different levels. These models serve as submodels coupled to the 2D hydrodynamic model to account for different effects of secondary flow on flow field. The major differences of them are summarized in Table 1, while L and B models can refer to the authors [45] for more details. Only NL model are briefly described as follows.

Table 1. Differences between L, B, and NL models.

	L Model	B Model	NL Model
Saturation effect	NO	NO	YES
Phase lag effect	NO	YES	YES
Wall boundary condition	-	no secondary flow produced	dispersion terms = 0
Velocity redistribution	YES	YES	YES

Based on linear models, the NL model is able to consider the feedback effects between secondary flow and main flow to reflect the saturation effect through a bend parameter β [1] (Equation (10)). However, the NL model proposed by Blanckaert and de Vriend [1] is limited to the centerline of the channel. Ottevanger [46] extended the model to the whole channel width through an empirical power law (f_w, Equation (9)). This method is as follows

$$\begin{aligned}S_\xi &= \frac{1}{J}\frac{\partial}{\partial\xi}(h_2T_{11})+\frac{1}{J}\frac{\partial}{\partial\eta}(h_1T_{12})+\frac{1}{J}\frac{\partial h_1}{\partial\eta}T_{12}-\frac{1}{J}\frac{\partial h_2}{\partial\xi}T_{22}\\ S_\eta &= \frac{1}{J}\frac{\partial}{\partial\xi}(h_2T_{12})+\frac{1}{J}\frac{\partial}{\partial\eta}(h_1T_{22})-\frac{1}{J}\frac{\partial h_1}{\partial\eta}T_{11}+\frac{1}{J}\frac{\partial h_2}{\partial\xi}T_{12}\end{aligned} \quad (5)$$

$T_{i,j}$ (i, j = 1,2) is called dispersion terms [37]. When the L model is adopted as the linear model, $T_{i,j}$ is expressed as

$$\begin{array}{c} T_{11} = HUU\left(\frac{\sqrt{g}}{\kappa C}\right)^2 \quad T_{12} = T_{21} = HUV\left(\frac{\sqrt{g}}{\kappa C}\right)^2 + f_{sn}(\beta) f_w \frac{(HU)^2}{\kappa^2 r} \frac{\sqrt{g}}{\kappa C} FF1 \\ T_{22} = HVV\left(\frac{\sqrt{g}}{\kappa C}\right)^2 + f_{sn}(\beta) f_w \frac{2HUHV}{\kappa^2 r} \frac{\sqrt{g}}{\kappa C} FF1 + f_{nn}(\beta) f_w^2 \frac{(HU)^2 H}{\kappa^4 r^2} FF2 \end{array} \tag{6}$$

where κ = the Von Karman constant, 0.4; r = the channel centerline, m; f_w = the empirical power law equation over the channel width; $FF1$, $FF2$ = the shape coefficients related to the vertical profiles of velocity which can refer to Lien et al. [37] for details; and $f_{sn}(\beta)$ and $f_{nn}(\beta)$ are the nonlinear correction coefficients expressed as Equations (7) and (8) [47], which directly reflect the saturation effect of secondary flow [17].

$$f_{sn}(\beta) = 1 - \exp\left(-\frac{0.4}{\beta(\beta^3 + 0.25)}\right) \tag{7}$$

$$f_{nn}(\beta) = 1.0 - \exp\left(-\frac{0.4}{1.05\beta^3 - 0.89\beta^2 + 0.5\beta}\right) \tag{8}$$

$$f_w = \left[1 - \left(\frac{2n}{w}\right)^{2n_p}\right] \tag{9}$$

$$\beta = C_f^{-0.275}(H/R)^{0.5}(1+\alpha)^{0.25} \tag{10}$$

$$\alpha_s = [w\partial U_s/\partial n/U_s]_{n_c} \tag{11}$$

β = the bend parameter which is a control parameter distinguishing the linear and nonlinear models; α_s = the normalized transversal gradient of the longitudinal velocity U at the centerline; and n_c = the position of channel centerline.

The phase lag effect of secondary flow is considered with the following transport equations [46].

$$\frac{1}{J}\left[\frac{\partial(h_2 HUY)}{\partial \xi} + \frac{\partial(h_1 HVY)}{\partial \eta}\right] = \frac{h\left|\vec{u}\right|}{\lambda}(Y_e - Y) \tag{12}$$

λ = the adaption length described by Johannesson and Parker [18]. Y = the terms referring to f_{sn}, f_{nn} in Equation (6), Y_e = the fully developed value of Y.

As L, B, and NL models serve as closure submodels in hydrodynamic Equations (1)–(4), the correction terms (S_ξ and S_η) are associated with the computed mean flow field, and the information on the relative variables of correction terms is available when solving these submodels. This is similar to the way to solve turbulence submodels. Detailed procedure for solving the NL model is shown in Figure 1. Equations (1)–(4) are solved first without considering the correction terms (S_ξ and S_η) for water depth and depth-averaged velocity. The nonlinear parameters in Equations (7)–(11) have been calculated next. Afterwards, the transport Equation (12) is solved for evaluating dispersion terms ($T_{i,j}$, Equation (6)) and (S_ξ and S_η) (Equation (5)). The correction terms (S_ξ and S_η) are then included in Equations (1)–(4), which are solved again to get new information on the mean flow field. The procedure continues until no significant variations in the magnitude of depth, velocity, and other variables in the model (Figure 1).

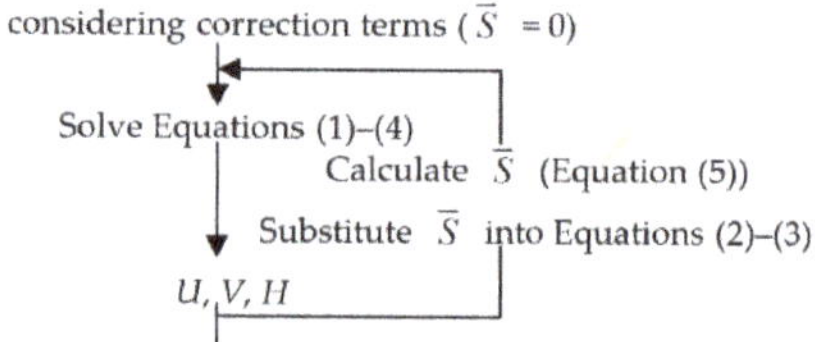

Figure 1. Solution procedure.

2.3. Cohesive Sediment Transport Equations

The cohesive sediment transport equation is similar to the noncohesive sediment transport equation [48], except the method to calculate the net exchange rate ($D_b - E_b$) [14].

$$\frac{\partial(HC)}{\partial t} + \frac{1}{J}\left[\frac{\partial(h_1 HCU)}{\partial \xi} + \frac{\partial(h_1 HCV)}{\partial \eta}\right] - \frac{H\varepsilon}{J}\left[\frac{\partial}{\partial \xi}\left(\frac{h_2}{h_1}\frac{\partial C}{\partial \xi}\right) + \frac{\partial}{\partial \eta}\left(\frac{h_1}{h_2}\frac{\partial C}{\partial \xi}\right)\right] + D_b - E_b = 0 \quad (13)$$

where C = the sediment concentration, kg·m^{-3}; D_b and E_b = the erosion and deposition rate respectively, kg·m^{-2}·s^{-1}, which are calculated [14] as follows

$$D_b = \alpha \omega_s C \quad (14)$$

where α = deposition coefficient calculated by Equation (15); ω_s = settling velocity, m·s^{-1}.

$$\alpha = \begin{cases} 1 - \frac{\tau_b}{\tau_{cd}}, \tau_b \le \tau_{cd} \\ 0, \tau_b > \tau_{cd} \end{cases} \quad (15)$$

where τ_b = the bed shear stress, Pa; τ_{cd} = the critical shear stress for deposition, Pa.

$$E_b = M\left(\frac{\tau_b}{\tau_{ce}}\right)^n \tau_b > \tau_{ce} \quad (16)$$

where n = an empirical coefficient; M = the erosion coefficient, kg·m^{-2}·s^{-1}; and τ_{ce} = the critical shear stress for erosion, Pa.

Most of the model parameters used for cohesive sediment calculation (Table 2) have been calibrated and validated by the sedimentation process of the Three Gorges Reservoir on Yangtze River [48], where the study area of this paper is located. A larger value of settling velocity is chosen from measurements by Li et al. [11,12] because only the medium diameter of the sediment is considered in this study.

Table 2. Model parameters used for cohesive sediment calculation.

Related Variables	Values
Settling velocity ω_s	2.1 mm·s^{-1}
Critical shear stresses τ_{cd}, τ_{ce}	0.41 Pa
Erosion coefficient M	1.0×10^{-8} kg·m^{-2}·s^{-1}
Empirical coefficient n	2.5

The morphological evolution due to cohesive sediment transport is calculated by the net sediment exchange rate ($D_b - E_b$), in the same way as noncohesive suspended sediment calculation does. The flow and sediment modules are solved in an uncoupled way. Details of the numerical method can be found in Wang et al. [42]. Central difference explicit scheme is applied to Equation (5), and

Equations (12) and (13) are solved by QUICK (Quadratic Upstream Interpolation for Convective Kinematics) finite difference scheme.

3. Study Case

The Hunghuacheng reach (HHC, Figure 2), located 364 km upstream from the Three Gorges Project (TGP), is approximately 13 km long, consisting two sharply curved bends with a center bar named "Huanghuacheng" splitting the reach into two branches. It belongs to the back water zone of TGP. The large mean annual discharge (32,000 $m^3 \cdot s^{-1}$) makes it a mega river reach [49]. Measurements of bed topographies and bed material size are taken at nine cross-sections from S201 to S209 twice each year. Due to huge amount of cohesive sediment siltation, the left branch of this reach has been blocked in September 2010 [13]. Secondary flow models are applied to this reach because the secondary flow caused by the upstream bend of this reach plays an important role in channel morphodynamics [50,51]. Also, it has been shown that similar models perform well in confluence [38] and braided rivers [25], which justify the application of these models in this reach.

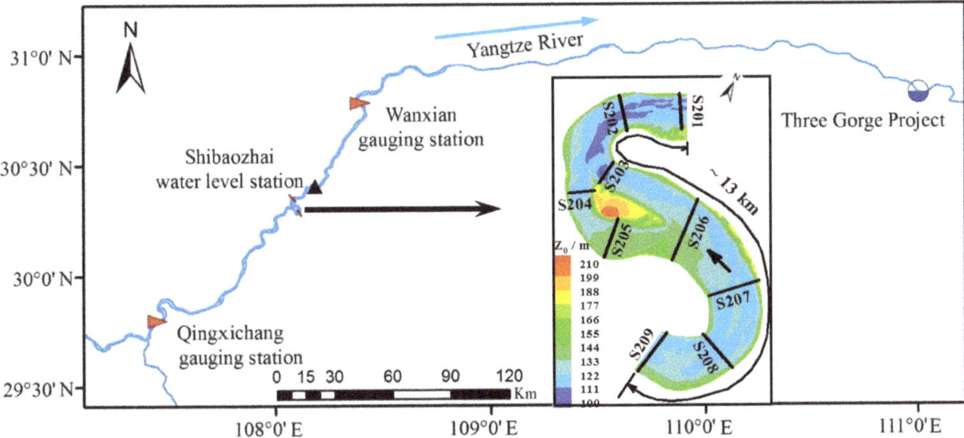

Figure 2. Planform geometry, bed elevation (Z_0) on March 2012 and nine cross-sections measured in HHC reach (S209 and S201 are the inlet and outlet boundaries, respectively; incoming flow discharge and sediment concentration used as inlet boundaries are interpolated from Qingxichang and Wanxian gauging station, located upstream 476.46 km and 291.38 km from TGP, respectively; and the outlet boundary applies the water stage measured at Shibaozhai station, located upstream 341.35 km from the TGP).

The year 2012 is selected to study the secondary flow effects on bed morphology variation in this reach because of the record amount of deposition that year. The inlet and outlet boundaries are S209 and S201, respectively (Figure 2). The observed flow and sediment discharges at Qingxichang (QXC) and Wanxian (WX) gauging stations have been depicted in Figure 3a,b, respectively. It clearly illustrates that the flow and sediment hydrographs are synchronous with each other at the two stations after the sediment discharges at WX station have been moved forward by one day. Considering the differences of hydrographs between the two stations and the contributions of tributary inflows are small, the interpolation method has been applied to calculate the incoming flow boundary condition at the HHC reach. The incoming suspended sediment concentration (SSC) boundary condition should be calculated through Equation (17). As the distance ratio of QXC-HHC to HHC-WX is equal to the ratio of the amount of deposition at QXC-HHC to that at HHC-WX in 2012, approximately 3:2 [52], and the flow discharges at the two stations are nearly the same, the interpolation method can be applied to approximate the SSC at the inlet boundary as well. The RMSE (Root Mean Square Error) value of the calculated SSC through the above two methods is 0.05 kg/m^3, which is acceptable for sediment deposition is negligible when the SSC is less than 0.1 kg/m^3. Besides, the SSC propagation is supposed to delay for one day from QXC station to the HHC reach. The water stage measured at Shibaozhai station is used as the outlet boundary condition (Figure 2). Only the flood season from May to November is simulated instead of a whole year because most sediment is transported during this period (Figure 4b), similar to the method applied by Fang and Rodi [53] to study the sedimentation of near dam region after TGP impoundment. This duration has been divided into six periods based on the water stage process (Figure 4a). It should be noted that the water stage rising during the last period of this process is resulted from the operation of TGP and the water stage and bed elevation data are both based on Wusong base level.

$$S_{\_HHC} = \left[0.4(Q_{S\_QXC} - Q_{S\_WX}) + Q_{S\_WX}\right]/Q_{\_HHC} = \left(0.4Q_{S\_QXC} + 0.6Q_{S\_WX}\right)/Q_{\_HHC} \approx 0.4S_{\_QXC} + 0.6S_{\_WX} \quad (17)$$

where $Q_S = Q \times S$, kg/s; Q = flow discharge, m^3/s; and S = sediment discharge, kg/s; 0.6 and 0.4 represent percentage of amount of sediment deposition at QXC-HHC and HHC-WX, respectively.

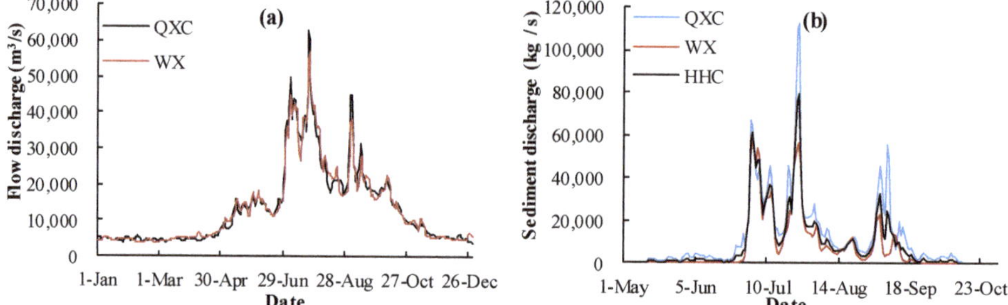

Figure 3. (**a**) Hydrograph at Qingxichang (QXC) and Wanxian (WX) gauging stations. (**b**) Sediment discharge (Q_S) measured at QXC and WX and calculated at HHC (The Q_S at WX station has been moved forward by one day).

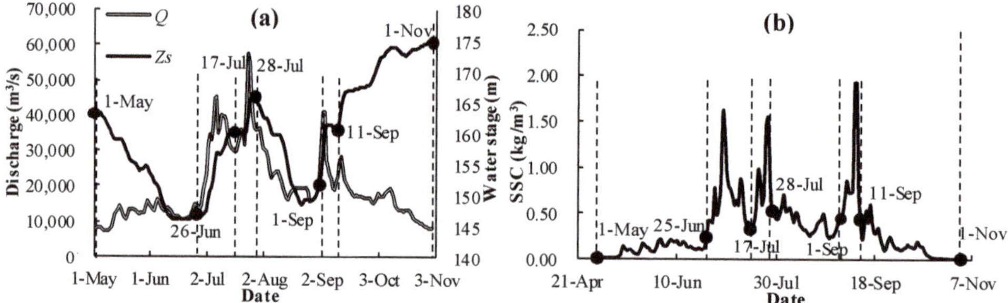

Figure 4. (**a**) Hydrograph and water stage from May 1 to November 1 (Q and Z_S represent discharge and water stage, respectively); the black filled circles divide the duration into several periods descripted clearly by the vertical black dash lines. (**b**) Suspended sediment concentration (SSC) as the inlet boundary in this duration.

A median size of 0.008 mm is used to represent the inflow cohesive sediment composition of this reach [13]. A flood event on 16 July 2012 is chosen as a verification case for this river reach simulation. Table 3 lists parameters and conditions of it. Because the radius to width ratio (r/w) is in the range of 0.8 to 2.0 (Table 3), this river reach belongs to sharply curved bends. The computation domain of the river reach is divided into 211 × 41 grids in longitudinal and transverse directions, with time steps of 1.0 s and 60.0 s for flow and sediment calculation, respectively.

Table 3. Channel dimensions and flow condition of HHC reach.

Study Case	Discharge Q ($m^3\ s^{-1}$)	Depth H (m)	Width w (km)	Bend Radius r (km)	r/w	H/r	Adaption Length λ
HHC	30,200	16–67	0.7–2.0	>0.4	0.8–2.0	0.001–0.066	0.001–0.2

4. Results

The flow simulation results of L, B, and NL models are verified for the discharge of 30,200 $m^3{\cdot}s^{-1}$, and the model with best performances has been selected. The preferable model L and the reference model N are used to predict cohesive sediment deposition during an annual hydrograph. The basic parameters, such as eddy viscosity coefficient and roughness of flow module, and parameters of sediment module are calibrated in N model first and then applied to the other models.

4.1. Verifications

4.1.1. Flow

Figure 5 shows simulated water stage at the right bank and the depth-averaged velocities across the channel width of the HHC reach. It can be seen that the results of the L model are more reasonable than those of the other models. The velocity shift due to secondary flow can be well predicted by the L model at the end of the bends (S202 and S206), especially at the exit of the second bend (S202), in contrast to other models. In addition, as the high velocity core shifts to the right bank at the end of the first bend (S206), velocity of the left branches (S205) has been reduced. That explains why the velocities predictions by B and L models are lower than those by N and NL models at S205 (Figure 5c). Overall, the differences among B, N and NL models are small, while the L model is preferable according to the flow simulation results of the HHC reach.

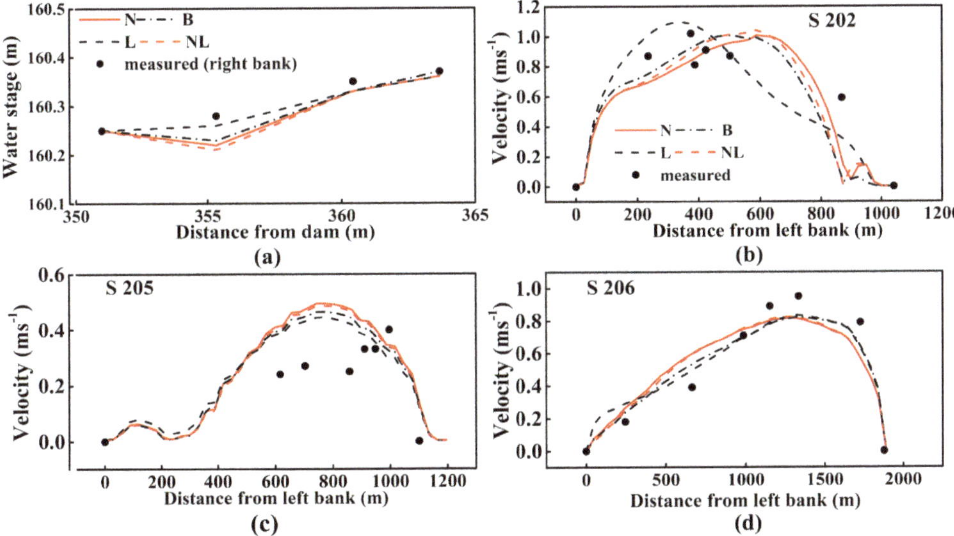

Figure 5. (**a**) Water stage of the right bank (downstream view). (**b**–**d**) Depth-averaged velocity distribution measured and predicted by N, B, L, and NL models at three cross-sections for discharge 30,200 m^3/s.

To quantitatively assess the performances of different models in flow simulation of the HHC reach, the RMSE of water stage and velocities of different models at typical cross-sections are listed in Table 4.

The L model with the smallest RMSE results outperforms the other models at the discharge of 30,200 m^3/s.

Table 4. The RMSE of water stage (rows 1–3) and velocities (rows 4–9) of different models.

RMSE	N	L	B	NL
Left bank	0.049	**0.049**	0.054	0.051
Right bank	0.032	**0.015**	0.027	0.037
Mean	0.041	**0.037**	0.043	0.045
S202	0.204	**0.173**	0.242	0.252
S203	**0.243**	0.249	0.236	0.233
S204	0.127	**0.093**	0.112	0.120
S205	0.151	**0.121**	0.133	0.145
S206	0.147	**0.110**	0.111	0.142
Mean	0.179	**0.160**	0.177	0.186

4.1.2. Sediment

Based on the above flow simulation results, the L model has been applied to the HHC reach to investigate the secondary flow effects on cohesive sediment deposition. The results of N model serve as references.

The deposition module is verified by field measurements (Figure 6a) in terms of planar distribution of deposition (Figure 6b,c), bed elevation (Figure 7), and amounts of deposition. Figure 6a–c show that the simulated planar distribution of deposits by the L and N models agree with field measurements qualitatively, with the maximum thickness of deposits found at the convex bank of the first bend, and the majority of deposits located at the right bank of the inlet and the left branch of the reach. The predicted thickness of sediment deposits by the L model is approximately 1 m thicker than that by N model on the concave bank of the first bends (region 1, Figure 6d), which is much closer to the measurement 5–7 m (Figure 6a). Bed elevations simulated by the two models matches well with measurements at S204–S206 (Figure 7). Predictions of total amounts of deposition from S206 to S203 are 8.33×10^6 m^3 and 8.0×10^6 m^3 by the N and L models respectively, while the field measurement during the same period is 8.18×10^6 m^3 [13]. The relative error is around 2%, which qualify the sediment module used in this paper. In general, the L model performs better than the N model in predicting the planar distribution of cohesive sediment deposition.

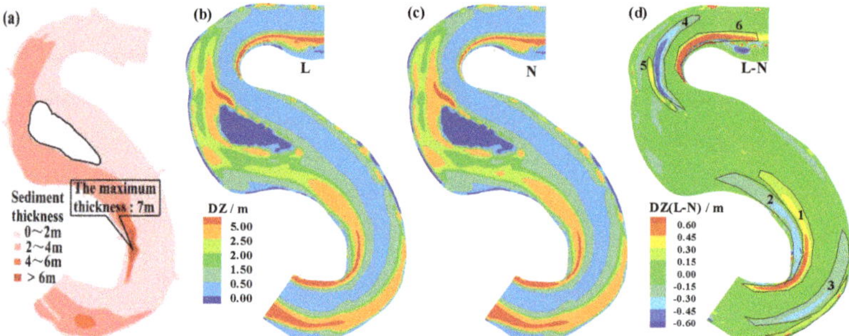

Figure 6. (**a**) Planar distribution of sediment thickness measured, the maximum is 7 m from March to August, 2012. (**b**) Sediment thickness simulated by the L model (**c**) and N model. (**d**) The difference between the L and N models.

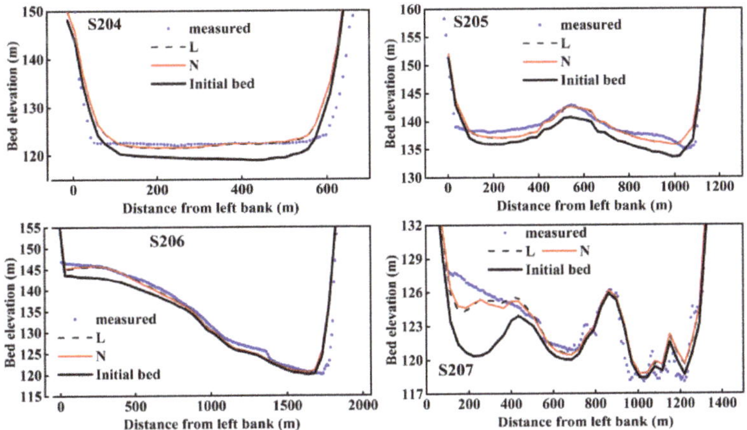

Figure 7. Comparison of bed elevation at cross-sections between measurements and predictions.

4.2. Secondary Flow Effects on Cohesive Sediment Deposition

The differences in planar distribution and amounts of deposition predicted by the L and N models have been illustrated in Figures 6d and 8, respectively, which clearly suggest the secondary flow effects on cohesive sediment deposition. Due to its impacts, high velocity core shifts from the convex to the concave bank of the bend, leading to the redistribution of bed shear stress and the consequent morphological changes [9]. Shifts of high velocities predicted by the L model result in the more deposition in region 1, 5, and 6 and less deposition in regions 3 and 4. The increase of sediment deposits in region 1 reduces sediment transported to region 2, resulting in less deposition here. The difference of predicted amount of deposition between the two models is about 0.31×10^6 m^3 from 11 September to 1 November, as is clearly shown in Figure 8. This difference is small compared to the total amount of deposition during the whole year, approximately 8.0×10^6 m^3. However, this difference can accumulate if the water stage keeps rising due to the impoundment of TGP. In general, secondary flow effects on cohesive sediment deposition become more obvious in the last period of the annual hydrograph when the sediment load is low and water stage is high (Figure 4).

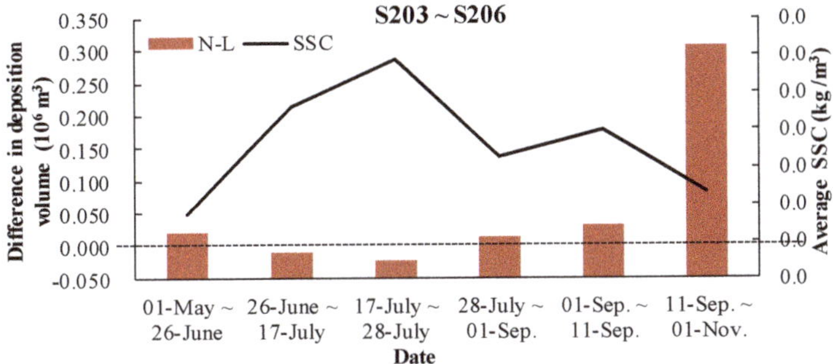

Figure 8. Differences in deposition volume during different periods (average SSC means the average suspended sediment concentration during each period).

The total deposition volume is calculated from S203–S206 during different periods of this year, because this part of the reach is seldom affected by the inlet and outlet boundaries. Deposition of this part is greatly impacted by the velocity redistribution at S206 (e.g., Figure 5c,d), which is controlled by the secondary flow produced in the upstream bend and the bed topography (transverse bed slope) there. In addition, the sediment load plays an important role in the deposition of this part. Therefore, the average of suspended sediment load during different periods has been shown in Figure 8 as well. When the sediment load is low, the velocity redistribution plays a dominate role resulting in more sediment transport downstream and less deposition due to the shift of high velocities to the right branch. Otherwise, the situation is just reversed, and more deposits can occur in the left branch resulting from the huge amount of sediment transported, despite of the fact that the velocities are higher in the right branch. These can qualitatively explain the difference in predicted amounts of deposition during different periods except the fifth period (1–11 September). In that period, the transverse bed slope at S206 is high enough to strengthen velocity redistribution further, thus surpasses the effects of higher sediment load and result in less predicted deposition by the L model than the N model. During the last period (11 September–1 November), the significant difference of predicted deposition volume is resulted from both the low sediment load transport and the large transverse bed slope.

Figure 9 shows the predicted depth gradients (a) and velocity distributions (b) by the L and N models at S206 on 5 June and 18 September (as typical days of the first and last periods), respectively, illustrating the effects of bed topography. It clearly reveals that the velocity redistribution on 18 September is resulted from the bed topography effects as the sediment load on the two days is ~0.1–0.3 kg·m^{-3}. In all, the low sediment load and the velocity redistribution induced by secondary flow produced by upstream bend and the bed topography result in the difference deposition predictions by the two models.

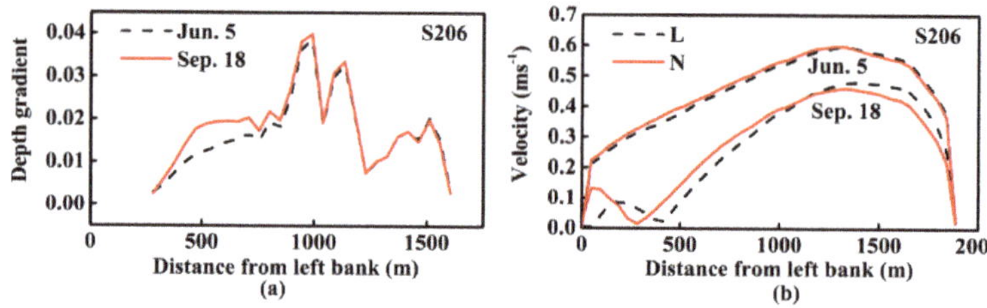

Figure 9. (**a**) Depth gradient (represents bed topography effects). (**b**) Velocity distribution predicted by the N and L models at S206 on typical days of the first and last period, respectively.

5. Discussion

One of the most important physical processes in meandering rivers is the outward shifting of main flow velocity caused by secondary flow, which is driven by channel curvature or point bars bed topography [3]. The latter one is called topography steering [9], which plays a significant role in meander dynamics [3]. Whether and how the correction terms representing the secondary flow effects quantify this process and the performances of these models in meandering channels of different scales will be discussed in this part. Besides, secondary flow effects on the total amount of deposition of the aforementioned part of this reach (S203–S206) are controlled by the properties of cohesive sediment, which will be investigated as well.

5.1. Secondary Flow Effects on Flow Field

5.1.1. Topography Effects

Equation (6) clearly reveals that the correction terms of the three models are directly proportional to the gradients of water depth (H). Due to the effects of bed topography, the longitudinal and transverse gradient of water depth in HHC reach is in the range of 0.01 to 0.001 and 0.01 to 0.1, respectively. Therefore the magnitudes of correction terms follow the same tendency as that of the gradients of water depths, in other words, the correction terms are able to reflect the topography effects. This finding has been justified by Lane [54] who pointed out that correction terms represent the gradients of the transport of momentum. Figure 10 depicts the distributions of (a) S_ξ and (b) S_η of the L, B and NL models along the channel. The orders of magnitude of them are within 0.01 to 0.001 in the longitudinal direction, which is the same as the longitudinal gradient of water depth. In the transverse direction, the order of magnitude of the L model is 0.01–0.1, which is consistent with the transverse gradient of water depth, while those of the B and NL models are approaching to zero and in the range of 0.01 to 0.001, respectively. The smaller orders of magnitude of the two models are resulted from the methods of them. As to the B model [35], it only considers the longitudinal correction. As to the NL model [1], the sharpness of the HHC reach limits the growth of the secondary flow. Since the L model considers the corrections in both directions and has larger correction values than the other two models, it outperforms the other models in the flow simulation as shown in Figure 5. In addition, the simulation results shown in Figure 5 clearly indicate that 2D depth-averaged model that include secondary flow effects (e.g., the L and B models) should be given first priority when it comes to sharp meandering channels with bed topography, such as the HHC reach. This has been confirmed by de Vriend [55] who found that his mathematical model with considering secondary flow effects worked better for curved bend flow simulation over developed bed.

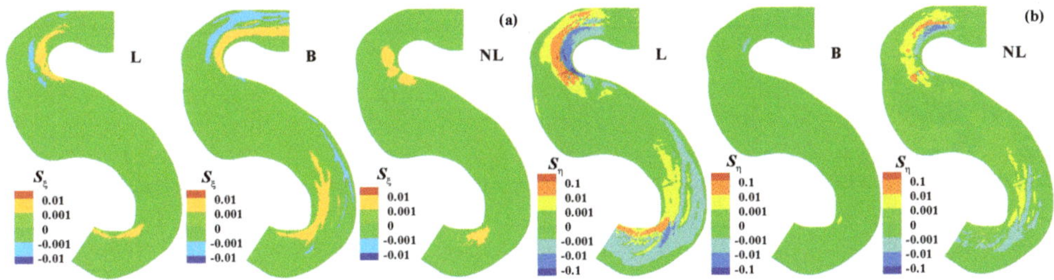

Figure 10. Correction terms (**a**) S_ξ and (**b**) S_η distributions of the L, B, and NL models along the channel.

5.1.2. Applicability of Different Secondary Flow Models

The differences among these models are listed in Table 1, which mainly lie in whether considering the effects of phase lag (B and NL models), sidewall boundary conditions (B model), and bend sharpness (NL model). As the HHC reach is sharply curved bends, the saturation effect considered by the NL model has weakened the secondary flow effects, which result in the minor differences of simulation results between the NL and N models (Figure 5). The depth to width ratio (H/w) distinguishes between meandering channels of different scales. It is approximately 0.001–0.06 in the HHC reach at the discharge of 30,200 m^3/s, while that in the laboratory bend channels and small meandering rivers are in the range of 0.05 to 0.25 [45] and 0.06 to 0.1 [56], respectively. Therefore, the effects of wall boundary conditions and phase lag have been reduced for such small value of H/w. Although B model has taken the bed topography effects into account in a similar way as the L model does, its correction terms only focus on the longitudinal direction. Consequently, the flow simulation results of the L model are better than that of the B model in the HHC reach. Overall, L model is preferable to flow simulation in meandering channels of mega scale, such as HHC reach. However, for laboratory scale curved bends with flat bathymetry, the B model obtains better results [45]. And for sharply curved bends of laboratory and small meandering rivers scales, the advantages of the NL model have been exhibited according to the flow simulation results by Blanckaert [1,2] and Ottevanger [57]. The H/w may play an important role, while the main reasons remain to be further investigated.

5.2. Secondary Flow Effects on Deposition Amounts

According to the deposition simulation results, secondary flow effects on the total deposition volume are small during an annual hydrograph (Figure 8). However, these effects vary with the changes of the cohesive sediment properties, such as settling velocity and critical shear stresses of cohesive sediment, which depend on the flow conditions and the process of bed consolidation. Series of numerical experiments are designed to investigate secondary flow effects on the deposition volume of cohesive sediment with different properties; these effects are reflected by the relative difference in deposition amounts (RD) predicted by the N and L models. Numerical experiments are conducted under the same flow condition (Table 3) to keep the strength of secondary flow constant in the HHC reach. The calculation time for each experiment is 33 days. Different properties of cohesive sediment (Table 5) are represented by the variation of settling velocity (ω_s) and the critical shear stress for deposition (τ_{cd}). Other parameters used in sediment module are the same as that of HHC reach.

Table 5. Settling velocity (ω_s) and critical shear stress for deposition (τ_{cd}) in numerical experiments and results.

ω_s (m/s)	RD [1] (%)	τ_{cd} (Pa)	RD [1] (%)
2.1	0.92	0.41	0.92
1.5	3.80	0.44	0.03
1.0	6.38	0.80	−9.36
0.5	9.01	1.00	−10.61

[1] The relative difference in deposition amounts (RD) predicted by N and L models.

Calculated RD values are listed in Table 5. It is obtained by calculating the difference of the predicted amounts of deposition by L and N models, and then divided by the N model predictions. The negative value of it means the amount of deposition simulated by the L model is smaller than that by the N model. The relationships of RD against ω_s and τ_{cd} are shown in Figure 11. RD is in reverse linear proportion to ω_s, which means the secondary flow effects on the deposition volume increase with the decrease of settling velocity of cohesive sediment. For τ_{cd} is ~0.44 Pa, RD is approaching zero. It implies that secondary flow nearly has no effect on the total deposition volume while its effects on planar distribution can still exit (Figure 6d). As the τ_{cd} increases, the secondary flow impacts on deposition become greater. In general, RD varies with the settling velocity and critical shear stress for deposition of cohesive sediment and the magnitudes of RD are within 11% based on the parameter values used here.

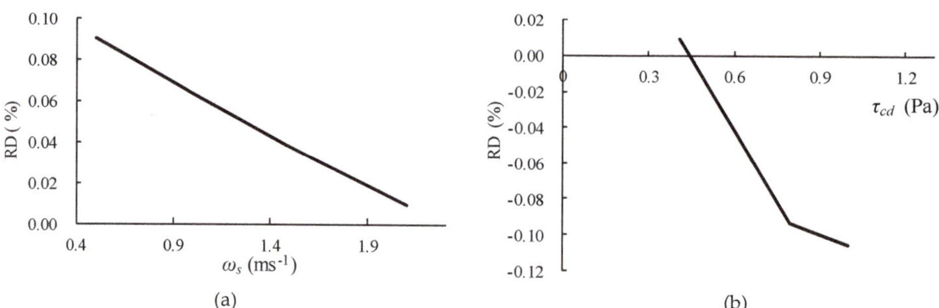

Figure 11. The relationships of relative difference in deposition volume (RD) predicted by the L and N models against (**a**) settling velocity (ω_s) and (**b**) critical shear stress (τ_{cd}).

5.3. Future Reseach Directions

1. As the study case is a reach of Yangtze River, which is classed as a mega river, secondary flow effects on bed morphology of meandering channels of different scales (natural rivers with different width to depth ratio) should be investigated. Besides, as the bank of HHC reach is nonerosional, the evolutions of natural rivers with floodplain consisting of cohesive sediment should be simulated by the 2D model developed here. In addition, long-term simulations, such as decadal timescales, should be considered in the future to research the cumulative effects of secondary flow.

2. As to the cohesive sediment transport, the values of parameters play important parts in the distributions and amounts of sediment deposition (Figure 11). The roles they played should be compared with that of secondary flow in bed morphology variations. More importantly, the erosion processes should be studied as these processes cannot be reflected obviously in the HHC reach.

6. Conclusions

In order to investigate secondary flow effects on cohesive sediment deposition in a meandering reach of the Yangtze River, a 2D depth-averaged model (N model) has been improved to consider different impacts of secondary flow and cohesive sediment transport. The improved 2D model includes three different submodels, that is, the Lien (L) model [37], with a wide application in literature; the Bernard (B) model [35], considering the phase lag effect and sidewall boundary conditions of secondary flow; and a nonlinear (NL) model [1] accounting for the saturation effect of secondary flow in sharp bends. All of the models can reflect velocity redistribution caused by secondary flow to a certain degree. A module for cohesive sediment transport has been coupled into the N model as well. The simulation results are as follows.

1. In flow calculations, the secondary flow effects on water stage and velocity distribution are well predicted. Velocity redistribution has been reproduced fairly well by the L model in the HHC reach, which means the improved 2D depth-averaged model is able to predict the secondary flow impacts on flow field in meandering channels of such mega scale. A previous study by the authors [45] pointed out that the B model is preferable in flow simulations of laboratory meandering channels with flat bathymetry. Further analyses found that secondary flow correction submodels can reflect the bed topography effects and the transverse bed topography, which is neglected by the B model, is more important than the longitudinal one. This explains why the L model performs better than the B model for curved flow simulation over bed topography. In addition, the NL model does not exhibit its advantages in field mega scale meandering reach with high curvatures as that in sharply curved bends of laboratory and small river scales, although the importance of their nonlinear effects on flow simulations have been emphasized by Blanckeart [1,2] and Ottevanger [57]. The reasons need to be further analyzed. In cohesive sediment deposition simulations, the L model performs better than the N model in planar distribution of deposition, due to more sediment deposit on the concave banks of the bends, which is resulted from the velocity redistribution caused by secondary flow.
2. The difference in predicted amounts of deposition between the L and N models is evident during the last period of an annual hydrograph when the sediment load is low and the velocity redistribution caused by bed topography is obvious in this reach. This implies that the secondary flow effects on the cohesive sediment deposition vary in an annual hydrography and temporal influence of secondary flow should be considered. This result is similar to that has been found by Guan et al. [28] who conducted a 2D depth-averaged model simulation with secondary flow correction in a natural meandering river dominated by bed load.
3. Secondary flow effects on predicted amounts of deposition vary with the settling velocity and critical shear stress for deposition of cohesive sediment, and the relative difference of predicted total amounts of deposition by the L and N models is within 11% based on the parameter values used here.

Author Contributions: Conceptualization, C.Q. and X.S.; methodology, C.Q.; software, C.Q.; validation, C.Q.; formal analysis, C.Q.; investigation, C.Q.; resources, Y.X.; data curation, Y.X.; writing—original draft preparation, C.Q.; writing—review and editing, X.S.; visualization, C.Q.; supervision, X.S.; project administration, X.S.; funding acquisition, X.S.

Funding: This research was funded by the National Key R&D Program of China (2018YFC0407402-01).

Acknowledgments: The authors would like to thank Baozhen Jia for her helpful advice and discussion about this paper and Yanjun Wang for the formation of the ideas about this paper.

Conflicts of Interest: The authors declare no conflict of interest.

References

1. Blanckaert, K.; de Vriend, H.J. Nonlinear modeling of mean flow redistribution in curved open channels. *Water Resour. Res.* **2003**, *39*. [CrossRef]
2. Blanckaert, K.; de Vriend, H.J. Meander dynamics: A nonlinear model without curvature restrictions for flow in open-channel bends. *J. Geophys. Res.* **2010**, *115*. [CrossRef]
3. Camporeale, C.; Perona, P.; Porporato, A.; Ridolfi, L. Hierarchy of models for meandering rivers and related morphodynamic processes. *Rev. Geophys.* **2007**, *45*. [CrossRef]
4. Duan, J.G.; Julien, P.Y. Numerical simulation of the inception of channel meandering. *Earth Surf. Process. Landf.* **2005**, *30*, 1093–1110. [CrossRef]
5. Jang, C.-L.; Shimizu, Y. Numerical Simulation of Relatively Wide, Shallow Channels with Erodible Banks. *J. Hydraul. Eng.* **2005**, *131*, 565–575. [CrossRef]
6. Sun, J.; Lin, B.-L.; Kuang, H.-W. Numerical modelling of channel migration with application to laboratory rivers. *Int. J. Sediment Res.* **2015**, *30*, 13–27. [CrossRef]
7. Odgaard, A.J.; Bergs, M.A. Flow Processes in a Curved Alluvial Channel. *Water Resour. Res.* **1988**, *24*, 45–56. [CrossRef]
8. Abad, J.D.; Garcia, M.H. Experiments in a high-amplitude Kinoshita meandering channel: 2. Implications of bend orientation on bed morphodynamics. *Water Resour. Res.* **2009**, *45*. [CrossRef]
9. Dietrich, W.E.; Smith, J.D. Influence of the Point Bar on Flow through Curved Channels. *Water Resour. Res.* **1983**, *19*, 1173–1192. [CrossRef]
10. Elina, K.; Petteri, A.; Matti, V.; Hannu, H.; Juha, H. Spatial and temporal distribution of fluvio-morphological processes on a meander point bar during a flood event. *Hydrol. Res.* **2013**, *44*, 1022–1039. [CrossRef]
11. Li, W.; Wang, J.; Yang, S.; Zhang, P. Determining the Existence of the Fine Sediment Flocculation in the Three Gorges Reservoir. *J. Hydraul. Eng.* **2015**, *141*, 05014008. [CrossRef]
12. Li, W.; Yang, S.; Jiang, H.; Fu, X.; Peng, Z. Field measurements of settling velocities of fine sediments in Three Gorges Reservoir using ADV. *Int. J. Sediment Res.* **2016**, *31*, 237–243. [CrossRef]
13. Xiao, Y.; Yang, F.S.; Su, L.; Li, J.W. Fluvial sedimentation of the permanent backwater zone in the Three Gorges Reservoir, China. *Lake Reserv. Manag.* **2015**, *31*, 324–338. [CrossRef]
14. Krone, R.B. *Flume Studies of the Transport of Sediment in Estuarial Shoaling Processes*; Hydraulic Engineering Laboratory, University of California: Berkeley, CA, USA, 1962.
15. De Vriend, H.J. Velocity redistribution in curved rectangular channels. *J. Fluid Mech.* **1981**, *107*, 423–439. [CrossRef]
16. Johannesson, H.; Parker, G. Velocity Redistribution in Meandering Rivers. *J. Hydraul. Eng.* **1989**, *115*, 1019–1039. [CrossRef]
17. Blanckaert, K. Saturation of curvature-induced secondary flow, energy losses, and turbulence in sharp open-channel bends: Laboratory experiments, analysis, and modeling. *J. Geophys. Res.* **2009**, *114*. [CrossRef]
18. Johannesson, H.; Parker, G. Secondary Flow in Mildly Sinuous Channel. *J. Hydraul. Eng.* **2015**, *115*, 289–308. [CrossRef]
19. Nicholas, A.P.; Ashworth, P.J.; Sambrook Smith, G.H.; Sandbach, S.D. Numerical simulation of bar and island morphodynamics in anabranching megarivers. *J. Geophys. Res. Earth Surf.* **2013**, *118*, 2019–2044. [CrossRef]
20. Ashworth, P.J.; Best, J.L.; Roden, J.E.; Bristow, C.S.; Klaassen, G.J. Morphological evolution and dynamics of a large, sand braid-bar, Jamuna River, Bangladesh. *Sedimentology* **2000**, *47*, 533–555. [CrossRef]
21. Hackney, C.R.; Darby, S.E.; Parsons, D.R.; Leyland, J.; Aalto, R.; Nicholas, A.P.; Best, J.L. The influence of flow discharge variations on the morphodynamics of a diffluence-confluence unit on a large river. *Earth Surf. Process. Landf.* **2018**, *43*, 349–362. [CrossRef]
22. Parsons, D.R.; Best, J.L.; Lane, S.N.; Orfeo, O.; Hardy, R.J.; Kostaschuk, R. Form roughness and the absence of secondary flow in a large confluence–diffluence, Rio Paraná, Argentina. *Earth Surf. Process. Landf.* **2007**, *32*, 155–162. [CrossRef]

23. Szupiany, R.N.; Amsler, M.L.; Hernandez, J.; Parsons, D.R.; Best, J.L.; Fornari, E.; Trento, A. Flow fields, bed shear stresses, and suspended bed sediment dynamics in bifurcations of a large river. *Water Resour. Res.* **2012**, *48*. [CrossRef]
24. Szupiany, R.N.; Amsler, M.L.; Parsons, D.R.; Best, J.L. Morphology, flow structure, and suspended bed sediment transport at two large braid-bar confluences. *Water Resour. Res.* **2009**, *45*. [CrossRef]
25. Nicholas, A.P. Modelling the continuum of river channel patterns. *Earth Surf. Process. Landf.* **2013**, *38*, 1187–1196. [CrossRef]
26. Rhoads, B.L.; Johnson, K.K. Three-dimensional flow structure, morphodynamics, suspended sediment, and thermal mixing at an asymmetrical river confluence of a straight tributary and curving main channel. *Geomorphology* **2018**, *323*, 51–69. [CrossRef]
27. Abad, J.D.; Buscaglia, G.C.; Garcia, M.H. 2D stream hydrodynamic, sediment transport and bed morphology model for engineering applications. *Hydrol. Process.* **2008**, *22*, 1443–1459. [CrossRef]
28. Guan, M.; Wright, N.G.; Sleigh, P.A.; Ahilan, S.; Lamb, R. Physical complexity to model morphological changes at a natural channel bend. *Water Resour. Res.* **2016**, *52*, 6348–6364. [CrossRef]
29. Iwasaki, T.; Shimizu, Y.; Kimura, I. Sensitivity of free bar morphology in rivers to secondary flow modeling: Linear stability analysis and numerical simulation. *Adv. Water Resour.* **2016**, *92*, 57–72. [CrossRef]
30. Yang, H.; Lin, B.; Sun, J.; Huang, G. Simulating Laboratory Braided Rivers with Bed-Load Sediment Transport. *Water* **2017**, *9*, 686. [CrossRef]
31. Kasvi, E.; Alho, P.; Lotsari, E.; Wang, Y.; Kukko, A.; Hyyppä, H.; Hyyppä, J. Two-dimensional and three-dimensional computational models in hydrodynamic and morphodynamic reconstructions of a river bend: Sensitivity and functionality. *Hydrol. Process.* **2015**, *29*, 1604–1629. [CrossRef]
32. Kang, T.; Kimura, I.; Shimizu, Y. Responses of Bed Morphology to Vegetation Growth and Flood Discharge at a Sharp River Bend. *Water* **2018**, *10*, 223. [CrossRef]
33. Reesink, A.J.H.; Ashworth, P.J.; Sambrook Smith, G.H.; Best, J.L.; Parsons, D.R.; Amsler, M.L.; Hardy, R.J.; Lane, S.N.; Nicholas, A.P.; Orfeo, O.; et al. Scales and causes of heterogeneity in bars in a large multi-channel river: Río Paraná, Argentina. *Sedimentology* **2014**, *61*, 1055–1085. [CrossRef]
34. Begnudelli, L.; Valiani, A.; Sanders, B.F. A balanced treatment of secondary currents, turbulence and dispersion in a depth-integrated hydrodynamic and bed deformation model for channel bends. *Adv. Water Resour.* **2010**, *33*, 17–33. [CrossRef]
35. Bernard, R.S. *STREMR: Numerical Model for Depth-Averaged Incompressible Flow*; Hydraulics Laboratory (Waterways Experiment Station), U.S. Army Corps of Engineers: Vicksburg, MS, USA, 1993.
36. Duan, J.G. Simulation of Flow and Mass Dispersion in Meandering Channels. *J. Hydraul. Eng.* **2004**, *130*, 964–976. [CrossRef]
37. Lien, H.C.; Hsieh, T.Y.; Yang, J.C.; Yeh, K.C. Bend-Flow Simulation Using 2D Depth-Averaged Model. *J. Hydraul. Eng.* **1999**, *125*, 1097–1108. [CrossRef]
38. Song, C.G.; Seo, I.W.; Kim, Y.D. Analysis of secondary current effect in the modeling of shallow flow in open channels. *Adv. Water Resour.* **2012**, *41*, 29–48. [CrossRef]
39. Jin, Y.-C.; Steffler, P.M. Predicting Flow in Curved Open Channels by Depth-Averaged Method. *J. Hydraul. Eng.* **1993**, *119*, 109–124. [CrossRef]
40. Abad, J.D.; Garcia, M.H. Experiments in a high-amplitude Kinoshita meandering channel: 1. Implications of bend orientation on mean and turbulent flow structure. *Water Resour. Res.* **2009**, *45*. [CrossRef]
41. Wu, W.; Wang, S.Y. Depth-Averaged 2-D Calculation of Flow and Sediment Transport in Curved Channels. *Int. J. Sediment Res.* **2004**, *19*, 241–257.
42. Wang, H.; Zhou, G.; Shao, X. Numerical simulation of channel pattern changes Part I: Mathematical model. *Int. J. Sediment Res.* **2010**, *25*, 366–379. [CrossRef]
43. WL|DelftHydraulics. *Delft3D-FLOW User Manual (Version: 3.15.30059): Simulation of Multi-Dimensional Hydrodynamic Flows and Transport Phenomena, Including Sediments*; Deltares: Rotterdamseweg, The Netherlands, 2013; pp. 226–230.
44. Hosoda, T.; Nagata, N.; Kimura, I.; Michibata, K.; Iwata, M. A Depth Averaged Model of Open Channel Flows with Lag between Main Flows and Secondary Currents in a Generlized Curvilinear Coordinate System. In Proceedings of the Advances in Fluid Modeling and Turbulence Measurements, Tokyo, Japan, 4–6 December 2001; pp. 63–70.

45. Qin, C.; Shao, X.; Zhou, G. Comparison of Two Different Secondary Flow Correction Models for Depth-averaged Flow Simulation of Meandering Channels. *Procedia Eng.* **2016**, *154*, 412–419. [CrossRef]
46. Ottevanger, W. Modelling and Parameterizing the Hydro-and Morphodynamics of Curved Open Channels. Ph.D. Thesis, Delft University of Technology, Delft, The Netherlands, 2013.
47. Gu, L.; Zhang, S.; He, L.; Chen, D.; Blanckaert, K.; Ottevanger, W.; Zhang, Y. Modeling Flow Pattern and Evolution of Meandering Channels with a Nonlinear Model. *Water* **2016**, *8*, 418. [CrossRef]
48. Lin, Q.; Wu, W. A one-dimensional model of mixed cohesive and non-cohesive sediment transport in open channels. *J. Hydraul. Res.* **2013**, *51*, 506–517. [CrossRef]
49. Latrubesse, E.M. Large rivers, megafans and other Quaternary avulsive fluvial systems: A potential "who's who" in the geological record. *Earth-Sci. Rev.* **2015**, *146*, 1–30. [CrossRef]
50. Dargahi, B. Three-dimensional flow modelling and sediment transport in the River Klarälven. *Earth Surf. Process. Landf.* **2004**, *29*, 821–852. [CrossRef]
51. Kleinhans, M.G.; Jagers, H.R.A.; Mosselman, E.; Sloff, C.J. Bifurcation dynamics and avulsion duration in meandering rivers by one-dimensional and three-dimensional models. *Water Resour. Res.* **2008**, *44*. [CrossRef]
52. YRWC. *Analyses of Flow-Sediment Characteristics, the Reservoir Siltation and the River Erosion in the Lower Reaches of the Three Gorges Reservoir in 2012*; Yangtze Water Resources Commission: Wuhan, China, 2013.
53. Fang, H.-W.; Rodi, W. Three-dimensional calculations of flow and suspended sediment transport in the neighborhood of the dam for the Three Gorges Project (TGP) reservoir in the Yangtze River. *J. Hydraul. Res.* **2010**, *41*, 379–394. [CrossRef]
54. Lane, S.N. Hydraulic modelling in hydrology and geomorphology: A review of high resolution approaches. *Hydrol. Process.* **2015**, *12*, 1131–1150. [CrossRef]
55. De Vriend, H.J. A Mathematical Model Of Steady Flow In Curved Shallow Channels. *J. Hydraul. Res.* **1977**, *15*, 37–54. [CrossRef]
56. Miao, W.; Blanckaert, K.; Heyman, J.; Li, D.; Schleiss, A.J. A parametrical study on secondary flow in sharp open-channel bends: Experiments and theoretical modelling. *J. Hydro-Environ. Res.* **2016**, *13*, 1–13.
57. Ottevanger, W.; Blanckaert, K.; Uijttewaal, W.S.J. Processes governing the flow redistribution in sharp river bends. *Geomorphology* **2012**, *163–164*, 45–55. [CrossRef]

Article

River Model Calibration Based on Design of Experiments Theory. A Case Study: Meta River, Colombia

Guillermo J. Acuña [1,2,*], Humberto Ávila [1] and Fausto A. Canales [2]

1 Department of Civil and Environmental Engineering, Instituto de Estudios Hidráulicos y Ambientales, Universidad del Norte, Km.5 Vía Puerto Colombia, Barranquilla 081007, Colombia

2 Department of Civil and Environmental, Universidad de la Costa, Calle 58 #55-66, Barranquilla 080002, Atlántico, Colombia

* Correspondence: guillermo.acunar@gmail.com; Tel.: +57-350-9509 (ext. 4236)

Received: 25 May 2019; Accepted: 29 June 2019; Published: 5 July 2019

Abstract: Numerical models are important tools for analyzing and solving water resources problems; however, a model's reliability heavily depends on its calibration. This paper presents a method based on Design of Experiments theory for calibrating numerical models of rivers by considering the interaction between different calibration parameters, identifying the most sensitive parameters and finding a value or a range of values for which the calibration parameters produces an adequate performance of the model in terms of accuracy. The method consists of a systematic process for assessing the qualitative and quantitative performance of a hydromorphological numeric model. A 75 km reach of the Meta River, in Colombia, was used as case study for validating the method. The modeling was conducted by using the software package MIKE-21C, a two-dimensional flow model. The calibration is assessed by means of an Overall Weighted Indicator, based on the coefficient of determination of the calibration parameters and within a range from 0 to 1. For the case study, the most significant calibration parameters were the sediment transport equation, the riverbed load factor and the suspended load factor. The optimal calibration produced an Overall Weighted Indicator equal to 0.857. The method can be applied to any type of morphological models.

Keywords: calibration; river modeling; design of experiments; MIKE-21C model; Meta River

1. Introduction

Numerical models have become an essential tool for researching and developing engineering solutions related to water resources problems [1]. Models enable complex underlying processes to be captured, and facilitate the analysis of interrelationships between variables in cases with limited data [2]. Modeling has major applications in fields such as hydrology, maritime and coastal studies [3,4], river hydraulics [5] and water quality [6].

Calibration is one of the most important activities within the modeling process, because the model's credibility strongly depends on it [7]. The calibration process can be defined as adjusting the parameter values of a model in order to reproduce the real-world response within an accuracy range defined in the performance criteria (i.e., an acceptable level of adjustment between model and reality) [8,9]. The importance and impact of calibration on hydrological and hydraulics models has been assessed and confirmed by several authors [10–13].

According to Troy et al. [14], the process for calibrating the parameters required in a model can be classified in three categories: trial and error (manual) calibration, automatic optimization (generally through computer programs) and multistep automatic methods that take advantage of combining the strengths of manual and automatic calibration. The trial and error approaches provide more control to the modeler; however, manual adjustment may become too complex when the parameters to calibrate are numerous and correlated. The automatic optimization methods are based on three elements: an objective function, an optimization algorithm (that usually includes a set of constraints) and a convergence criterion. This type of calibration has been widely used in recent works related to hydrological and hydrodynamic modeling [10,11,15–17]. However, this type of calibration faces a numerical problem, the equifinality of the problem (i.e., that there might exist a set of different combinations of parameters for which the objective function returns the same value) [18]. It is also possible that the combination of values obtained in the calibration process is inconsistent with real world phenomena that the model tries to represent.

In general, river models consist of a combination of two or three of the following components: hydrodynamics, sediments and morphology. A river model calibration is usually focused on the hydrodynamic component, whose usual indicators are water surface levels and discharge under steady flow conditions [19]. The most common indicator for sedimentological calibration is the sediment transport rate [20,21]; however, this information is not always available and, furthermore, it may include high levels of uncertainty depending on the measuring or estimation techniques employed to obtain the data. Morphology calibration typically involves comparing field data on bathymetry as well as erosion and sediment transport rates, with results obtained from simulation. This type of calibration requires a set of parameters that allows sensitivity analysis to be performed on the simulated river form, in order to emulate field data conditions from an initial state to a final state [22]. Morphology calibration is complex and rarely performed, because topobathymetric data is usually unavailable [3]. Many numerical models are unable to effectively reproduce the physical processes related to the morphological evolution of the river channel [1], and understanding the relationship between the river channel morphology and erosion processes is the subject of many recent studies [23–25] aiming for a better understanding and forecasting of the river evolution.

Because of the difficulties commonly caused by the lack of sufficient and accurate data, as well as by poor model performance under certain conditions, the calibration of a hydromorphological model is usually complicated due to the interaction of its calibration parameters; consequently, the sequential adjustment of each parameter value might not be the most efficient calibrating method [26]. This may be even more relevant for rivers whose morphology is heavily affected by hydrological and sedimentological variations, because the slightest modification in the parameters of one of the three components (hydrodynamics, sediments or morphology) would possibly affect the adjustment of the other two. For example, a significant change in the river cross-section could influence flow velocities, water surface levels and shear stress, which in turn would impact on erosion and sediment transport processes. In such cases, a common calibration approach used by most consultants and researchers is to follow recommendations given by software developers and more experienced users. Nevertheless, these recommendations do not represent a structured method, and they are also hard to replicate when different conditions arise [27].

Within this context, the main contribution of this paper is to present a calibration method based on Design of Experiments (DOE) theory that allows: calibration of the model considering the interplay between the calibration parameters; identification of the most sensitive parameters within the calibration process; and determination of a value or a range of values for which the calibration parameters produce an adequate performance of the model in terms of accuracy. The method consists of a systematic process for assessing the qualitative and quantitative performance of a hydromorphological numeric model based on calibration parameters (riverbed level changes, velocity vectors, sediment transport rates, etc.).

The combination of sensitivity analysis and optimization techniques for a better model calibration has been described by van Waveren et al. [27] as a good modeling practices. The DOE approach presented in this paper incorporates these good practices while also allowing the modelers to get a better understanding of the effect of the calibration parameters on the adjustment indicators, which is a useful feature, especially for beginners in modelling. Using DOE to define the number of simulations might even reduce the computational times when compared to other heuristic approaches. Because it is based on DOE theory, the method aims at easy implementation and adjustment.

To validate the method, this paper presents a case study evaluating a 75 km-long reach of the Meta River, in Colombia. The modeling was conducted with MIKE-21C, a two-dimensional flow model which can analyze spatial and temporal variation in depth, bed level, velocity, and shear stress during extended time intervals [28]. Nevertheless, the method described in this paper can be coupled with similar hydromorphological models, many of them briefly described in [29].

The remainder of this paper has the following structure: Section 2 reports the set of parameters required for calibrating the model, and descriptions of the method and case study; Section 3 presents the model setup and the main results obtained by applying the method, followed by the corresponding discussion. Section 5 lists the main conclusions.

2. Materials and Methods

The method is summarized as follows: (i) definition of the model objective and possible simplifications; (ii) selection the calibration parameters and indicators used for hydrodynamic, sedimentologic and morphologic components; (iii) experiment design and calibration. After presenting the method, the river reach taken as the case study is described.

2.1. Modeling Objective and Simplifications

Some of the typical modeling purposes include: understanding the hydrodynamics, sediment transport or morphological development of a specific segment of a river; aiding in the design of ports, bridges and other hydraulic and navigation structures; assessing river restoration and management plans; analyzing aquatic ecosystems; etc. Defining the model objective is perhaps the most important step of the process. Once the objective is defined, and before calibrating the model, it is necessary to reduce the number of calibration parameters, in order to have a better understanding and control over the process. Depending on the spatiotemporal resolution of the simulations, the use that will be given to the results and the characteristics of the river in consideration, some parameters may become less significant. For example, the river hydrodynamic conditions may be such that the eddy viscosity calibration becomes unnecessary, possibly because it is not a sensitive parameter in the hydrodynamic adjustment process [30]. Another example is the existence of hydraulic structures or marginal protections providing riverbank erosion control [31], which in turn make the erosion rate parameter expendable.

2.2. Parameters Used as Hydrodynamic, Sedimentologic and Morphological Indicators

For every parameter considered during the calibration process, and in order to determine the goodness-of-fit of the model, a set of qualitative and quantitative indicators is defined for comparing measurements with the results from simulations. Qualitative indicators refer to variables difficult to compare at a specific time or place, or variables whose interpretation requires modeler experience. For this method, the goodness-of-fit of the calibration based on the quantitative indicators is measured by means of the coefficient of determination, also known as R-squared (R^2), a statistical measure of how close the data fit the regression line, in this case, comparing whether the simulations match real-data. The method admits other statistical measures for goodness-of-fit, like the mean squared error (MSE) and the mean absolute error (MAE).

Based on suggestions by Matte et al. [32] and some common outputs from river models, a set of quantitative indicators were chosen to compose an Overall Weighted Indicator (OWI) for assessing calibration. The OWI is expressed as follows:

$$\mathrm{OWI} = \beta_1 \cdot \mathrm{WL} + \beta_2 \cdot \mathrm{QL} + \beta_3 \cdot \mathrm{FD} + \beta_4 \cdot \mathrm{MP} + \beta_5 \cdot \mathrm{SST} + \beta_6 \cdot \mathrm{SBT} + \beta_7 \cdot \mathrm{ST} + \beta_8 \cdot \mathrm{BL} + \beta_9 \cdot \mathrm{BE}, \quad (1)$$

where $\sum_{i=1}^{n} \beta_i = 1$, and:

- WL = coefficient of determination (R^2) related to water level;
- QL = coefficient of determination (R^2) related to flow rate at control cross-sections;
- FD = coefficient of determination (R^2) related to flow distribution through island branches;
- MP = coefficient of determination (R^2) related to depth through the 2D domain;
- SST = coefficient of determination (R^2) related to suspended-load sediment transport;
- SBT = coefficient of determination (R^2) related to bed load sediment transport;
- ST = coefficient of determination (R^2) related to total load sediment transport;
- BL = coefficient of determination (R^2) related to bed level through the 2D sector;
- BE = coefficient of determination (R^2) related to bank erosion rate.

The different values of β_i correspond to the relative weight of each indicator on the OWI. These weights are usually chosen by consensus among the modeling team based on experience or based on the uncertainty related to the variables of each indicator. For beginners, the authors of this paper recommend using the same value for all β_i, as explained by Chaves and Alipaz [33] based on Shannon's principle of maximum entropy, which warrants that the probabilities of underestimating or overestimation the parameters would be the same, considering that the indicators and the parameters are random variables (with also random distributions).

It is worth mentioning that the OWI can be subdivided in terms of the components of the river model. Elements WL, QL, FD and MP are calibration parameters of the hydrodynamic component (OWI_{HD}); SST, SBT and ST correspond to the sedimentological component (OWI_{ST}); while BL and BE are calibration parameters accounting for the morphological component (OWI_{MF}).

The input parameters can be categorized into two types: those that can be determined from field measurements or obtained from literature or other reliable sources; and those that must be determined through calibration [8]. Table 1 shows some of the most common parameters required for modeling the hydrodynamic, sedimentological and morphological characteristics of a river. It is worth noticing that the calibration parameters listed in Table 1 emphasize the ones used by MIKE-21C, the computer model used as the main software tool in the present work.

Table 1. Typical calibration parameters. (Source: Adapted from [28]).

Component	Parameter	Description
Hydrodynamics	• Bed resistance to flow	This is related to the riverbed's resistance to flow. For MIKE-21C, the value at each cell is determined as a function of water depth and Manning's or Chézy's coefficient. The model uses as input a constant value or maps corresponding to their spatial distribution.
	• Eddy viscosity—K	Parameter related to the calibration of the velocity vector and depends on the turbulence state. For MIKE-21C, usually between 0.2 m^2/s and 5.0 m^2/s.
Sedimentology/ Morphology	• Sediment transport equation	The selection of sediment transport equation defines the parameters to be used in the model, and it mainly depends on granulometry (i.e., particle size and shape). A complete description of the transport formulae can be found in the MIKE-21C user's guide [28].
	• Bed load factor—Kb • Suspended load factor—Ks	The total sediment load is estimated through a sediment transport equation (MIKE-21C includes a set of options for this purpose).
	• Longitudinal slope coefficient—LSC • Transverse slope coefficient—TSC • Transverse slope power—TSP	These parameters modify sediment transport rates in the stream current, considering morphological changes affecting transverse and longitudinal slopes.
	• Helical flow calibration constant—HL	In MIKE-21C this parameter defines the intensity of the helical flow on the riverbed due to the secondary currents. It is usually assumed to be equal to 1; however and ranges from 0.4 to 1.2, according to [34].
	• Erosion rate on riverbanks	According to MIKE-21C, the erosion rate on the riverbanks can be calibrated using three different parameters, α: Transversal slope of the riverbank. ψ: Fraction of sediment transport rate near the bank. γ: Erosion constant which does not depend on the hydraulic condition.

Depending on the river characteristics or the chosen computational model (or numerical approach), the model might require more or less parameters. For instance, MIKE-21C was employed by de Villiers [35] for studying cohesive sediment transport in a shallow reservoir, including within this analysis some additional calibration parameters such as Critical Shear Stress for deposition (τ_{de}), Critical Shear Stress for erosion (τ_{er}), Erosion constant (E_0) and Exponent of the erosion. Beck and Basson [36] assessed the hydraulic and sedimentological behavior of the Klein river estuary by means of MIKE-21C, including as calibration parameters: Hydrodynamic time step, Morphological time step, Flooding Depth, Drying depth, Median grain diameter, Mass density of sediment and Porosity, as well as the calibration parameters shown in Table 1. Therefore, if additional calibration parameters are required for a more accurate representation of the process, the authors of the present paper recommend to the reader to conduct a literature review in order to define the range of feasible values for each calibration parameter. For example, the critical shear stress for deposition (τ_{de}) and the critical shear stress for erosion (τ_{er}) are very variable parameters which have a great impact on sediment transport mechanisms, but some authors [25,35] have performed sensitivity analysis on these calibration parameters, that could be used as baseline for setting the range of feasible values.

In this paper, two types of visual analysis are defined: (i) a comparison between simulated and measured (ADCP) velocity vectors, and (ii) an assessment of the simulated and historical morphological behavior of the river. Low performance model setups are easily detected by visual analysis before any numerical analysis.

2.3. Experiment Design and Calibration

The aim of DOE is to maximize the information obtained from a minimum number of experiments. It also helps to determine which factors might affect the performance of the model [37].

Once the calibration parameters and goodness-of-fit statistics have been chosen, the proposed method considers a screening design of experiment (definitive screening type) that allows identifying which calibration parameters and interactions have a significant effect on the overall accuracy of the model [38,39]. For this screening design, authors propose using a 2^k factorial type of experiment, which evaluates k calibration parameters, each of them having two alternatives or levels that must be defined according to its corresponding physical or numerical meaning [37,38,40]. In consequence, goodness-of-fit and OWI calculations must be performed for each possible combination defined in the 2^k experiment.

A second DOE aims at optimizing the calibration parameters whose effects (or the interactions with other parameter) were found as significant from the screening DOE. The present authors recommend a 3^k type of experiment, that means, evaluating these parameters for at least three different representative levels from their previously defined range of possible values [38]. However, following this recommendation depends on available resources. As in the previous DOE, goodness-of-fit and OWI calculations must be performed for each possible combination defined in the 3^k experiment.

Depending on the results from the second DOE, it is possible to obtain a regression equation expressing the relationship between calibration parameters and the model response, as well as the corresponding goodness-of-fit statistic, allowing optimizing a specific component of the OWI. The calibration process proposed in this paper is represented, for a better understanding, in the flowchart shown in Figure 1.

The process begins with hydrodynamic calibration. For this component, the main calibration parameter is the roughness coefficient, since water depth and flow velocity are function of this parameter, and it also has influence on flow distribution between the islands. If the river topology presents a cross-section contraction or sudden change of direction, it is likely that the eddy viscosity coefficient will be another significant calibration parameter [41].

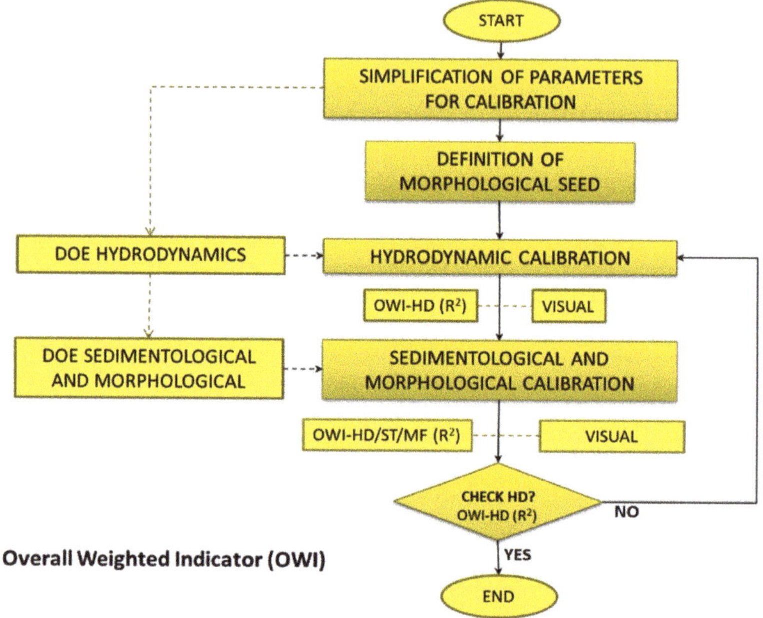

Figure 1. Flowchart describing the calibration process.

The present method is suitable for calibrating and modeling rivers with a significant relationship between hydrodynamics and morphology. Therefore, it is necessary to include the effect of river morphological changes when conducting hydrodynamic calibration. For the method explained in this document, it is proposed the use of a morphological seed or preliminary values for morphological calibration parameters. This would allow a preliminary calculation of the morphological component of the model, during hydrodynamics calibration. The selection of the morphological seed can be done through a conceptual analysis, by using reference values available in the literature or based on known values used in other sections of the same river or other rivers with similar characteristics.

Once the hydrodynamic calibration is complete, the process continues with the sedimentological and morphological calibration. As explained before, such calibration is performed by an experimental design which can be used to define the significant parameters, the interactions between them, and the range of values for the calibrating parameters that provide a good fit between the results of the model and the observed data. When assessing the calibration of the morphological (OWI_{MF}) and sedimentological (OWI_{ST}) components of the Overall Weighted Indicator, it is important to maintain the calibration of the hydrodynamic component and its corresponding parameters.

It is worth mentioning that the method can be used for unidimensional or multidimensional models (2D or 3D). A greater quantity of calibration parameters and indicators increases the required computational processing time, but it does not guarantee more accuracy.

2.4. Case Study: A Reach of the Meta River, Colombia

The Meta River is one of the major tributaries of the Orinoco river. The methods and results presented in this paper were produced from a project called "Update of studies and designs for navigation between Cabuyaro (K804) and Puerto Carreño (K0)" [42], conducted by Universidad del Norte (UNINORTE) in 2013, for the Colombian National Roads Institute (INVIAS) and the Colombian Ministry of Transport. From the field measurements made, the length of the Meta River was calculated as 1002 km, from its source near the town of Guamal, in the Meta Department, to its mouth in the Orinoco river. The Meta River watershed is approximately 99,500 km^2 in area.

Recovering and improving navigability in Colombian rivers has been one of the main goals of the Colombian government over the last years [43]. For this reason, the aforementioned project was commissioned in order to analyze and model the Meta River, aiming at identifying the actions required to improve its navigability conditions.

The main data for this study consists of water depth, velocity and flow rate measurements, taken at the Orinoco river mouth (K0) and near the town of Cabuyaro (K796). The measurements were taken in two field campaigns that occurred between August 2012 and January 2013. Besides these on-site measurements, hydrological records from the Institute for Hydrology, Meteorology and Environmental Studies of Colombia (IDEAM) and Landsat (NASA) satellite imagery were other sources of information used for this case study.

The navigable channel of the Meta River is very unstable, and the flow and water depth variations through the year are significant and often abrupt, which causes some boats to become stranded. At the same time, the river presents high erosion and sedimentation rates [44]. From historical records between 1983 and 2010 at the IDEAM gauging station "Aceitico" (located 127 km upstream the mouth of the Orinoco), it was found that the average hydrograph presents maximum flow rates of around 10,323 m^3/s, and minimum of around 755 m^3/s. This ratio of 13.67 between maximum and minimum average indicates wide variability in flow rate in any given year. Furthermore, a unimodal trend was found in all hydrometric records; with high flow rates between May and August, and low flow rates between early December and late February.

Sedimentological measurements at the IDEAM stations in Puerto Texas (K669), Aguaverde (K360) and Aceitico (K127) show that sand and a small fraction of silt are the main materials transported by the river. The average grain size in the bed (D50-bed) is approximately 0.35 mm and the average grain size of the suspended particles (D50-susp) ranges between 0.057 and 0.17 mm. It is also worth noting that the flow rate in the river presents strong correlation with the sediment transport rate. According to the IDEAM records, this transport rate ranges between 419,680 and 4057 ton/day, based on data from Aceitico station for maximum and minimum average flow rates.

The results from the geomorphological studies carried out during the project showed that the Meta River morphology could be described as: 34% tabular, 26% sinuous / meandering, 24% straight, 10% braided and 6% anastomosed, based on satellite images from August and September 2012. Although the Meta River does not hold a particular morphological form, the selected satellite images, as well as aerial photographs taken between years 1986 and 2012 evidenced a high lateral and frontal mobility of the channel. From the analysis of the satellite images, it was also found that the floodplain of the river ranges from 3 to 9 km wide.

Based on the previous characterization and due to the technical and financial unfeasibility to model the whole river, a representative reach 75 km long was selected, between abscissas K235 (6°06′18.81″ N/69°12′55.04″ W) and K310 (6°02′50.38″ N/69°44′40.11″ W) (See Figure 2). The selection of this analysis section was based on:

1. Morphological Stability: this section presented a low riverbank variability between 1980 and 2012.
2. Hydraulic and sedimentological stability: this section is downstream of the last significant tributary of the Meta River, thus, variations in the hydraulic and sedimentological regime up to its mouth are not significant.
3. Morphological typology: presents a mix of braided, straight and sinuous reaches, indicating associated morphological response patterns along much of the river.

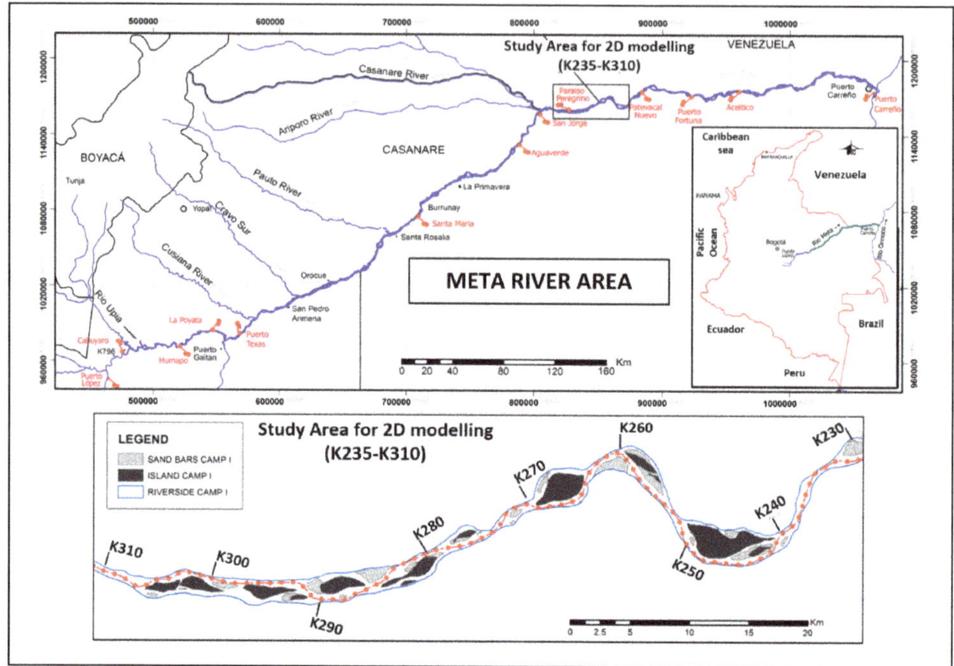

Figure 2. Representative section of the river, between K235–K310.

2.4.1. Indicators for the Meta River Model

As described in the previous section, the calibration process begins by selecting the available parameters that can be used to describe the river behavior. The rating curves for Meta River were obtained from the unidimensional model presented by UNINORTE [42], and this information was used as boundary conditions for hydrodynamic calibration.

The following comparable elements, based on the measurements obtained in the aforementioned study, were used as indicators for the calibration process of the Meta River model:

- **Adjustment of water level series (WL):** Coefficient of determination (R^2) between the hydrograph created from the water level series and the water level simulated by interpolation for the calibration period. The comparison point was set at the abscissa K310 corresponding to the upstream boundary condition.
- **Adjustment of flow distribution (FD):** Coefficient of determination (R^2) between the percentages of flow distribution through river branches around islands, measured on-site and those calculated by the 2D model.
- **Visual comparison of velocity vectors (Visual Indicator):** A visual comparison was performed between the simulated velocity vectors and those recorded by the ADCP. It was considered that the average velocity value in the measured columns would occur close to $0.6 \times H$ [45], where H is the height of the water column.
- **Adjustment of suspended sediment transport rate (SST):** The IDEAM station "Aceitico" (K127) holds information regarding suspended sediment transport, estimated through regression models from data between 1996 and 2010. Based on the information available, the following equation was used in this work for estimating suspended sediment transport rate:

$$Q_S = 0.0096 \cdot {Q_L}^{1.774}, \quad (2)$$

where Q_S is suspended sediment transport rate (ton/day) and Q_L is the river discharge (m^3/s). To calculate R^2, comparison was made between the Q_S values obtained by using discharge measurements and the ones calculated by using the results from the model.

- **Bed Level Adjustment (BL):** The coefficient of determination (R^2) was calculated between the mesh of riverbed levels created from the bathymetric surveys from field campaigns and those simulated by MIKE-21C.
- **Adequate morphological evolution (Qualitative Indicator):** This visual indicator refers to the morphological evolution of the riverbed and riverbanks. The modeler shall verify if the morphologic changes through time are coherent with the numerical capabilities of the model, for example, no sudden and intense sedimentation or erosion phenomena occurred during the time span considered in the simulation.

For this case study, the OWI described in Equation (1) was divided into two components: The weighted indicator of hydrodynamic adjustment (OWI_{HD}) and the weighted indicator of sedimentological and morphological adjustment ($OWI_{ST, MF}$). The β_i coefficients were defined by the team of modelers participating of this project, based on agreed confidence levels for each variable measured during the field campaigns. Lower β_i coefficients were assigned to variables with higher uncertainty levels.

$$OWI_{HD} = 0.60 \text{ WL} + 0.40 \text{ FD} \tag{3}$$

$$OWI_{ST, MF} = 0.30 \text{ SST} + 0.70 \text{ BL} \tag{4}$$

By considering that both components have the same weight, the global OWI can be expressed as:

$$OWI = 0.30 \text{ WL} + 0.20 \text{ FD} + 0.15 \text{ SST} + 0.35 \text{ BL} \tag{5}$$

2.4.2. Parameters for Modeling the Meta River

For this case study, the modeling objective was to assess navigation capabilities in the Meta River. Based on this objective and the river characteristics, the modelers defined that:

- For the river reach under consideration, the Chézy coefficient is used for quantifying bed resistance. In MIKE-21C, this value is estimated as a function of water depth, following the approach explained by Talmon [46]:

$$\text{Chézy} = C \cdot h^{0.17} \tag{6}$$

  where Chézy is given in $m^{1/2}$/s and h is the water depth in meters. Coefficient C (in $m^{1/3}$/s) is the parameter to calibrate, and corresponds to the reciprocal of the Manning's roughness coefficient [47], considered constant by the authors for this study. According to Talmon [46], this approach allows to incorporate small bathymetric irregularities (due to dunes, ripples, etc.) as bed resistance. The adequate calibration of the bed resistance directly impacts on the accuracy of the model to correctly estimate flow distribution and morphologic evolution. For example, if the resistance over a shallow region is too high, too much flow will be deflected and there will be a greater tendency towards developing sandbars [35].
- According to the sedimentological information supplied by IDEAM, the median grain size in the riverbed (D50-bed) is 0.35 mm. Based on this grain size, García [48] suggests using Engelund-Hansen [49], Yang's [50] and Van Rijn's [51] equations for estimating sediment transport rates. From modeling studies carried out in 2003 Hidroconsultas LTDA [52] on the same river, it was observed that the sediment transport rates estimated by using Yang's equation showed a good fit when compared to values from measurements. Based on this and aiming to reduce the number of variables in the study, the modelers decided to use only Yang's and Van Rijn's equations.
- From the available imagery, it was found that riverbank variations in time intervals shorter than three years were not significant for the representative section. Islands were defined as covered

with vegetation, which favors stability. The erosion rate at the riverbank and the corresponding coefficients were not considered as calibration parameters.

- Due to absence of significant patterns and/or phenomena affecting the morphology, the helical flow coefficient HL was simplified to its default value of 1.00.

Based on the previous considerations, the calibration parameters for this case study can be summarized as:

- Chézy Roughness coefficient as a function of depth; where C is the calibration parameter.
- Sediment Transport Equation;
- Riverbed Load Factor (Kb);
- Suspended Load Factor (Ks);
- Transverse Slope coefficient (TSC);
- Transverse Slope power (TSP).

3. Results

This section describes the main results of the calibration process.

3.1. Hydrodynamic Calibration

To initiate the hydrodynamic calibration, it was necessary to define a sedimentological and morphological seed given as:

• Sediment Transport Equation:	Van Rijn
• Riverbed Load Factor (Kb):	0.100
• Suspended Load Factor (Ks):	0.300
• Transverse Slope coefficient (TSC):	0.625
• Transverse Slope power (TSP):	0.500

The hydrodynamic calibration was focused on the adjustment of coefficient C in the roughness equation (Equation (6)). As seen in Tables 2 and 3, three coefficient C values were evaluated for this purpose: 50, 55 and 60. As described in the previous section, their accuracy was assessed through the measured water levels and percentages of flow distribution through river branches as indicators. The best fit was obtained using $C = 55$, which results in an $OWI_{HD} = 0.9062$, as shown in Table 3.

Table 2. Flow distribution through river branches around islands of the river reach.

Abscissa (km)	Arm	Observed	$C = 50$	$C = 55$	$C = 60$
204	Right	0.7076	0.6875	0.6915	0.6962
	Left	0.2924	0.3125	0.3085	0.3038
289	Right	0.1598	0.1757	0.1694	0.1514
	Left	0.8402	0.8243	0.8306	0.8486
267	Right	0.2580	0.2718	0.2453	0.2516
	Left	0.7420	0.7282	0.7547	0.7484
248	Right	0.0560	0.0527	0.0436	0.0327
	Left	0.9440	0.9473	0.9564	0.9673

Table 3. Hydrodynamic calibration results.

C	R^2		OWI_{HD}
	Water Level (WL)	Flow Distribution (FD)	
50	0.7893	0.9979	0.8727
55	0.8450	0.9981	0.9062
60	0.8272	0.9984	0.8957

Once the hydrodynamic calibration has been conducted, the behavior of the velocity vectors is visually assessed, basically to verify if the model adequately represents this feature. An example of the model results for these visual comparisons is presented in Figure 3.

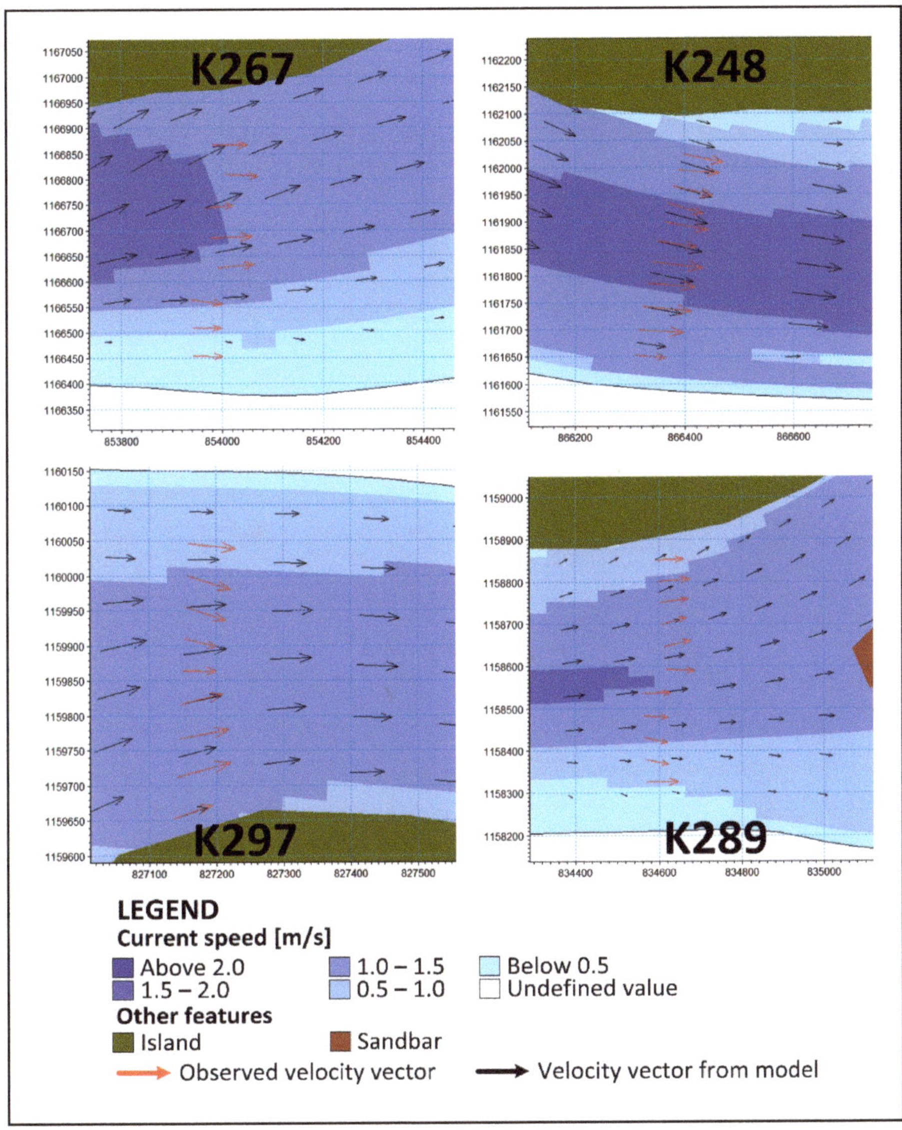

Figure 3. Velocity vectors from the model for hydrodynamic calibration.

3.2. *Screening Design—2^k Experiment*

The sedimentological and morphological components are calibrated using two consecutive experimental designs. The first design (2^k) has the function of detecting the statistically significant factors. The parameters and levels at which the first design was evaluated are shown in Table 4.

Table 4. Factors (Calibration parameters) for 2^k experimental design.

Level	Sediment Transport Equation (A) *	Suspended Load Factor—Ks (B) *	Riverbed Load Factor Kb (C) *	Transverse Slope Coeff.—TSC (D) *	Transverse Slope Power—TSP (E) *
High	Yang	0.1	0.1	0.625	0.5
Low	Van Rijn	0.9	0.5	1.250	1.0

* The letters in parentheses will be used to abbreviate the name of the corresponding calibration parameters.

To determine which factors are statistically significant, it is necessary to perform an Analysis of Variance (ANOVA) using the results from simulation. As previously stated, this experiment does not include replicates, and consequently, the degrees of freedom of error within the ANOVA are few. This could induce to errors when estimating each parameter significance [38]. Therefore, before performing the ANOVA, it is necessary to identify which parameters have a significant effect on the model response and the probability of statistical noise, which might be interpreted as natural variability (error). To do so, a normality test was performed on the standardized effects of each parameter and its corresponding interactions, shown in Figure 4.

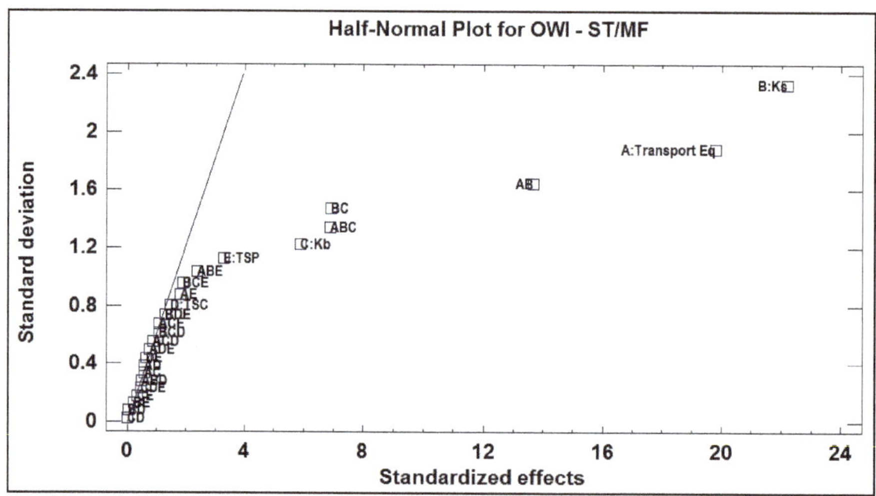

Figure 4. Normal Probability of Effects for the weighted indicator of sedimentological and morphological adjustment ($OWI_{ST, MF}$) in experimental design 2^k.

Once the most significant parameters and interactions have been identified, the ANOVA table is presented in Table 5. The R^2 statistic indicates that the adjusted model explains 96.91% of the variability in $OWI_{ST, MF}$.

Table 5. Analysis of Variance (ANOVA) results for the 2^k experiment.

Source	Sum of Squares	Df	Mean Square	*F*-Ratio	*P*-Value
A: Transport Eq	0.084400	1	0.084400	245.66	0.0000
B: Ks	0.106300	1	0.106300	309.35	0.0000
C: Kb	0.007400	1	0.007400	21.76	0.0001
AB	0.040200	1	0.040200	117.23	0.0000
BC	0.010200	1	0.010200	29.90	0.0000
ABC	0.010200	1	0.010200	29.67	0.0000
Total error	0.008200	24	0.000340		
Total (corrected)	0.267200	31			

The statistical analysis for the 2^k experiment shall be completed with the graphic verification of assumptions of Homoscedasticity, Normality and Independence. These charts are presented in Figure 5.

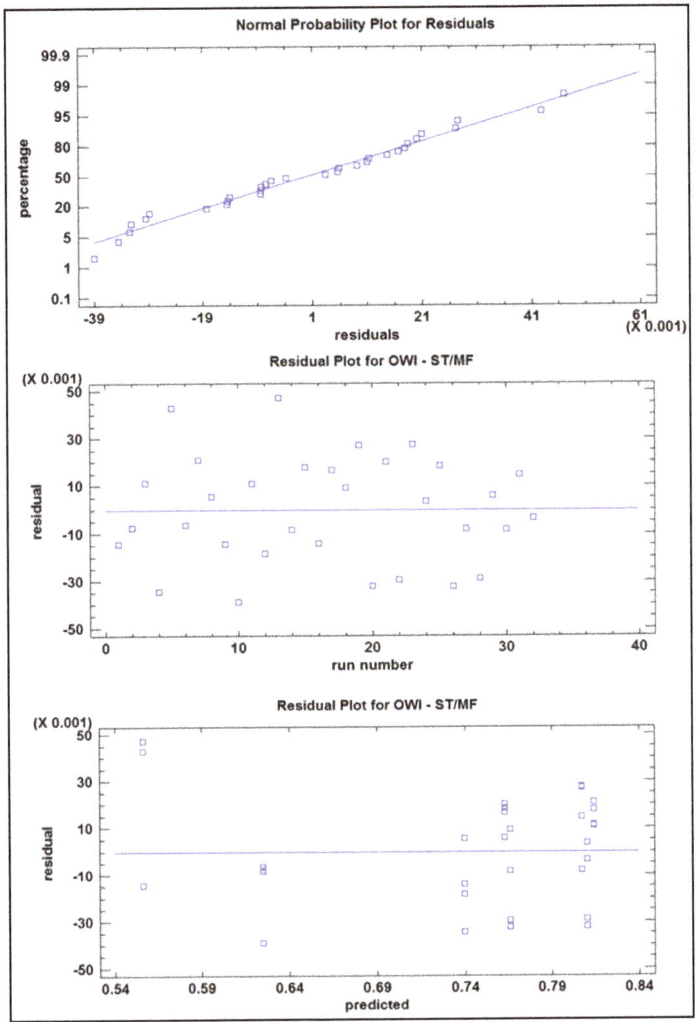

Figure 5. Graphic verification of Homoscedasticity, Normality, and Independence—Experiment 2^k.

3.3. Fit Design—3^k Experiment

This second DOE, of the 3^k type, allows specifying the calibration value of the parameters. The parameters and levels at which the second design was evaluated are shown in Table 4. However, since some 2nd and 3rd order significant interactions were found at Table 5, they cannot be excluded from the experiment. Nevertheless, if all these interactions are considered, the necessary degrees of freedom in the residues will not be enough to perform the analysis of variance. Therefore, the modelers decided to vary the TSP as a replica generator, because it is one of the original main parameters whose effect was recognized as not significant at Figure 4. As previously defined at Table 4, the TSP was varied within the range from 0.5 to 1.

Based on the previous consideration, the parameters and levels at which this second design was evaluated are shown in Table 6, and the ANOVA results for this experiment are shown in Table 7, where all F-ratios are based on the mean square of residual error. The R^2 statistic indicates that this fit model explains 97.69% of the variability in $OWI_{ST, MF}$. As in the 2^k experiment, verification tests were performed for Homoscedasticity, Normality and Independence assumptions. Homoscedasticity and Independence charts are presented in Figure 6.

Table 6. Factors (Calibration parameters) for 3^k experimental design.

Level	Sediment Transport Equation (A)*	Suspended Load Factor—Ks (B)*	Riverbed Load Factor Kb (C)*
High	Yang	0.1	0.1
Mid	-	0.5	0.3
Low	Van Rijn	0.9	0.5

Table 7. ANOVA results for the 3^k experiment.

Source	Sum of Squares	Df	Mean Square	F-Ratio	P-Value
A: Transport Eq	0.1033940	1	0.1033940	91.53	0.0000
B: Ks	0.0666866	2	0.0333433	29.52	0.0000
C: Kb	0.0196977	2	0.0098489	8.72	0.0016
AB	0.0281542	2	0.0140771	12.46	0.0002
BC	0.0134977	2	0.0067489	5.97	0.0085
ABC	0.0231324	4	0.0057831	5.12	0.0045
Total error	0.0248521	22	0.0011296		
Total (corrected)	0.2794150	35			

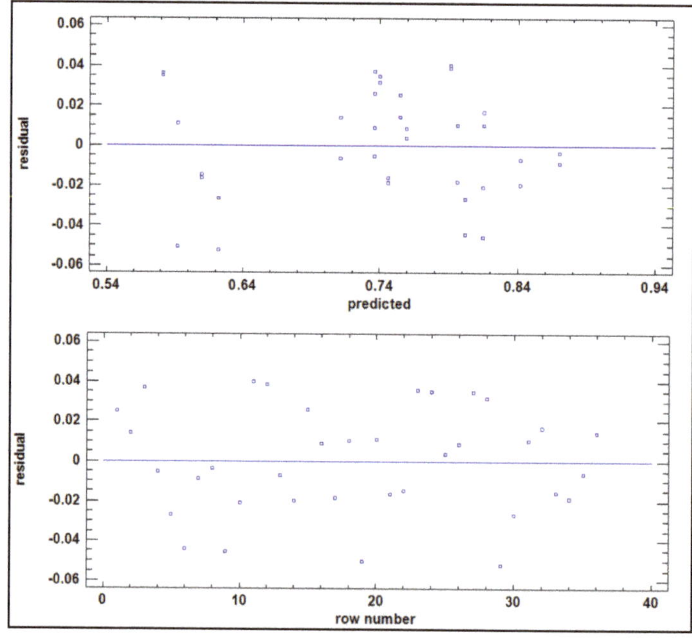

Figure 6. Graphic verification of Homoscedasticity and Independence—Experiment 3^k.

From the analysis of the individual behavior and interactions of the main parameters of this case study, the modelers observed that the higher $OWI_{ST, MF}$ values were obtained when using the Van Rijn Transport equation.

Optimal calibration has as its objective to maximize $OWI_{ST, MF}$ within the boundaries of the experimental region. From this, the best $OWI_{ST, MF}$ result found was 0.864, associated with the following conditions: Van Rijn transport equation; Ks = 0.5; Kb = 0.1.

As described in the method section and summarized in Figure 1, once the morphological and sedimentological calibration has been performed and quantified by means of the $OWI_{ST, MF}$, the hydrodynamic component of OWI has to be checked in order to determine if its calibration remains valid.

For this case study, the water level (WL) and flow distribution (FD) were calculated by the model using the conditions previously mentioned for optimizing $OWI_{ST, MF}$. An OWI_{HD} of 0.90 was obtained from using those results. The qualitative calibration parameters related to velocity vectors and morphological evolution were also evaluated for this optimal case. For two sections of the river reach used as case study, Figure 7 displays the comparison between observed and modeled velocity vectors, where a good fit was observed both for magnitude and direction.

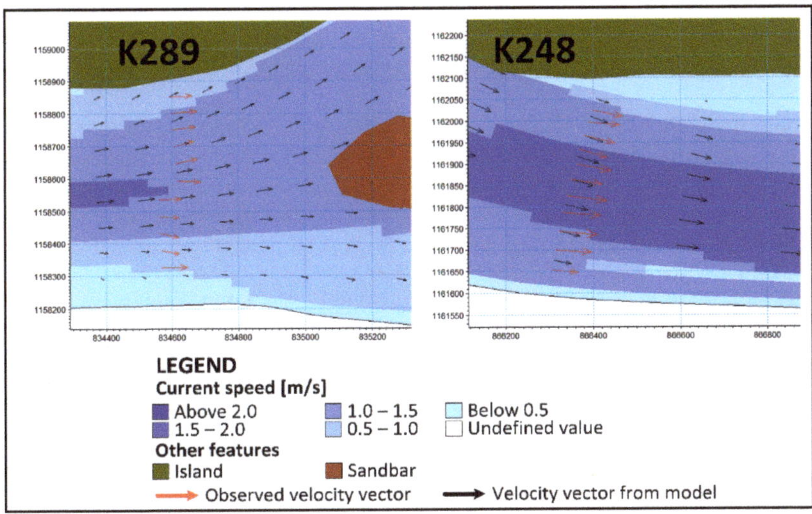

Figure 7. Comparison between Velocity Vectors in two sections of the reach for the optimal case.

The morphological evolution was assessed and verified, with no significant instabilities or bathymetric changes found.

4. Discussion

4.1. Hydrodynamic Calibration

For the three different options of C, the error found between the simulated and observed values did not exceed 1.7%. Consequently, for this case study, the model sensitivity to this parameter is low, because in general, all the evaluated cases achieved an excellent adjustment when estimating water levels and flow distribution through river branches.

The visual inspection showed a good adjustment of the direction and magnitude of the average vectors extracted from the ADCP measurements with respect to the velocity vectors with respect to the simulated ones. The morphological behavior developed properly. There were no significant instability or abrupt changes in the bottom morphology of the river.

4.2. Screening Design—2^k Experiment

Based on Figure 4, the most distant effects from the normal distribution trend were: B (Ks), A (Transport Equation), AB, BC, ABC, C (Kb). These were taken to perform the ANOVA analysis, allowing to dissect $OWI_{ST, MF}$ variability and to evaluate the statistical significance of each effect by comparing its mean square against an estimate of the experimental error.

The results of the ANOVA Table concluded that the 6 evaluated effects include a P-value less than 0.05, indicating that they are significantly different from zero with a confidence level of 95.0%. From Figure 5 it can be observed that the graph of Residual vs. Predicted Values does not present any unusual structure. From the normality graph, it was observed that residues are correctly fitted to this type of statistical distribution. Finally, the graph of Residual vs. Order did not show any trend regarding the structure of execution; instead, the behavior of the Residual is erratic and dispersed.

From this screening experiment, it was found that just three of the five calibration parameters and only three interactions have significant effects on $OWI_{ST, MF}$. Based on this, only those three parameters were used for the second 3^k type experiment design, or fit design.

4.3. Fit Design—3^k Experiment

From this fit design experiment, the ANOVA summarized in Table 7 indicates that all factors and interactions considered are significant. With all P values smaller than 0.05, all the calibration parameters have a statistically significant effect on $OWI_{ST, MF}$, when considering a confidence level of 95%.

The homoscedasticity plot in Figure 6 did not show any unusual structure. As for Normality verification, a chi-square normality test was performed, where the $OWI_{ST, MF}$ residues were distributed into 16 equally probable classes, and each class was then compared with the expected number of observations. Using Statgraphics software for this analysis, a P-value was obtained equal to 0.9273 for this normality test, which is greater than or equal to 0.05. This prevents rejecting the hypothesis that $OWI_{ST, MF}$ residues do not come from a normal distribution, with statistical confidence level of 95%.

The assumption of data independence was verified from the plot of Residues vs. Execution order in Figure 6. As in the 2^k experiment, the results shown in this figure indicate that the residuals did not follow a trend in terms of their execution order. The plot of residues in Figure 6 suggests that there is no autocorrelation among the residuals of the ANOVA.

After verifying the fit of the qualitative parameters, it can be summarized that the optimal calibration for this case study for the Meta River occurs at these experimental conditions: Van Rijn transport equation; Ks = 0.5 and Kb = 0.1. As result, the modelers obtained an $OWI_{ST, MF} = 0.8293$ and a global OWI = 0.8571.

4.4. Additional Remarks

As previously stated, the method proposed in this paper is based on DOE theory, and because of this, it shares some of its limitations:

- The domain of quantitative variables must be continuous;
- It is possible to discard a significant calibration parameter value, because of an inadequate choice of levels or alternatives during the screening design;
- Some combinations of calibration parameter values might generate numeric instability during the simulation;
- The lack of replicates (with different results) for the same configuration limits the degrees of freedom of the experiment.

It is also worth noting that the goodness of fit might considerably vary depending on the numeric model used and its capacity to recreate the hydrodynamic, sedimentologic and morphological processes under analysis.

5. Conclusions

By including principles of design of experiments (DOE), the method presented in this paper improves the reliability of the calibration process of hydromorphological models. This becomes even more relevant for rivers whose hydrodynamics depends on morphological processes. Within this context, this paper presented a calibration method based on DOE, which consists of a systematic approach for evaluating the qualitative and quantitative performance of a hydromorphological numerical model based on its corresponding calibration parameters, with the aim of allowing the modelers to get a better understanding and of the effects of these parameters on the adjustment indicators.

Using a 75 km reach of the Meta River as case study, and the MIKE-21C model, the method and its results were described and discussed, along with some of its limitations. The calibration parameters for this case study were the roughness coefficient, the sediment transport equation, the riverbed and suspended load factors, the transverse slope coefficient and the transverse slope power. An optimal overall weighted indicator, which can be defined as an integrated measure of the performance of the model was found for the example, with a value of 85.71%.

The method is versatile in terms of number of calibration parameters and selection of adjustment indicators, because their selection depends on the modeling objective and characteristics of the analyzed section. This method can be applied to unidimensional or multidimensional hydromorphological models.

Based on the findings of this study, some future research directions are:

- Compare the efficiency of the method and calibration performance by using other DOE types (Fractional Factorials, Taguchi, Latin Square, etc.) and statistical measures of goodness of fit (Nash-Sutcliffe's efficiency coefficient (NSE), P-Bias, etc.).
- Compare the performance of the method with other calibration approaches for different conditions: numerical models, river characteristics, quantity and type of calibration parameters, etc.

Author Contributions: Conceptualization, G.J.A. and H.A.; Methodology, G.J.A. and H.A.; Formal Analysis, G.J.A. and H.A.; Writing and Original Draft Preparation, G.J.A. and F.A.C.; Writing, Review and Editing, G.J.A. and F.A.C.; Funding acquisition, H.A.

Funding: This research was funded by the Universidad del Norte, Colombia.

Acknowledgments: The authors acknowledge the teamwork with the Institute of Hydraulic and Environmental Studies of the Universidad del Norte and data from the project "Update of the Studies and Designs for the Navigation of the Meta River between Cabuyaro (K804) and Puerto Carreño (K0)".

Conflicts of Interest: The authors declare no conflict of interest.

References

1. Church, M.; Ferguson, R.I. Morphodynamics: Rivers beyond steady state. *Water Resour. Res.* **2015**, *51*, 1883–1897. [CrossRef]
2. Yadav, B.; Eliza, K. A hybrid wavelet-support vector machine model for prediction of lake water level fluctuations using hydro-meteorological data. *Measurement* **2017**, *103*, 294–301. [CrossRef]
3. Zhu, Q.; Wang, Y.P.; Gao, S.; Zhang, J.; Li, M.; Yang, Y.; Gao, J. Modeling morphological change in anthropogenically controlled estuaries. *Anthropocene* **2017**, *17*, 70–83. [CrossRef]
4. Pascolo, S.; Petti, M.; Bosa, S. On the wave bottom shear stress in shallow depths: The role ofwave period and bed roughness. *Water* **2018**, *10*, 1348. [CrossRef]
5. Logan, B.L.; McDonald, R.R.; Nelson, J.M.; Kinzel, P.J.; Barton, G.J. *Use of Multidimensional Modeling to Evaluate a Channel Restoration Design for the Kootenai River, Idaho*; Scientific Investigations Report 2010–5213; U.S. Geological Survey: Reston, VA, USA, 2011.
6. Stewart, G.; Anderson, R.; Wohl, E. Two-dimensional modelling of habitat suitability as a function of discharge on two Colorado rivers. *River Res. Appl.* **2005**, *21*, 1061–1074. [CrossRef]

7. Ouédraogo, W.; Raude, J.; Gathenya, J. Continuous modeling of the Mkurumudzi River catchment in Kenya using the HEC-HMS conceptual model: Calibration, validation, model performance evaluation and sensitivity analysis. *Hydrology* **2018**, *5*, 44. [CrossRef]
8. Refsgaard, J.C.; Henriksen, H.J. Modelling guidelines—Terminology and guiding principles. *Adv. Water Resour.* **2004**, *27*, 71–82. [CrossRef]
9. Kannan, N.; Santhi, C.; White, M.J.; Mehan, S.; Arnold, J.G.; Gassman, P.W. Some Challenges in hydrologic model calibration for large-scale studies: A case study of SWAT model application to Mississippi–Atchafalaya River basin. *Hydrology* **2019**, *6*, 17. [CrossRef]
10. Arsenault, R.; Brissette, F.; Martel, J.L. The hazards of split-sample validation in hydrological model calibration. *J. Hydrol.* **2018**, *566*, 346–362. [CrossRef]
11. Hernandez-Suarez, J.S.; Nejadhashemi, A.P.; Kropp, I.M.; Abouali, M.; Zhang, Z.; Deb, K. Evaluation of the impacts of hydrologic model calibration methods on predictability of ecologically-relevant hydrologic indices. *J. Hydrol.* **2018**, *564*, 758–772. [CrossRef]
12. Kavetski, D.; Kuczera, G.; Franks, S.W. Calibration of conceptual hydrological models revisited: 1. Overcoming numerical artefacts. *J. Hydrol.* **2006**, *320*, 173–186. [CrossRef]
13. Guerrero, M.; Di Federico, V.; Lamberti, A. Calibration of a 2-D morphodynamic model using water–sediment flux maps derived from an ADCP recording. *J. Hydroinforma.* **2013**, *15*, 813–828. [CrossRef]
14. Troy, T.J.; Wood, E.F.; Sheffield, J. An efficient calibration method for continental-scale land surface modeling. *Water Resour. Res.* **2008**, *44*, 1–13. [CrossRef]
15. Getirana, A.C.V. Integrating spatial altimetry data into the automatic calibration of hydrological models. *J. Hydrol.* **2010**, *387*, 244–255. [CrossRef]
16. Francés, F.; Vélez, J.I.; Vélez, J.J. Split-parameter structure for the automatic calibration of distributed hydrological models. *J. Hydrol.* **2007**, *332*, 226–240. [CrossRef]
17. Singh, S.; Bárdossy, A. Hydrological model calibration by sequential replacement of weak parameter sets using depth function. *Hydrology* **2015**, *2*, 69–92. [CrossRef]
18. Beven, K. Prophecy, reality and uncertainty in distributed hydrological modelling. *Adv. Water Resour.* **1993**, *16*, 41–51. [CrossRef]
19. Wright, K.A.; Goodman, D.H.; Som, N.A.; Alvarez, J.; Martin, A.; Hardy, T.B. Improving hydrodynamic modelling: An analytical framework for assessment of two-dimensional hydrodynamic models. *River Res. Appl.* **2017**, *33*, 170–181. [CrossRef]
20. Paarlberg, A.J.; Guerrero, M.; Huthoff, F.; Re, M. Optimizing dredge-and-dump activities for river navigability using a hydro-morphodynamic model. *Water* **2015**, *7*, 3943–3962. [CrossRef]
21. Wu, K.; Yeh, K.C.; Lai, Y.G. A combined field and numerical modeling study to assess the longitudinal channel slope evolution in a mixed alluvial and soft bedrock stream. *Water* **2019**, *11*, 735. [CrossRef]
22. Guan, M.; Liang, Q. A two-dimensional hydro-morphological model for river hydraulics and morphology with vegetation. *Environ. Model. Softw.* **2017**, *88*, 10–21. [CrossRef]
23. Kang, T.; Kimura, I.; Shimizu, Y. Responses of bed morphology to vegetation growth and flood discharge at a sharp river bend. *Water* **2018**, *10*, 223. [CrossRef]
24. Castro-Bolinaga, C.F.; Fox, G.A. Streambank erosion: Advances in monitoring, modeling and management. *Water* **2018**, *10*, 1346. [CrossRef]
25. Bosa, S.; Petti, M.; Pascolo, S. Numerical modelling of cohesive bank migration. *Water* **2018**, *10*, 961. [CrossRef]
26. Klein, A. Verification of Morphodynamic Models on Channels, Trenches, and Pits. Master's Thesis, TU Delft, Delft, The Netherlands, March 2004.
27. Van Waveren, R.H.; Groot, S.; Scholten, H.; van Geer, F.; Wösten, H.; Koeze, R.; Noort, J. *Good Modelling Practice Handbook*; RWS-RIZA: Lelystad, The Netherlands, 1999.
28. DHI. MIKE 21C Curvilinear model for river morphology—Scientific Documentation. Available online: http://manuals.mikepoweredbydhi.help/2017/Water_Resources/MIKE21C_Scientific_documentation.pdf (accessed on 11 November 2018).
29. Papanicolaou, A.N.T.; Krallis, G.; Edinger, J. Sediment transport modeling review—Current and future developments. *J. Hydraul. Eng.* **2008**, *134*, 1–14. [CrossRef]
30. Mueller, E.R.; Pitlick, J. Sediment supply and channel morphology in mountain river systems: 1. Relative importance of lithology, topography, and climate. *J. Geophys. Res. Earth Surf.* **2013**, *118*, 2325–2342. [CrossRef]

31. Sear, D.A.; Newson, M.D.; Thorne, C.R. *Guidebook of Applied Fluvial Geomorphology*; Thomas Telford Ltd: London, UK, 2010.
32. Matte, P.; Secretan, Y.; Morin, J. Hydrodynamic modeling of the St. Lawrence fluvial estuary. I: Model setup, calibration, and validation. *J. Waterw. Port Coast. Ocean Eng.* **2017**, *143*, 04017010. [CrossRef]
33. Chaves, H.M.L.; Alipaz, S. An integrated indicator based on basin hydrology, environment, life, and policy: The watershed sustainability index. *Water Resour. Manag.* **2007**, *21*, 883–895. [CrossRef]
34. DHI Water & Environment. *MIKE 21 Flow Model FM—User Guide: Sand Transport Module, incl. Shoreline Morphology*; DHI Water & Environment: Hørsholm, Denmark, 2017.
35. De Villiers, J. 2D Modelling of Turbulent Transport of Cohesive Sediments in Shallow Reservoirs. Master's Thesis, University of Stellenbosch, Stellenbosch, South Africa, 2006.
36. Beck, J.S.; Basson, G.R. Klein River estuary (South Africa): 2D numerical modelling of estuary breaching. *Water SA* **2008**, *34*, 33–38.
37. Jain, R. *The Art of Computer Systems Performance Analysis: Techniques for Experimental Design, Measurement, Simulation, and Modeling*, 1st ed.; John Wiley & Sons: Hoboken, NJ, USA, 1991.
38. Montgomery, D.C. *Design and Analysis of Experiments*, 9th ed.; John Wiley & Sons: Hoboken, NJ, USA, 2017.
39. Kleijnen, J.P.C. Experimental design for sensitivity analysis, optimization, and validation of simulation models. In *Handbook of Simulation*; Banks, J., Ed.; John Wiley & Sons: Hoboken, NJ, USA, 1998; pp. 173–223.
40. Lee, R. Statistical design of experiments for screening and optimization. *Chemie-Ingenieur-Technik* **2019**, *91*, 191–200. [CrossRef]
41. Dorfmann, C.; Knoblauch, H. ADCP measurements in a reservoir of a run-of-river Hydro Power Plant. In Proceedings of the 6th International Symposium on Ultrasonic Doppler Methods for Fluid Mechanics and Fluid Engineering, Prague, Czech Republic, 9–11 September 2008; pp. 45–48.
42. Universidad del Norte. *Actualización de los Estudios y Diseños para la Navegabilidad del río Meta entre Cabuyaro (K804) y Puerto Carreño (K0)*; Universidad del Norte: Barranquilla, Colombia, 2013.
43. *Plan Maestro Fluvial de Colombia 2015*; Ministerio de Transporte: Bogotá, Colombia, 2015.
44. *Caracterización del Transporte en Colombia—Diagnóstico y Proyectos de Transporte e Infraestructura*; Ministerio de Transporte: Bogotá, Colombia, 2005.
45. Pasternack, G.B.; Gilbert, A.T.; Wheaton, J.M.; Buckland, E.M. Error propagation for velocity and shear stress prediction using 2D models for environmental management. *J. Hydrol.* **2006**, *328*, 227–241. [CrossRef]
46. Talmon, A.M. Bed Topography of River Bends with Suspended Sediment Transport. Ph.D. Thesis, Delft University of Technology, Delft, The Netherlands, 1992.
47. DHI Water & Environment. *MIKE 21 Flow Model—User Guide: Hydrodynamic Module*; DHI Water & Environment: Hørsholm, Denmark, 2017.
48. García, M.H. Sediment transport and morphodynamics. In *Sedimentation Engineering*; American Society of Civil Engineers: Reston, VA, USA, 2008; pp. 21–163.
49. Engelund, F.; Hansen, E. *A monograph on sediment transport in alluvial streams*; Technical University of Denmark: Copenhagen, Denmark, 1967.
50. Yang, C.T. Incipient motion and sediment transport. *J. Hydraul. Div.* **1973**, *99*, 1679–1704.
51. Van Rijn, L.C. Sediment transport, part II: Suspended load transport. *J. Hydraul. Eng.* **1984**, *110*, 1613–1641. [CrossRef]
52. Hidroconsultas LTDA. *Estudios básicos en el río Meta para la línea base de ingeniería tendiente a definir el sistema más adecuado para el mantenimiento de un canal navegable, obras de encauzamiento y demás obras fluviales entre la desembocadura del río Casanare y Puerto Texas*; Hidroconsultas LTDA: Bogota, Colombia, 2003.

Article

Hydraulic Parameters for Sediment Transport and Prediction of Suspended Sediment for Kali Gandaki River Basin, Himalaya, Nepal

Mahendra B. Baniya [1,2,*], Takashi Asaeda [3,4], Shivaram K.C. [5] and Senavirathna M.D.H. Jayashanka [1]

1 Graduate School of Science and Engineering, Saitama University, 255 Shimo-okubo, Sakura-ku, Saitama 338-8570, Japan; jayasanka@mail.saitama-u.ac.jp

2 Provincial Government, Ministry of Physical Infrastructure Development, Gandaki Province, Pokhara 33700, Nepal

3 Institute for studies of the Global Environment, 7-1 Sophia University, Kioicho, Chiyoda, Tokyo 102-0094, Japan; asaeda@mail.saitama-u.ac.jp

4 Hydro Technology Institute, 4-3-1 Shiroyama Trust Tower, Tranomon, Minato, Tokyo 105-0001, Japan

5 Nepal Electricity Authority, Central Office, Durbarmarga, Kathmandu 44600, Nepal; kcramshiva@gmail.com

* Correspondence: baniyam57@gmail.com; Tel.: +81-048-858-9186

Received: 22 May 2019; Accepted: 10 June 2019; Published: 12 June 2019

Abstract: Sediment yield is a complex phenomenon of weathering, land sliding, and glacial and fluvial erosion. It is highly dependent on the catchment area, topography, slope of the catchment terrain, rainfall, temperature, and soil characteristics. This study was designed to evaluate the key hydraulic parameters of sediment transport for Kali Gandaki River at Setibeni, Syangja, located about 5 km upstream from a hydropower dam. Key parameters, including the bed shear stress (τ_b), specific stream power (ω), and flow velocity (v) associated with the maximum boulder size transport, were determined throughout the years, 2003 to 2011, by using a derived lower boundary equation. Clockwise hysteresis loops of the average hysteresis index of +1.59 were developed and an average of 40.904 ± 12.453 Megatons (Mt) suspended sediment have been transported annually from the higher Himalayas to the hydropower reservoir. Artificial neural networks (ANNs) were used to predict the daily suspended sediment rate and annual sediment load as 35.190 ± 7.018 Mt, which was satisfactory compared to the multiple linear regression, nonlinear multiple regression, general power model, and log transform models, including the sediment rating curve. Performance indicators were used to compare these models and satisfactory fittings were observed in ANNs. The root mean square error (RMSE) of 1982 kg s^{-1}, percent bias (PBIAS) of +14.26, RMSE-observations standard deviation ratio (RSR) of 0.55, coefficient of determination (R^2) of 0.71, and Nash–Sutcliffe efficiency (NSE) of +0.70 revealed that the ANNs' model performed satisfactorily among all the proposed models.

Keywords: sediment yield; bed shear stress; specific stream power; flow velocity; hysteresis index; suspended sediments

1. Introduction

It is important to understand the sediment transport and river hydraulics in river systems for a variety of disciplines, such as hydrology, geomorphology, and risk management, including reservoir management. The sediment yield from a catchment is dependent on several parameters, including the topography, terrain slope, rainfall, temperature, and soil type of the catchment area [1]. On the other hand, the yield of sediment fluxes is a combination effect of weathering, land sliding, glacial, and fluvial erosions [2]. Sediment yield from these effects is quite complex [3] and the sediment transport in rivers varies seasonally. The hydrology of Nepal is primarily dominated by the monsoons, characterized by higher precipitation during the summer monsoon from June to September, contributing about 80% of the total annual precipitation [4]. Dahal and Hasegawa [5] reported that about 10% of the total precipitation occurs in a single day and 50% of the total annual precipitation occurs within 10 days of the monsoon period, responsible for triggering landslides and debris flows. The main natural agents for triggering landslides in the Himalayas are the monsoon climate, extremities in precipitation, seismic activities, excess developed internal stress, and undercutting of slopes by streams [6]. The sediments are transported by mountain streams in the form of a suspended load, as well as a bedload [7], depending on the intensity of the rainfall and number of landslide events that occurred within the catchment area [8]. Dams constructed to regulate flood magnitudes limits the downstream transportation of all suspended sediments [9]. However, the annual deposition of sediment in reservoirs decreases the capacity of reservoirs, which compromises the operability and sustainability of dams [10]. Basin morphology and lithological formation governs the amount of sediment crossing a stream station at a certain timepoint, which is generally acted upon by both active and passive forces [11].

Outbursts of glaciers and the failure of moraine dams trigger flash floods [6,12–14], which is one of the main causes of large boulder transportation in high gradient rivers in mountain regions. Different hydraulic parameters, such as shear stress, specific stream power, and flow velocity, can be combined in different ways to form sediment transport predictors [15,16]. Shear stress is a well-known hydraulic parameter that can easily determine the ability of rivers to transport coarse bedload material [17,18]. Similarly, flow competence assessments of floods related to the largest particle size transported are described by the mean flow stress, specific stream power, and mean velocity [19,20]. A number of studies have demonstrated the relationships of shear stress [20–24], specific stream power [20,23,24], and flow velocity [20,21,23–26] of rivers with the size of the boulder movement in the river. It is important to perform this study in Kali Gandaki River as this river originates from the Himalayas and there is limited research on sediment transport by this river, which is crucial in Nepal due to differences in the terrain within a short distance.

In this study, relationships between the fluvial discharge and hydraulic parameters, such as the shear stress, specific stream power, and flow velocity, were generated to derive a lowest boundary equation for the maximum size of particles transported by fluvial discharge in the Kali Gandaki River at a point 5 km upstream of the hydropower dam. The equation was used to calculate the maximum size of particles transported by fluvial discharge during 2006 to 2011. Additionally, it explored the nature of hysteresis loops, developed a hysteresis index, quantified the annual suspended sediment load (ASSL) transport, developed different suspended sediment transport models for Kali Gandaki River, and applied them to predict the suspended sediment rate as well as the average ASSL transport.

2. Materials and Methods

2.1. Study Site Description

The Kali Gandaki River is a glacier-fed river originating from the Himalaya region, Nepal [27]. The basin has a complex geomorphology and watershed topography with rapid changes in elevation, ranging from about 529 m MSL to 8143 m MSL. It flows from north to south in the higher Himalayan region before flowing eastward through the lower Himalayan region, entering the Terai plains of Nepal and connecting with Narayani River, which ultimately merges with the Ganges River in India. The snowfall area is separated, with elevation ranges less than 2000 m MSL having no snow cover, 2000 to 4700 m MSL having seasonal snow, 4700 to 5200 m MSL having complete snow cover except for 1 or 2 months, and elevations greater than 5200 m MSL having permanent snow [4]. The Kali Gandaki catchment basin covers a 7060 km^2 area, comprised of elevations of 529~2000 m MSL covering 1317 km^2 (19% coverage); 2000~4700 m MSL covering 3388 km^2 (48% coverage); 4700~5200 m MSL covering 731 km^2 (10% coverage); and elevations greater than 5200 m MSL covering 1624 km^2 (23% coverage). Figure 1a shows the different altitude areas' coverage map showing river networks, with the locations of meteorological stations, created in ArcGIS 10.3.1 (ERSI Inc., Berkeley, CA, USA) software. The elevations of Kali Gandaki River decrease from 5039 m MSL in the higher Himalayas to 529 m MSL at Setibeni, 5 km upstream of the hydropower dam (Figure 1b). This encompasses a wide variation in mean rainfall, ranging from less than 500 mm year^{-1} in the Tibetan Plateau to about 2000 mm year^{-1} in the monsoon-dominated Himalayas [8].

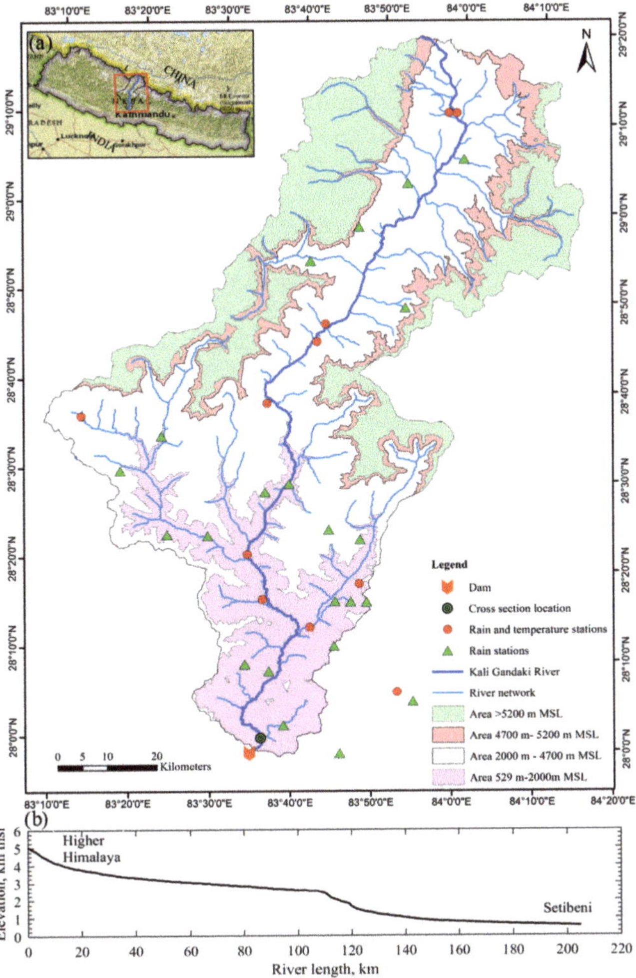

Figure 1. (**a**) Map of Kali Gandaki River catchment area; (**b**) longitudinal profile of Kali Gandaki River.

The main physiographic characteristics of the Kali Gandaki River basin at the hydropower station are shown in Table 1.

The discharge of this river varies seasonally and is dependent on the rainfall received by its tributaries' catchments in addition to the amount of snow melting from the Himalayas. A dam (27°58′44.88″ N, 83°34′49.68″ E) was constructed in 2002 for a hydropower project with a 144 MW power generation capacity, at Mirmi, Syangja.

Table 1. Main characteristics of the river basin.

Parameters	
Catchment area	7060 km^2
Length of river up to dam	210 km
Mean gradient of river	2.20%
Extreme discharge	3280 m^3 s^{-1} in 1975, 2824.5 m^3 s^{-1} in 2009
Elevation ranges	529 m MSL–8143 m MSL
Precipitation	Tibetan plateau <500 mm year^{-1}, monsoon dominated Himalayas~2000 mm year^{-1}

2.2. Data Collection and Acquisition

The department of Hydrology and Meteorology (DHM), Nepal established a gauge station (28°00′30″ N, 83°36′10″ E) in 1964 (www.dhm.gov.np) and it operated until 1995. The gauge station was not operated during the hydropower dam construction period (1997–2002). The bed level of the dam increased yearly due to the trapping of bedload as well as a suspended sediment load by the dam, which reduced the sediment load downstream. The cross-sectional areas of different years were calculated from area–discharge regression equations obtained from historical discharge rating DHM data (1964–1995). Sedimentation lowers the reservoir capacity of the dam annually.

2.3. Analysis of Shear Stress, Specific Power, and Flow Velocity

Historical discharge and cross profile elevations data sourced from Nepal Electricity Authority (NEA), Nepal were used to calculate the bed shear stress, specific power developed, and flow velocity by using the following common equations [28–30]:

$$\text{Bed shear stress, } \tau_b = \rho \times g \times R \times i. \quad (1)$$

The mean available power supply over a unit of bed area is calculated by:

$$\omega = \frac{\Omega}{w_t} = \frac{\rho \times g \times Q \times i}{w_t}, \quad (2)$$

where w_t represents the width of the flow, and Ω is the available stream power supply or the time rate of the energy supply to the unit length of the stream in w.m^{-1} and is given by:

$$\Omega = \rho \times g \times Q \times i. \quad (3)$$

The flow velocity is calculated by Manning's formula:

$$v = \frac{1}{n} R^{2/3} i^{1/2}, \quad (4)$$

where τ_b is the bed shear stress (N·m^{-2}), ρ is the density of water (1000 kg·m^{-3}), g is the acceleration due to gravity (9.81 m·s^{-2}), R is the hydraulic radius (m), i is the slope of the river bed (m·m^{-1}), ω is the mean available specific stream power per unit area (w·m^{-2}), Q is the observed discharge (m^3·s^{-1}), v is the flow velocity (m·s^{-1}), and n is Manning's constant. Manning's constant, n, in a steep natural channel is calculated by the equation proposed by Jarrett [31]:

$$n = 0.39 S^{0.38} (3.28R)^{-0.16}. \quad (5)$$

2.4. Development of Different Models for Suspended Sediment Predictions

The daily suspended sediment load transported by the river in the catchment area is a key indicator to visualize the sediment losses from the higher Himalayas and to assess the reservoir management in hydropower projects. Different researchers have developed multiple linear regression (MLR) and nonlinear multiple regression (NMLR), sediment rating curve (SRC), and artificial neural networks (ANNs) models for the prediction of the daily suspended sediment load [32,33].

2.4.1. Multiple Linear Regression

Multiple linear regression assumes that the sediment load transported by a river is in the linear form. A dependent variable, the suspended sediment load, Q_{s_t}, depends on two independent variables, the daily average discharge of a river (Q_{w_t}) and the average rainfall (R_t) of a catchment area, and the model is expressed in the form of a regression equation [32,33]:

$$Q_{s_t} = \beta_0 + \beta_1 Q_{w_t} + \beta_2 R_t. \tag{6}$$

The different linear models were created by considering Q_{w_t}, a day lag discharge; $Q_{w_{t-1}}$ and R_t, and a day lag rainfall, R_{t-1}, were the input variables and the performance of different models was also evaluated.

2.4.2. Nonlinear Multiple Regression

The suspended sediment transported by the river shows a dynamic state in a nonlinear form so that it is expressed in the form of a polynomial equation [32,33]:

$$Q_{s_t} = \beta_0 + \beta_1 Q_{w_t} + \beta_2 R_t + \beta_{11} {Q_{w_t}}^2 + \beta_{22} {R_t}^2 \tag{7}$$

Different nonlinear models were also created and their performance was evaluated separately.

2.4.3. Sediment Rating Curve

SRC is expressed [34] in the form of:

$$Q_{s_t} = a {Q_{w_t}}^b \tag{8}$$

where Q_{s_t} is the suspended sediment load (kg·s^{-1}), Q_{w_t} is the daily average discharge of river, and a and b are coefficients that depend on the characteristics of a river.

2.4.4. Artificial Neural Networks

An artificial neural network is capable of solving complex nonlinear relationships between input and output parameters, which consists of three different layers known as an input, hidden, and output layer, respectively [33]. MATLAB (R2016a) software was used to develop different artificial neural networks, where the input consisted of the average daily river discharge (Q_{w_t}), a day lag discharge ($Q_{w_{t-1}}$), and average daily rainfall (R_t), a day lag rainfall (R_{t-1}), where the output consisted of the average daily suspended sediment load (Q_{s_t}). Out of 2191 data sets, 70% of the data was used for training, 15% for validation, and 15% for testing in the ANNs.

2.5. Model Performance

The performance of different models was assessed in terms of the root mean square (RMSE), percent BIAS (PBIAS), RMSE-observations standard deviation ratio (RSR), coefficient of determination (R^2), and Nash–Sutcliffe efficiency (NSE) [32,35,36]:

$$RMSE = \sqrt{\frac{\sum_{i=1}^{n} \left(Q_{s_{o,i}} - Q_{s_{p,i}}\right)^2}{N}}. \tag{9}$$

The lower the *RMSE* value, the better the model's performance:

$$PBIAS = \left\{ \frac{\sum_{i=1}^{n} \{ Q_{s_{o,i}} - Q_{s_{p,i}} \}}{\sum_{i=1}^{n} Q_{s_{o,i}}} \right\} \times 100, \quad (10)$$

where the optimal PBIAS value is 0.0, and positive values indicate a model underestimation bias and negative values indicate a model overestimation bias:

$$RSR = \frac{RMSE}{STDEV_o} = \frac{\left\{ \sqrt{\sum_{i=1}^{n} \{ Q_{s_{o,i}} - Q_{s_{p,i}} \}^2} \right\}}{\left\{ \sqrt{\sum_{i=1}^{n} \{ Q_{s_{o,i}} - \overline{Q}_{s_{p,i}} \}^2} \right\}}. \quad (11)$$

The optimal value for RSR is 0.0; the lower the RSR, the lower the RMSE and the better the model's performance.

Coefficient of determination:

$$R^2 = \left\{ \frac{\sum_{i=1}^{n} \{ Q_{s_{o,i}} - \overline{Q}_{s_{o,i}} \} \{ Q_{s_{p,i}} - \overline{Q}_{s_{p,i}} \}}{\sqrt{\sum_{i=1}^{n} \{ Q_{s_{o,i}} - \overline{Q}_{s_{o,i}} \}^2 \sum_{i=1}^{n} \{ Q_{s_{p,i}} - \overline{Q}_{s_{p,i}} \}^2}} \right\}^2. \quad (12)$$

The optimal value for R^2 is 1.0; the higher the value of R^2, the better the model's performance:

$$NSE = \left\{ 1 - \frac{\sum_{i=1}^{n} \{ Q_{s_{o,i}} - Q_{s_{p,i}} \}^2}{\sum_{i=1}^{n} \{ Q_{s_{o,i}} - \overline{Q}_{s_{o,i}} \}^2} \right\}. \quad (13)$$

The optimal value for NSE is 1.0 and values range from $-\infty$ to 1. Values between 0.0 and 1.0 are taken as acceptable levels of performance whereas negative values indicate that the mean observed value is a better predictor than the predicted value, which indicates an unacceptable performance. Here, $Q_{s_{o,i}}$ and $Q_{s_{p,i}}$ are the observed and predicted suspended sediment and $\overline{Q}_{s_{o,i}}$ and $\overline{Q}_{s_{p,i}}$ are the average observed and average predicted suspended sediment, respectively.

3. Results and Discussion

The average discharge of Kali Gandaki River from 2003 to 2012 was 306 $m^3 \cdot s^{-1}$ with a minimum discharge of 40.73 $m^3 \cdot s^{-1}$ during winter in 2009 and a maximum discharge 2824.50 of $m^3 \cdot s^{-1}$ during summer in 2009. The maximum discharge showed a decreasing trend from 2003 to 2006 whereas an increasing trend from 2007 onwards was observed, as shown in Figure 2a. The yearly transported sediment load increased nearby upstream river bed level elevations of the reservoir (Figure 2b) and sediment deposited into the reservoir decreased the reservoir's capacity. The effects of climate change in the higher Himalayas appeared in the form of uneven patterns of increasing rainfall, glacial rate erosion, and permafrost degradation, resulting in an increase in landslides and debris flows [2], which also reflects the temporal and spatial variation of the water balance components in the Kali Gandaki basin [37]. The amount and intensity of rainfall around its catchment affected the discharge rating curve [27].

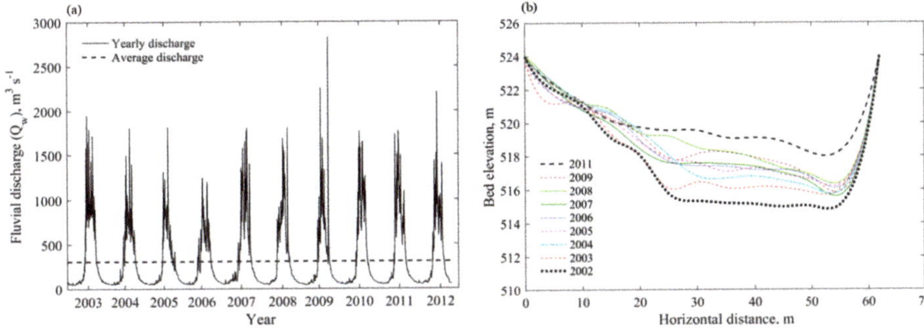

Figure 2. Yearly (**a**) discharge (NEA 200–2012) (**b**) cross profiles (2002–2011) of Kali Gandaki River.

3.1. Relationship of Shear Stress, Specific Stream Power, and Flow Velocity with Discharge

The calculated shear stress, specific stream power, and flow velocity of Kali Gandaki River at the discharge gauge station, which was about 5 km upstream from the dam within limited data from 2003 to 2011, was related as:

$$\tau_b = 3.143 \times Q^{0.621} \left(R^2 = 0.72\right), \tag{14}$$

$$\omega = 27.40 \times Q^{0.612} \left(R^2 = 0.65\right), \tag{15}$$

$$v = 0.108 \times Q^{0.519} \left(R^2 = 0.95\right). \tag{16}$$

The highest shear stress, specific stream power, and flow velocity were observed during 2008 whereas the lowest were observed during 2007. These parameters were directly related with the hydraulic radius in the case of shear stress and flow velocity, whereas the fluvial discharge in the case of specific power (Equations (1), (2), and (4)). The sedimentation process increased the bed level elevation, changing the cross geomorphology of the bed (Figure 2b). These parameters followed nearly the same trends during the remaining years. The shear stress, specific power, and flow velocity of the river increased the function of the fluvial discharge, as shown in Figure 3a,b and Figure 4a, respectively.

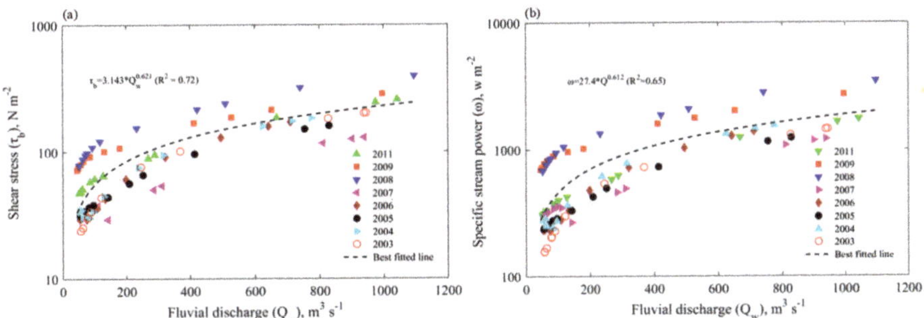

Figure 3. Relationship of fluvial discharge and (**a**) shear stress and (**b**) specific power.

3.2. Relationship of Particle Sizes and Fluvial Discharge

The hydraulic parameters were the shear stress, specific stream power, and flow velocity depict transportation of different particle sizes. When subjected to the same fluvial discharge, the specific power showed an increase of 327 mm to 2062 mm particle size whereas the flow velocity depicted an increase of 37 mm to 1794 mm. The shear stress exhibited an increase of 147 mm to 1492 mm particles, which covered the lowest maximum particle sizes compared to the specific power and flow velocity

(Figure 4b). These three parameters were derived from the fluvial discharge, as summarized in the lowest boundary equation form of the fluvial discharge as shown in Figure 4b:

$$d_{mm} = 0.4 \times Q^{1.093} \ (25 \text{ mm} \leq d \leq 840 \text{ mm}). \tag{17}$$

Equation (17) predicted that from the 2003 to 2011, the discharge during monsoons was capable of transporting an 840 mm particle size. Hydraulic parameters, such as the bed shear stress, specific stream power, and flow velocity, have gained wider acceptability among different researchers [20–26] regarding their useful contribution to the derivation of the relationship between particle sizes and hydraulic parameters. The shear stress and particle size relationship of this study was compared with Costa's [20] average of $\tau_b = 0.163d^{1.213}$, lower boundary of $\tau_b = 0.056d^{1.213}$ for 50 mm ≤ d ≤ 3290 mm; Komar's [21] $\tau_b = 0.164d^{1.21}$ for 50 mm ≤ d ≤ 5000 mm; Lenzi's [22] $\tau_b = 86.629d^{0.25}$ for 20 mm ≤ d ≤ 1000 mm; O'Connor's [23] average of $\tau_b = 0.0249\ d^{1.12}$ for 270 mm ≤ d ≤ 6240 mm; and Williams [24] lower boundary of $\tau_b = 0.17d^{1.0}$ for 10 mm ≤ d ≤ 3300 mm (Figure 5).

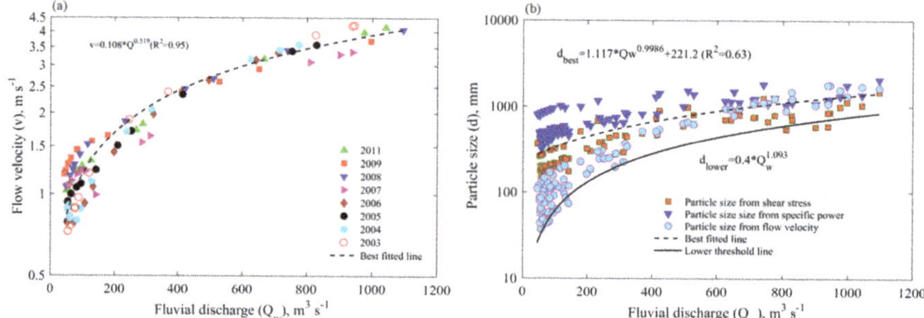

Figure 4. Relationship of the fluvial discharge, (**a**) flow velocity, and (**b**) particle size (boulder).

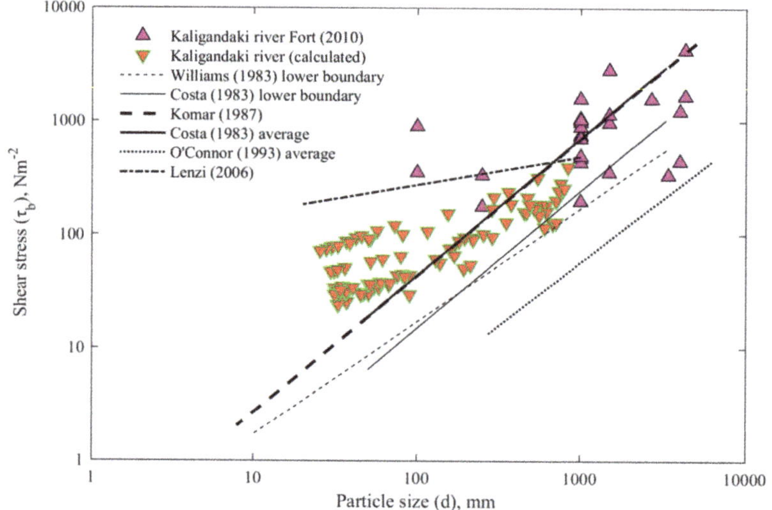

Figure 5. Relationship of shear stress and particle size (boulder) and a comparison with different researchers.

For a comparative study of the specific stream power, the particle size relationship of this river was compared with Costa's [20] average of $\omega = 0.030d^{1.686}$ and lower boundary of $\omega = 0.009d^{1.686}$ for 50 mm $\leq d \leq$ 3290 mm; O'Connor's [23] average of $\omega = 0.002d^{1.71}$ and lower boundary $\omega = 30 \times 1.00865d^{0.1d}$ for a particle size of 270 mm $\leq d \leq$ 6240 mm; and Williams' [24] lower boundary of $\omega = 0.079d^{1.3}$ for a particle size of 10 mm $\leq d \leq$ 1500 mm (Figure 6).

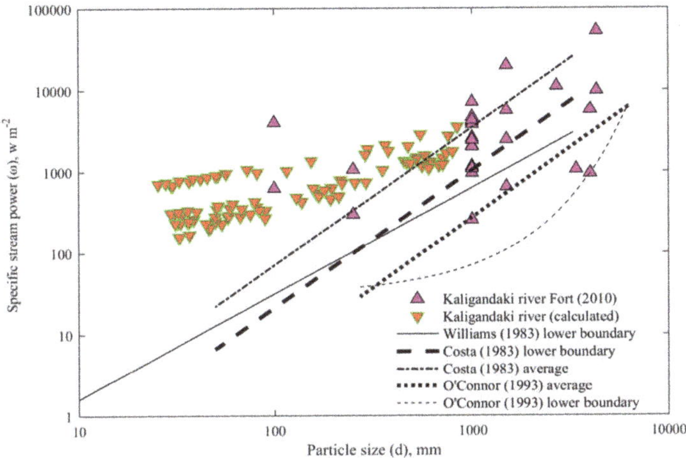

Figure 6. Relationship of the specific power and particle size (boulder) and a comparison with different researchers.

The flow velocity and particle size relationship of this study was compared with Costa's [20] average of $v = 0.20d^{0.455}$ and lower boundary of $v = 0.14d^{0.455}$ for 50 mm $\leq d \leq$ 3290; U.S.B.R.'s [20] $v = 0.187d^{0.50}$ for 1 mm $\leq$ d $\leq$ 600 mm; Komar's [21] $v = 0.197d^{0.46}$ for 8 mm $\leq$ d $\leq$ 5000 mm; O'Connor's [23] average of $v = 0.074d^{0.60}$ for 270 mm $\leq$ d $\leq$ 6240 mm; Williams' [24] lower boundary of $v = 0.065d^{0.50}$ for 10 mm $\leq$ d $\leq$ 1500 mm; Bradely and Mears' [25] $v = 0.163d^{0.5}$ for 50 mm $\leq d \leq$ 3290 mm; and Helley's [26] $v = 0.1545d^{0.499}$ for 1 mm $\leq d \leq$ 600 mm (Figure 7).

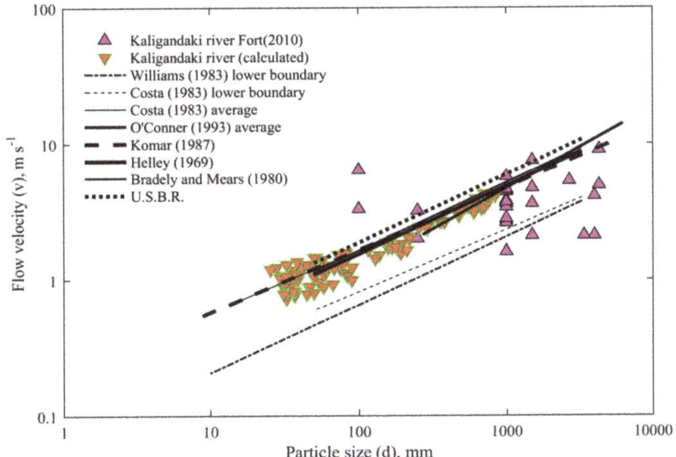

Figure 7. Relationship of the flow velocity and particle size (boulder) and comparison with different researchers.

The calculated values of the shear stress, specific stream power, and flow velocity were less than the observed values by Fort [6], who reconstructed the 1998 landslide dam located about 76 km upstream of the existing hydropower dam of Kali Gandaki River and estimated the hydraulic parameters with an exceptional dam breach discharge of 10,035 $m^3 s^{-1}$. This high discharge was responsible for the movement of a maximum boulder size of 4300 mm [6]. The higher shear stress, specific stream power, and flow velocity observed due to a higher fluvial discharge after the breaching of landslide dam were responsible for the transportation of larger sized boulders (Figures 5–7).

3.3. Estimation of the Return Period by Gumbel's Distribution

The flood return period from the historical data of DHM, Nepal can be forecasted by the Gumbel method [38] as $Q_T = \overline{Q} + k\sigma_n$, where $\overline{Q}$ is the mean discharge, k is the frequency factor, and σ_n is the standard deviation of the maximum instantaneous flow, respectively. The frequency factor is given by $k = \left(\frac{y_t - \overline{y}_n}{s_n}\right)$, where $\overline{y}_n$ is the mean and s_n is the standard deviation of Gumbel's reduced variate; y_t is given by $y_t = -ln\left[ln\left[\frac{T}{T-1}\right]\right]$. The observed highest flood in 1975 was 3280 $m^3 \cdot s^{-1}$. According to the Gumbel frequency of flood distribution, the highest flood will occur after a 40 year return period, as shown in Figure 8a, and the observed extreme discharge, as shown in Figure 8b.

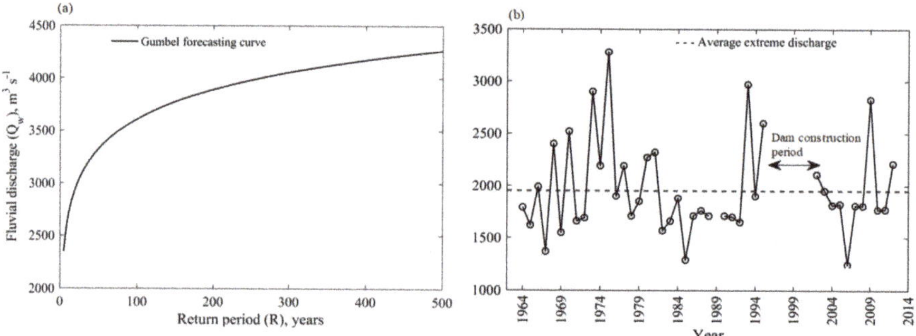

Figure 8. (**a**) Gumbel flood return period, (**b**) extreme fluvial discharge.

3.4. Boulder Movement Mechanisms in the Himalayas

High gradient river hydraulics are strongly influenced by large boulders, with the diameters on the same scale as the channel depth or even the width [39]. Williams [24] mentioned that five possible mechanisms of boulder transport by high gradient river are by ice, mudflow, water stepwise creep by periodic erosion, undermining of stream banks, and avalanches. The bed forming material remains immobile during typical flows, and larger bed forming particles in steep gradient channels typically become mobile only every 50 to 100 years during a hydrologic event [40]. After that, the gravel stocked in low energy sites during lower floods is mobilized and travels as the bedload [40].

The failure of the mountain slope of Kali Gandaki catchment in 1988, 1989, and 1998 was due to an evolved rock avalanche and caused the damming of the Kali Gandaki River [2]. The shockwaves after the massive 7.8 M_W Gorkha earthquake, Nepal on 25 April 2015 and its aftershocks on 23 May 2015 created cracks in the weathered rocks and weakened the mountain slopes of this catchment, which brought rocks, debris, and mud down into the river [41,42]. The river was blocked about 56 km upstream from the hydropower dam by a landslide on 24 May 2015 for 15 h [41] (Figure 9a,b). The downstream fluvial discharge after the blockage was almost zero and a flash flood occurred after an outburst of the natural landslide dam (Figure 9c,d). Extreme flooding during the monsoon period due to high rainfall and a flash flood (Figure 9b), generated by the overtopping of landslide dams [42], was responsible for the noticeable transport of large boulders in the river bed of Kali Gandaki River.

Figure 9. (**a**) Natural landslide dam formation on 24 May 2015 (~56 km upstream of dam), (**b**) lake formation after the blockage of the river, (**c**) downstream fluvial discharge after the blockage of the river, and (**d**) extreme fluvial discharge after breaching of the landslide dam on 25 May 2015 (Source: http://kathmandupost.ekantipur.com/news/2015-05-24/blocked-kali-gandaki-river-flows-again-with-photos.html).

The combination of fluid stress, localized scouring, and undermining of the stream banks may cause small near vertical displacements of large boulders [43]. Catastrophic events, such as natural dam breaks and debris flows, are responsible for larger translations of boulders in rivers [40,43].

3.5. Quantification and Prediction of the Suspended Sediment

3.5.1. Hysteresis Curve and Hysteresis Index (HI_{mid}) Analysis

The relationship between the suspended sediment concentration and fluvial discharge can be studied by the nonlinear relationship between them known as hysteresis [44]. Generally, a clockwise hysteresis loop is formed due to an increasing concentration of sediment that forms more rapidly during rising limb, which suggests a source of sediment close to the monitoring point and sediment depletion in the channel system. Conversely, an anticlockwise hysteresis loop shows a long gap between the discharge and concentration peak, which suggests that the source is located far from the monitoring point or bank collapse [45,46].

Clockwise hysteresis loops were developed, increasing the suspended sediment load on the rising limb of hysteresis from December to July, leading to a maximum value of the suspended sediment load of 10,691 $kg{\cdot}s^{-1}$ for a fluvial discharge of 1053 $m^3{\cdot}s^{-1}$ on August 2009. The suspended sediment load decreased on the falling limb of hysteresis from July/September to November. Overall, these six years were characterized by distinct clockwise hysteresis patterns (Figure 10a).

The HI_{mid} is a numerical indicator of hysteresis, which effectively shows the dynamic response of suspended sediment concentrations to flow changes during storm events [47].

The midpoint discharge was calculated by Lloyd [46] and Lawler [47]:

$$Q_{mid} = k(Q_{max} - Q_{min}) + Q_{min}, \tag{18}$$

where k is 0.5, Q_{max} is the peak discharge, and Q_{min} is the starting discharge of an event.

The hysteresis index value was calculated by Lloyd [46] and Lawler [47]:

$$HI_{mid} = \left(\frac{Q_{sRL}}{Q_{sFL}}\right) - 1 \text{ for a clockwise loop}, \tag{19}$$

$$HI_{mid} = \left(-1/\left(\frac{Q_{sRL}}{Q_{sFL}}\right)\right) + 1 \text{ for an anticlockwise loop}, \tag{20}$$

where Q_{sRL} and Q_{sFL} are the suspended sediment on the rising and falling limb, respectively.

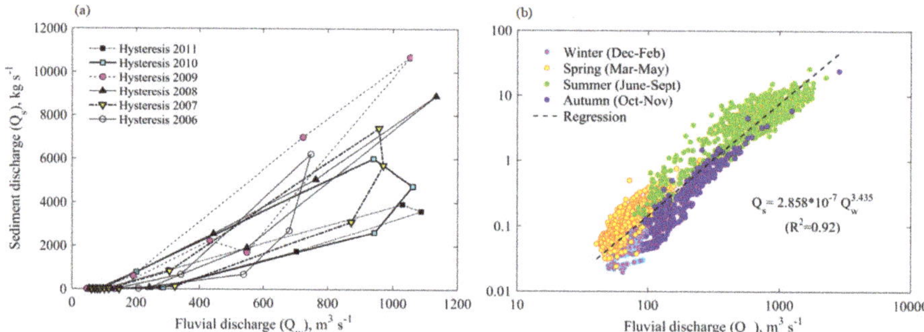

Figure 10. (**a**) Seasonal hysteresis loop of the sediment load. (**b**) Suspended sediment–discharge rating curve.

3.5.2. Yearly Suspended Sediment Yield and Prediction by Different Models

A regression equation derived from the observed data (2006 to 2011) of the suspended sediment versus the discharge of the river shown in Figure 10b is given by:

$$Q_s = 2.858 \times 10^{-7} \times Q_w^{3.435}\left(R^2 = 0.92\right). \quad (21)$$

The total suspended sediment yield from the catchment is given by:

$$Y_s = \int_{t=0}^{T} C_{wi} Q_{wi} dt = \sum_{i=1}^{365} C_{wi} Q_{wi} \times 10^{-3} \times (t_{i+1} - t_i), \quad (22)$$

where Y_s is the total annual sediment yield from the catchment, C_{wi} is the suspended sediment concentration in mg·L^{-1}, Q_{wi} is the fluvial discharge in m^3·s^{-1}, dt is the time interval, t_i and t_{i+1} are the preceding and succeeding time in seconds, respectively.

This study showed that the median ASSL transported by KaliGandaki River in the hydropower reservoir was 0.003 Mt during winter, increased to 0.026 Mt during spring, was 41.405 Mt during the summer season, and decreased 0.175 Mt during the autumn season (Figure 11a). Compared to the seasonal transport of suspended sediment, more than 96% of the suspended sediment was transported during the summer season. This depicts a wide seasonal variability of the suspended sediment caliber, which was nearly 14,000 times higher than the winter season (Figure 11a). The maximum observed ASSL transported by the river was 58.426 Mt in 2009, and after that it decreased (Figure 11b).

The $HI_{mid} \approx 0$ indicated a weak hysteresis loop whereas $HI_{mid} > 0$ indicated a clockwise hysteresis loop, and $HI_{mid} < 0$ an anticlockwise hysteresis loop. Moreover, the maximum HI_{mid} developed was +2.64 in 2006, depicting the higher sediment transport rate in the rising limb but lower sediment transport rate in the falling limb (Figure 10a). The minimum HI_{mid} developed was +0.53 in 2008, depicting the nearly same paths of the rising and falling limb and indicating a weak hysteresis loop (Figures 10a and 11b).

Different types of MLR, NLMR, general power, log transform linear, and ANNs models, including inputs of the fluvial discharge and average rainfall of the catchment, were developed to select the most suitable model and the results are shown in Tables 2–6, respectively. The performance parameters of MLR and NLMR were satisfactory but predicted negative sediment values for low fluvial discharges and low rainfall, thus these models are unacceptable.

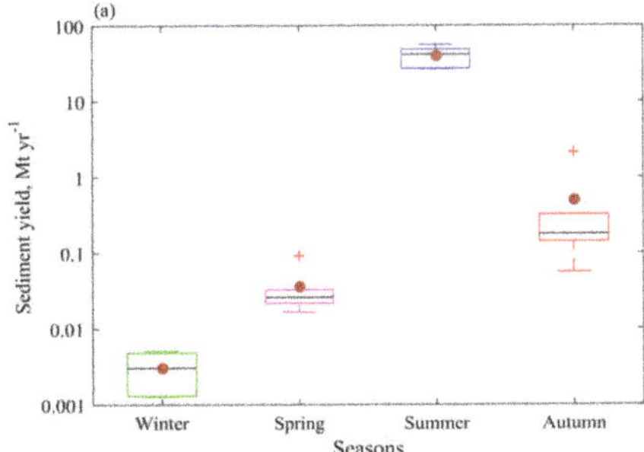

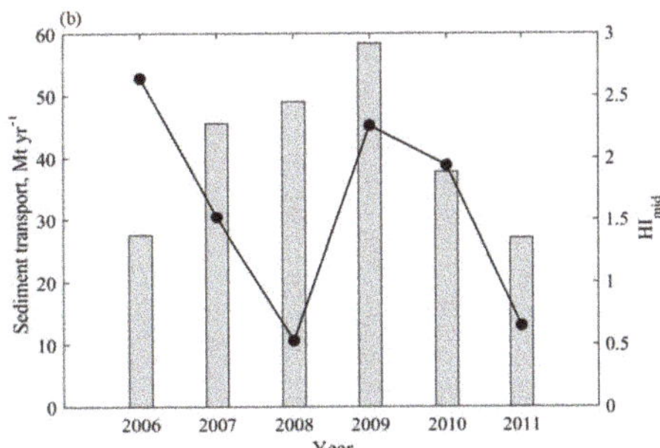

Figure 11. (**a**) Seasonal suspended yield from the catchment. Central lines indicate the median, bottom and top edges of the box indicate the 25th and 75th percentiles respectively. The whiskers extend to the most extreme data points not considered outliers, the '+' symbol represents outliers (1.5 fold interquartile range), the circle shows the mean value. (**b**) Yearly suspended sediment transport and hysteresis index (HI_{mid}).

Table 2. MLR models.

Model Scenario	RMSE ($kg{\cdot}s^{-1}$)	PBIAS	RSR	R^2	NSE	Model Equation
Q_t	2498	+0.47	0.66	0.53	+0.56	$Q_s = 7.12Q_t - 920.70$
R_t	2729	+0.34	0.73	0.44	+0.47	$Q_s = 199.54R_t - 229.07$
Q_tR_t	2442	+0.22	0.64	0.55	+0.59	$Q_s = 5.31Q_t + 71.0R_t - 897.03$
Q_tR_{t-1}	2494	+0.35	0.66	0.53	+0.56	$Q_s = 6.68Q_t + 18.12R_{t-1} - 920.66$
$Q_tQ_{t-1}R_tR_{t-1}$	2339	+0.29	0.59	0.59	+0.65	$Q_s = 13.47Q_t - 8.02Q_{t-1} - 14.02R_t + 64.44R_{t-1} - 784.15$

Table 3. NLMR models.

Model Scenario	RMSE ($kg{\cdot}s^{-1}$)	PBIAS	RSR	R^2	NSE	Model Equation
Q_t	2314	+0.33	0.57	0.59	+0.67	$Q_s = 5.02 \times 10^{-3}Q_t^2 + 0.71Q_t - 111.61$
R_t	2697	+0.66	0.71	0.46	+0.49	$Q_s = 1.30R_t^2 + 138.75R_t - 36.72$
Q_tR_t	2280	+0.15	0.56	0.61	+0.68	$Q_s = 4.04 \times 10^{-3}Q_t^2 + 0.74Q_t + 0.57R_t^2 + 24.10R_t - 188.70$
Q_tR_{t-1}	2303	+0.32	0.57	0.59	+0.67	$Q_s = 5.14 \times 10^{-3}Q_t^2 - 0.17Q_t - 0.024R_{t-1}^2 + 30.46R_{t-1} - 93.99$
$Q_tQ_{t-1}R_tR_{t-1}$	2250	+0.43	0.55	0.62	+0.69	$Q_s = 3.73 \times 10^{-3}Q_t^2 - 8.10 \times 10^{-4}Q_{t-1}^2 + 4.97Q_t - 3.02Q_{t-1} + 8.18 \times 10^{-2}R_t^2 + 0.91R_{t-1}^2 + 8.27R_t + 0.28R_{t-1} - 272.04$

Table 4. General power model.

Model Scenario	RMSE ($kg{\cdot}s^{-1}$)	PBIAS	RSR	R^2	NSE	Model Equation
General power model 1 Q_t	2039	+3.81	0.56	0.67	+0.68	$Q_s = 1.027 \times 10^{-3}Q_t^{2.238}$
General power model 2 Q_t	2039	+0.22	0.56	0.67	+0.68	$Q_s = 0.847 \times 10^{-3}Q_t^{2.263} + 71.08$

Table 5. Log transform models.

Model Scenario	RMSE ($kg{\cdot}s^{-1}$)	PBIAS	RSR	R^2	NSE	Model Equation
Linear model (SRC) $logQ_t$	4451	−21.65	1.23	0.59	−0.51	$logQ_s = 3.435 \log Q_t - 6.544$ $Q_s = 2.858 \times 10^{-7} Q_t^{3.435}$
General power model 2 $logQ_t$	4039	−17.50	1.12	0.59	−0.25	$logQ_s = 3.915 logQ_t^{0.931} - 7.131$
Linear model $logQ_t logR_t$	3715	−15.47	1.03	0.61	−0.05	$logQ_s = 3.112 \log Q_t + 0.10 logR_t - 5.714$

Table 6. ANN models.

Model Scenario	RMSE ($kg{\cdot}s^{-1}$)	PBIAS	RSR	R^2	NSE	Model Equation
$logR_t$ $1-10-1-1$	2768	+54.07	0.77	0.45	+0.41	Levenberg-Marguardt
$logQ_t$ $1-10-1-1$	2070	+14.91	0.57	0.67	+0.66	Levenberg-Marguardt
$logQ_t logR_t$ $2-10-1-1$	2052	+15.99	0.56	0.71	+0.68	Levenberg-Marguardt
$logQ_t logR_{t-1}$ $2-10-1-1$	2123	+22.95	0.59	0.69	+0.66	Levenberg-Marguardt
$logQ_t logQ_{t-1} logR_t logR_{t-1}$ $4-10-1-1$	1982	+14.26	0.55	0.71	+0.70	Levenberg-Marguardt

The RMSE, PBIAS, RSR, R^2, and NSE values of the general power model, log transform models, and ANNs are shown in Tables 4–6. In general, the model simulation can be judged as "satisfactory" if NSE > 0.50, and RSR ≤ 0.70, and if the PBIAS value is ±25% for the stream flow and the PBIAS value is ±55% for the sediment [35]. In this study, the predicted values from ANNs (4−10−1−1) showed an RMSE value of 1982 $kg{\cdot}s^{-1}$, PBIAS value of +14.26, RSR value of 0.55, R^2 value of 0.71, and an NSE value of +0.70, which indicates that the ANNs model's performance was satisfactory. Figure 12a–d show the comparison between the model's predicted transport rates of the suspended sediment discharge in $kg{\cdot}s^{-1}$ of the SRC, log transform power model, log transform linear models, and ANNs and the observed suspended sediment values respectively.

Among the SRC, power, log transform, and ANN models, the best median ASSL predicted by the ANN model was 37.611 Mt for the period of 2006 to 2011, whereas the observed median ASSL was 41.678 Mt. The mean ASSL transported by the river to the hydropower reservoir was 40.904 ± 12.453 Mt for 2006 to 2011 and the ANNs' predicted mean value was 35.190 ± 7.018 Mt (Figure 13). Struck [8] reported that the average annual suspended sediment transported by this river was 36.9 ± 10.6 Mt.

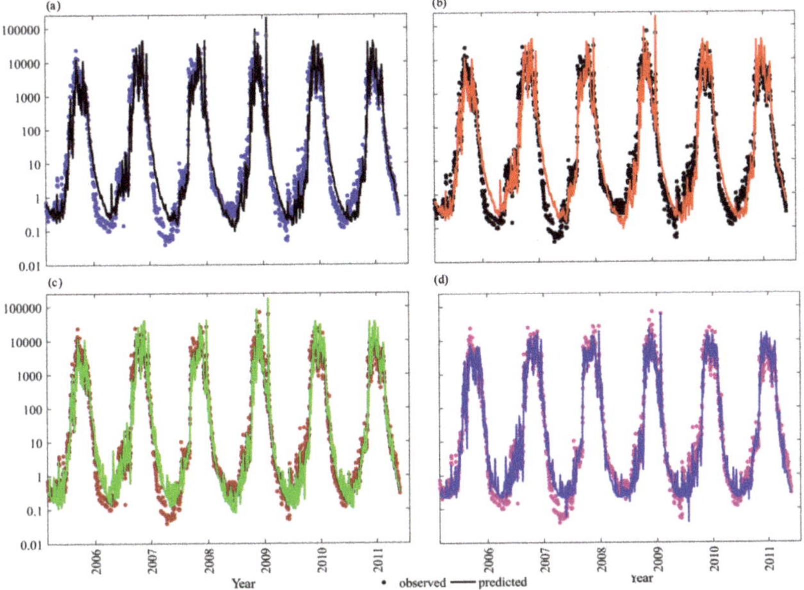

Figure 12. Observed and predicted sediment (**a**) SRC (Q_W and Q_s) model, (**b**) power model (Q_W), (**c**) log transform linear model (Q_W and R_t), and (**d**) ANN model.

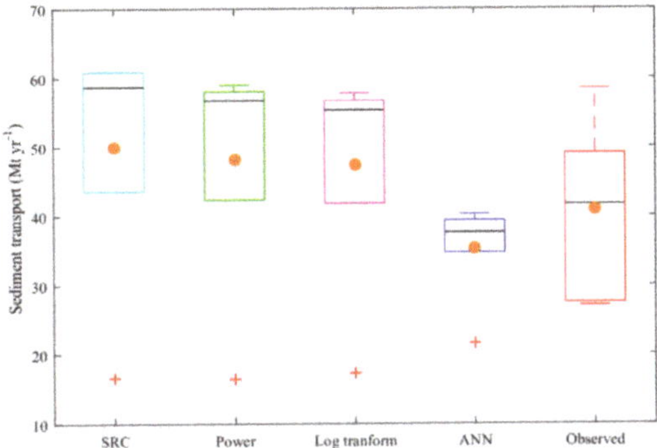

Figure 13. Comparison of different models' predicted and observed yearly total suspended sediment transport. Central lines indicate the median, and bottom and top edges of the box indicate the 25th and 75th percentiles, respectively. The whiskers extend to the most extreme data points not considered outliers, the '+' symbol represents outliers (1.5-fold interquartile range), and the circle shows the mean value.

4. Conclusions

Shear stress, specific stream power, and flow velocity are important key hydraulic parameters to describe sediment transport in river systems. The monsoon fluvial discharge and landslide dam outburst flood (LDOF) were responsible for boulder movements in Kali Gandaki River, Nepal. The lower boundary equation derived from a broad range of observed and calculated data sets estimated that a maximum particle size of 840 mm was transported by the monsoon fluvial discharge from 2003 to 2011. The ASSL transported by KaliGandaki River in the hydropower reservoir increased from winter to pre-monsoon to monsoon, respectively, and decreased in the post-monsoon period. It was estimated that 40.904 ± 12.453 Mt suspended sediment is lost annually from the higher Himalayas. Additionally, the ANN model provided satisfactory results for the prediction of the suspended sediments' transport rate in Kali Gandaki River, where the annual predicted mean ASSL value was 35.190 ± 7.018 Mt. These parameters are important for visualizing sediment loss from the higher Himalayas to the sea and also for monitoring the dead storage volume of reservoirs for hydroelectric power generation.

Author Contributions: For the conceptualization and methodology, M.B.B. and T.A.; software, validation, formal analysis, and writing—original draft preparation, M.B.B.; writing—review and editing T.A. and S.M.D.H.J.; supervision T.A.; data curation S.K.C.

Funding: This research received no external funding.

Acknowledgments: The authors would like to give thanks to Saitama University for managing the research environment. We would also like to thank to Nepal Electricity Authority for providing fluvial and sediment data and Ministry of Energy, Water Resources and Irrigation, Department of Hydrology and Meteorology, Nepal for providing historical fluvial and climate data.

Conflicts of Interest: The authors declare no conflict of interest.

References

1. Gudino-Elizondo, N.; Biggs, T.W.; Bingner, R.L.; Langendoen, E.J.; Kretzschmar, T.; Taguas, E.V.; Taniguchi-Quan, K.T.; Liden, D.; Yuan, Y. Modelling Runoff and Sediment Loads in a Developing Coastal Watershed of the US-Mexico Border. *Water* **2019**, *11*, 1024. [CrossRef]
2. Fort, M. Sedimentary fluxes in Himalaya. In *Source-to-Sink Fluxes in Undisturbed Cold Environments*; Beylich, A.A., Dixon, J.C., Zwolinski, Z., Eds.; Cambridge University Press: Cambridge, UK, 2016; pp. 326–350.
3. Fort, M.; Cossart, E. Erosion assessment in the middle Kali Gandaki (Nepal): A sediment budget approach. *J. Nepal Geol. Soc.* **2013**, *46*, 25–40.
4. Mishra, B.; Babel, M.S.; Tripathi, N.K. Analysis of climatic variability and snow cover in the Kaligandaki River Basin, Himalaya, Nepal. *Theor. Appl. Climatol.* **2014**, *116*, 681–694. [CrossRef]
5. Dahal, R.K.; Hasegawa, S. Representative rainfall thresholds for landslides in the Nepal Himalaya. *Geomorphology* **2008**, *100*, 429–443. [CrossRef]
6. Fort, M.; Cossart, E.; Arnaud-Fassetta, G. Hillslope-channel coupling in the Nepal Himalayas and threat to man-made structures: The middle Kali Gandaki valley. *Geomorphology* **2010**, *124*, 178–199. [CrossRef]
7. Lenzi, M.A.; D'Agostino, V.; Billi, P. Bedload transport in the instrumented catchment of the Rio Cordon: Part I: Analysis of bedload records, conditions and threshold of bedload entrainment. *Catena* **1999**, *36*, 171–190. [CrossRef]
8. Struck, M.; Andermann, C.; Hovius, N.; Korup, O.; Turowski, J.M.; Bista, R.; Pandit, H.P.; Dahal, R.K. Monsoonal hillslope processes determine grain size-specific suspended sediment fluxes in a trans-Himalayan river. *Geophys. Res. Lett.* **2015**, *42*, 2302–2308. [CrossRef]
9. Asaeda, T.; Baniya, M.B.; Rashid, M.H. Effect of floods on the growth of Phragmites japonica on the sediment bar of regulated rivers: A modelling approach. *Int. J. River Basin Manag.* **2011**, *9*, 211–220. [CrossRef]
10. Guillén-Ludeña, S.; Manso, P.; Schleiss, A. Multidecadal Sediment Balance Modelling of a Cascade of Alpine Reservoirs and Perspectives Based on Climate Warming. *Water* **2018**, *10*, 1759. [CrossRef]
11. Megnounif, A.; Terfous, A.; Ouillon, S. A graphical method to study suspended sediment dynamics during flood events in the Wadi Sebdou, NW Algeria (1973–2004). *J. Hydrol.* **2013**, *497*, 24–36. [CrossRef]
12. Fort, M. Sporadic morphogenesis in a continental subduction setting: An example from the Annapurna Range, Nepal Himalaya. *Z. Geomorphol.* **1987**, *63*, 36.
13. Costa, J.E.; Schuster, R.L. The formation and failure of natural dams. *Geol. Soc. Am. Bull.* **1988**, *100*, 1054–1068. [CrossRef]
14. O'Connor, J.E.; Costa, J.E. Geologic and hydrologic hazards in glacierized basins in North America resulting from 19th and 20th century global warming. *Nat. Hazards* **1993**, *8*, 121–140. [CrossRef]
15. Zhang, G.; Liu, Y.; Han, Y.; Zhang, X.C. Sediment Transport and Soil Detachment on Steep Slopes: I. Transport Capacity Estimation. *Soil Sci. Soc. Am. J.* **2009**, *73*, 1291. [CrossRef]

16. Ali, M.; Sterk, G.; Seeger, M.; Boersema, M.; Peters, P. Effect of hydraulic parameters on sediment transport capacity in overland flow over erodible beds. *Hydrol. Earth Syst. Sci.* **2012**, *16*, 591–601. [CrossRef]
17. Baker, V.R.; Ritter, D.F. Competence of rivers to transport coarse bedload material. *Geol. Soc. Am. Bull.* **1975**, *86*, 975–978. [CrossRef]
18. Lotsari, E.; Wang, Y.; Kaartinen, H.; Jaakkola, A.; Kukko, A.; Vaaja, M.; Hyyppä, H.; Hyyppä, J.; Alho, P. Geomorphology Gravel transport by ice in a subarctic river from accurate laser scanning. *Geomorphology* **2015**, *246*, 113–122. [CrossRef]
19. Komar, P.D.; Carling, P.A. Grain sorting in gravel-bed streams and the choice of particle sizes for flow-competence evaluations. *Sedimentology* **1991**, *38*, 489–502. [CrossRef]
20. Costa, J.E. Paleohydraulic reconstruction of flash-flood peaks from boulder deposits in the Colorado Front Range. *Geol. Soc. Am. Bull.* **1983**, *94*, 986–1004. [CrossRef]
21. Komar, P.D. Selective Gravel Entrainment and the Empirical Evaluation of Flow Competence. *Sedimentology* **1987**, *34*, 1165–1176. [CrossRef]
22. Lenzi, M.A.; Mao, L.; Comiti, F. When does bedload transport begin in steep boulder-bed streams? *Hydrol. Process.* **2006**, *20*, 3517–3533. [CrossRef]
23. O'Connor, J.E. *Hydrology, Hydraulics, and Geomorphology of the Bonneville Flood*; Geological Society of America: Boulder, CO, USA, 1993; Volume 274, ISBN 0813722748.
24. Williams, G.P. Paleohydrological methods and some examples from Swedish fluvial environments: I cobble and boulder deposits. *Geogr. Ann. Ser. A Phys. Geogr.* **1983**, *65*, 227–243. [CrossRef]
25. Bradley, W.C.; Mears, A.I. Calculations of flows needed to transport coarse fraction of Boulder Creek alluvium at Boulder, Colorado. *Geol. Soc. Am. Bull.* **1980**, *91*, 1057–1090. [CrossRef]
26. Helley, E.J. *Field Measurement of the Initiation of Large Bed Particle Motion in Blue Creek Near Klamath, California*; US Government Printing Office: Washington, DC, USA, 1969.
27. Bhusal, J.K.; Subedi, B.P. Effect of climate change on suspended sediment load in the Himalayan basin: A case study of Upper Kaligandaki River. *J. Hydrol.* **2015**, *54*, 1–10.
28. Kale, V.S.; Hire, P.S. Effectiveness of monsoon floods on the Tapi River, India: Role of channel geometry and hydrologic regime. *Geomorphology* **2004**, *57*, 275–291. [CrossRef]
29. Mao, L.; Uyttendaele, G.P.; Iroumé, A.; Lenzi, M.A. Field based analysis of sediment entrainment in two high gradient streams located in Alpine and Andine environments. *Geomorphology* **2008**, *93*, 368–383. [CrossRef]
30. Wicher-Dysarz, J. Analysis of Shear Stress and Stream Power Spatial Distributions for Detection of Operational Problems in the Stare Miasto Reservoir. *Water* **2019**, *11*, 691. [CrossRef]
31. Jarrett, R.D. Hydraulics of high-gradient streams. *J. Hydraul. Eng.* **1984**, *110*, 1519–1539. [CrossRef]
32. Uca; Toriman, E.; Jaafar, O.; Maru, R.; Arfan, A.; Ahmar, A.S. Daily Suspended Sediment Discharge Prediction Using Multiple Linear Regression and Artificial Neural Network. *J. Phys. Conf. Ser.* **2018**, *954*, 012030.
33. Ulke, A.; Tayfur, G.; Ozkul, S. Predicting suspended sediment loads and missing data for Gediz River, Turkey. *J. Hydrol. Eng.* **2009**, *14*, 954–965. [CrossRef]
34. Glysson, G.D. *Sediment-Transport Curves*; Open-File Report 87-218; US Geological Survey: Reston, VA, USA, 1987.
35. Moriasi, D.N.; Arnold, J.G.; Van Liew, M.W.; Bingner, R.L.; Harmel, R.D.; Veith, T.L. Model evaluation guidelines for systematic quantification of accuracy in watershed simulations. *Trans. ASABE* **2007**, *50*, 885–900. [CrossRef]
36. Pandey, M.; Zakwan, M.; Sharma, P.K.; Ahmad, Z. Multiple linear regression and genetic algorithm approaches to predict temporal scour depth near circular pier in non-cohesive sediment. *ISH J. Hydraul. Eng.* **2018**. [CrossRef]
37. Bajracharya, A.R.; Bajracharya, S.R.; Shrestha, A.B.; Maharjan, S.B. Climate change impact assessment on the hydrological regime of the Kaligandaki Basin, Nepal. *Sci. Total Environ.* **2018**, *625*, 837–848. [CrossRef] [PubMed]
38. Onen, F.; Bagatur, T. Prediction of Flood Frequency Factor for Gumbel Distribution Using Regression and GEP Model. *Arab. J. Sci. Eng.* **2017**, *42*, 3895–3906. [CrossRef]
39. Grant, G.E.; Swanson, F.J.; Wolman, M.G. Pattern and origin of stepped-bed morphology in high-gradient streams, Western Cascades, Oregon. *Bull. Geol. Soc. Am.* **1990**, *102*, 340–352. [CrossRef]
40. Montgomery, D.R.; Buffington, J.M. Channel-reach morphology in mountain drainage basins. *Bull. Geol. Soc. Am.* **1997**, *109*, 596–611. [CrossRef]

41. Marahatta, S. Earthquake and Ramche Landslide Dam in Kali Gandaki. Post April 25, 2015. Available online: http://coramnepal.org/wp-content/uploads/2017/09/Kali-Gandaki-Landslide-Dam-Outburst-Flood.pdf (accessed on 18 May 2019).
42. Bricker, J.D.; Schwanghart, W.; Adhikari, B.R.; Moriguchi, S.; Roeber, V.; Giri, S. Performance of Models for Flash Flood Warning and Hazard Assessment: The 2015 Kali Gandaki Landslide Dam Breach in Nepal. *Mt. Res. Dev.* **2017**, *37*, 5–15. [CrossRef]
43. Griffiths, G.A. Origin and transport of large boulders in mountain streams. *J. Hydrol.* **1977**, *16*, 1–6.
44. Williams, G.P. Sediment concentration versus water discharge during single hydrologic events in rivers. *J. Hydrol.* **1989**, *111*, 89–106. [CrossRef]
45. Bača, P. Hysteresis effect in suspended sediment concentration in the Rybárik basin, Slovakia/Effet d'hystérèse dans la concentration des sédiments en suspension dans le bassin versant de Rybárik (Slovaquie). *Hydrol. Sci. J.* **2008**, *53*, 224–235. [CrossRef]
46. Lloyd, C.E.M.; Freer, J.E.; Johnes, P.J.; Collins, A.L. Using hysteresis analysis of high-resolution water quality monitoring data, including uncertainty, to infer controls on nutrient and sediment transfer in catchments. *Sci. Total Environ.* **2016**, *543*, 388–404. [CrossRef] [PubMed]
47. Lawler, D.M.; Petts, G.E.; Foster, I.D.L.; Harper, S. Turbidity dynamics during spring storm events in an urban headwater river system: The Upper Tame, West Midlands, UK. *Sci. Total Environ.* **2006**, *360*, 109–126. [CrossRef] [PubMed]

Article

A Combined Field and Numerical Modeling Study to Assess the Longitudinal Channel Slope Evolution in a Mixed Alluvial and Soft Bedrock Stream

Kuowei Wu [1,*], Keh-Chia Yeh [1] and Yong G. Lai [2]

[1] Department of Civil Engineering, National Chiao Tung University, 1001 University Road, Hsinchu City 30010, Taiwan; kcyeh@mail.nctu.edu.tw
[2] Technical Service Center, U.S. Bureau of Reclamation, Bldg. 67, P.O. Box 25007, Denver, CO 80225-007, USA; ylai@usbr.gov
* Correspondence: kuowei@wrap.gov.tw; Tel.: +886-4-2330-2980

Received: 26 February 2019; Accepted: 7 April 2019; Published: 9 April 2019

Abstract: A study approach is developed to assess the longitudinal channel slope under the equilibrium condition as well as the transient evolution of a mixed alluvial-soft-bedrock stream. Both the historical field data and 2D mobile-bed numerical modeling are adopted. The proposed approach is applied to a 14 km reach downstream of the Ji-Ji Weir, Chuo-Shui River, Taiwan, where continuous maintenance works have been carried out to stabilize this reach. In this study, the temporal evolution of the longitudinal channel profile is assessed numerically with three spatial scales: The large (the entire study reach), the medium (four sub-reaches), and the local (cross-sections) scale. The large scale analysis is the approach for purely alluvial streams and is shown to be difficult to use to characterize mixed alluvial-bedrock streams. The local scale analysis shows that the soft-bedrock incision has a widely fluctuating slope, reflecting the compound environmental forcing and complex riverbed setting. With the medium scale analysis, the longitudinal channel profile is found to follow a predictive trend if the reach is partitioned into four distinctive sub-reaches. Characteristics of the dynamic channel slope evolution in different spatial scales are computed and presented. The study results can be used to select the proper locations and types of the engineering stabilizing structures in a mixed alluvial and soft bedrock stream.

Keywords: soft bedrock; SRH-2D; head-cutting; mixed transition; knickpoint; spatial scales; river morphology

---

1. Introduction

Rivers in Taiwan are very steep and subject to a large amount of sediment transport. In the last two decades, most lower and even middle river reaches in west-central Taiwan have experienced much increased degradation and deposition caused by earthquake induced riverbed uplift, construction of in-stream structures, and mining activities. In many locations, armor layers were washed away and underlying bedrock was exposed. In particular, significant degradation occurred downstream of constructed structures, such as dams and weirs. Downstream degradation has disrupted the normal operation of the structures and caused safety concerns at several major locations [1–3].

Most geomorphological studies focused on alluvial rivers (e.g., [4]). A few were reported concerning bedrock channels (e.g., [5–7]). Little work, however, has been done with mixed bedrock-alluvial systems (e.g., [8,9]). A unique feature of the degradation in Taiwan is that the exposed bedrock consists primarily of soft bedrock, which is even less resistant to shear and abrasive erosions [10]. In general, the exposed bedrock in west-central Taiwan is mainly composed of sandstone, siltstone, mudstone, and shale in a monotonous alternating sequence and belongs to Pleistocene and Pliocene formation [3].

Field and laboratory tests showed that the unconfined compressive strength (UCS) of the exposed bedrock was under 25 MPa [11,12], under the upper limit of the soft rock strength [13,14]. As a result, bedrock-exposed reaches experience degradation characteristics different from the widely known channel profile evolution of alluvial streams. Channel stability becomes an important issue for limited water resources and flood prevention in such dynamic rivers. To reduce the failure risk of in-stream and cross-stream infrastructures from the continued channel profile adjustment due to soft bedrock incision and knickpoint (abrupt change in gradient) propagation, continuous engineering measures have been carried out; their designs, however, are mostly based on design guidelines derived from alluvial rivers. These traditional design guidelines were found inadequate for the mixed alluvial-soft-bedrock reach and repeated and continuous maintenance has been carried out over the years. There is an urgent need in Taiwan to understand the longitudinal channel profile evolution under purely soft-bedrock or mixed alluvial-soft-bedrock conditions. Such an understanding would help to plan and develop engineering solutions to protect upstream infrastructure.

This objectives of this study are: (1) To assess a transient stable longitudinal channel profile evolution for river management; (2) to evaluate the optimum spatial scales in characterizing the channel evolution processes for mixed alluvial-soft-bedrock reaches; and (3) to propose and demonstrate an approach applicable to the analysis of mixed alluvial-soft-bedrock reaches. The proposed procedure uses both field data and numerical modeling to gain insight into the morphdynamic characteristics and longitudinal profile evolution processes. The present study contributes to the knowledge about the morphological processes of mixed alluvial and soft bedrock rivers. In this study, the mixed channel is the focus and a study approach is developed to assess the hydraulic geometry as well as the transient evolution of the mixed alluvial-soft-bedrock stream. Specifically, a 2D mobile-bed model, SRH-2D (Sedimentation and River Hydraulics, Two-Dimensional model), is used to replicate and predict the channel slope evolution.

2. Study Site and Background

Cho-Shui River drains from a basin of 3157 km^2 and is the longest river in Taiwan (187 km) with an average longitudinal channel slope of 0.018. The upstream originates from Mt. He-Huan and is very steep; the downstream reach is relatively flatter, forming an alluvial fan of 1800 km^2 near the Er-Shui Pass. The average annual runoff is about 6100 million m^3. The Ji-Ji Weir is located in the middle reach; the 100-year discharge at the weir is about 20,500 m^3/s. The largest flood recorded since 1980 was 12,582 m^3/s in August 2009—close to the 10-year return period discharge. The weir construction started in 1993 and was completed in 2001. It has a width of 353 m, a height of 15 m, with 18 spillway gates and four sluice gates.

The river reach downstream of the Ji-Ji Weir on the Cho-Shui River is the primary study site of this work. The Ji-Ji weir diverts water in the Cho-Shui River for agriculture and industrial use. Downstream of the weir, the channel has experienced incision of up to 20 m since 1998. At present, the reach is undergoing continued degradation, which is impacting the weir operation and causing safety concerns [12,15]. Studies suggested that the channel incision was intensified once the alluvial cover was eroded and the soft bedrock was exposed [16,17]. The study area, as shown in Figure 1, has tectonic structures across the channel: Chu-Suang and Che-Long-Pu Faults. There are two distinctive bed features. The reach upstream of the weir and that downstream of Min-Zhu Bridge are primarily alluvial channels with a grain size between 10 to 100 mm. The in-between reach consists primarily of the exposed soft bedrock with an only thin layer of fine sediments. The exposed bedrock is composed of sedimentary rocks from the Pliocene Cho-Lan Formation and poorly cemented with low erosion resistance [11,12].

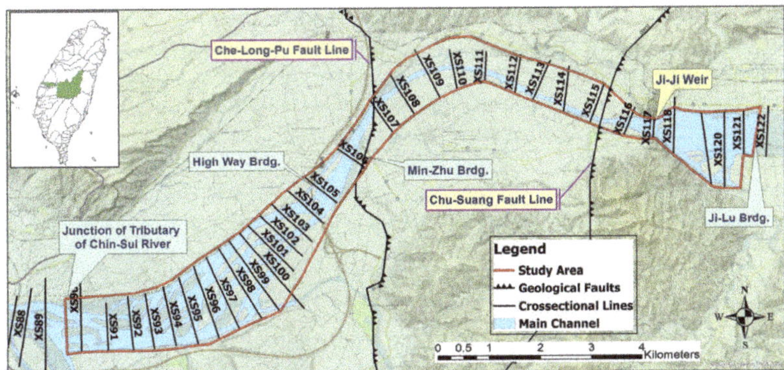

Figure 1. A map of the study site at the Ji-Ji Weir (background map source: maps.nlsc.gov.tw web map service).

2.1. Factors Impacting Channel Stability

Chen et al. [18] investigated the important factors that may impact channel stability in Taiwan. These included geological factors (e.g., earthquake), flow factors (precipitation, flood and erosion), and human factors (e.g., manmade cross-stream structures). At the preset study site, a key factor is the soft bedrock incision along with the downstream alluvial channel morphology. Other important factors at the site include the earthquake, cross-stream weir, and sand mining further downstream. The co-seismic fault movement, which occurred in 1999, caused a riverbed uplift of about 7 m near the Min-Zhu Bridge. The earthquake event altered channel slope continuity and increased the sediment supply upstream. The cross-stream weir disrupted further the existing longitudinal channel profile, limited the lateral channel migration, and blocked the coarse grain sediment supply from the weir upstream. In addition, sand mining was active between the early 1970s and late 1980s; it was estimated that about 30 to 60 million tons were removed per year downstream of the Min-Zhu Bridge [19,20]. The mining altered the bed sediment gradation and sediment rate.

2.2. Channel Longitudinal Evolution

A large amount of data have been collected by the Water Resources Agency, Taiwan (WRA) since 1984, including field surveying, photogrammetry, and laser LiDAR (Light Detection and Ranging) data. The chronological topographies (1984–2015) are used to generate the longitudinal channel profiles. The specific longitudinal profiles are plotted as Figure 2; the time period is chosen so that the impacts of mining, earthquake uplift, and cross-stream weir operation may be assessed. The historical channel morphological characteristics at the study reach shows that the channel evolution may be divided into four stages: Stage 1 before 1984, stage 2 between 1984 and 1998, stage 3 between 1998 and 2004, and stage 4 from 2004 to the present.

Stage 1 was the natural channel development period during which there were no significant interferences from human activities and natural hazards. The longitudinal channel maintained at an almost constant slope of 0.0068 for several decades. The channel plan form was braided for the whole reach. Stage 2 witnessed the impacts of human activities in the 1980s, primarily extensive gravel mining in the lower reach of the study area. Channel degradation occurred in the lower reach and migrated upstream (Figure 2a). The upstream sediment supply, however, was not altered. Abundant sediment supply, in particular the availability of large size sediments, kept the downstream erosion moderate at this stage. Stage 3 saw the outcome of the in-stream and cross-stream structures as well as the co-seismic channel uplift. Large disturbances were introduced, with channel slope discontinuity, into the system. The upstream sediment supply was reduced and sediment composition was altered. Three sections were identified: Upstream of the weir, downstream of the channel uplift, and the in-between middle section. The channel slope at the beginning of this stage was similar to those of stages 1 and 2 (Figure 2b). Finally, stage 4 corresponded to the evolution processes after weir operation and was controlled primarily by the exposed soft bedrock incision. At the early period of the stage (Figure 2c), the channel slope is quite different in the alluvial and the soft bedrock section. The slope of the alluvial section followed the historical dynamic slope exhibited in stage 1. The soft bedrock section, however, became narrower and deeper with the decreased slope. At the later period of stage 4 (Figure 2d), deposition occurred in the upper part of the alluvial section, which tried to recover to the historical dynamic slope of alluvial channel; degradation in the soft bedrock section, however, continued.

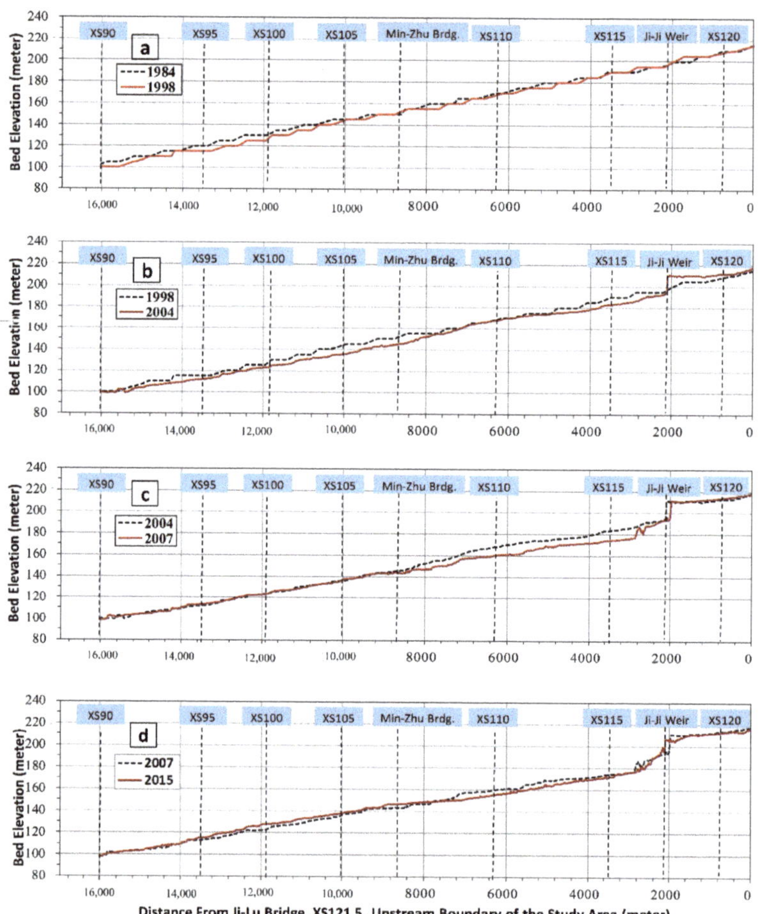

Figure 2. Longitudinal profiles during four time periods corresponding to the four stages: (**a**) period 1984–1998; (**b**) period 1998–2004; (**c**) period 2004–2007; (**d**) period 2007–2015.

2.3. Bed Material Changes

The size of the bed materials and the distribution are important parameters to assess the channel hydraulic geometry. The variation of the bed material grain size is highly correlated to the river morphological and sedimentation processes. Periodic bulk sample and sieve analysis of bed material gradation in Cho-Shui River has been performed by the WRA (Water Resources Agency, Taiwan) since 1983. In general, the grain size increases with the distance from the river mouth for most rivers. For the study reach, the variations of the bed material size in three different years are plotted in Figure 3.

In the figure, abscissa is the cross sectional station numbered from the river mouth, ordinate is the representative grain size (D_{50}).

The longitudinal distribution of the bed materials in 1983 is shown as the filled blue diamonds; this year was the situation before the existence of the geological uplift and the cross-stream weir. It can be seen that the median grain size ranged from 55 to 116 mm with an average of 88 mm. The size distribution in 2004 is plotted with the square red symbols; it was after the impacts of the uplift and weir operation. The grain size decreased from upstream to downstream consistently, in response to the geological uplift and the cross-stream weir. Finally, the latest longitudinal distribution in 2016 was also displayed using the filled black triangles. The median grain size in the lower reach has increased and was in the process of returning to the pre-disturbance condition in 1983. Upstream of the weir, however, the sediment size remains low, far from the 1983 condition. The data implies that the channel slope adjustment in the mixed alluvial-soft-bedrock stream is still progressing.

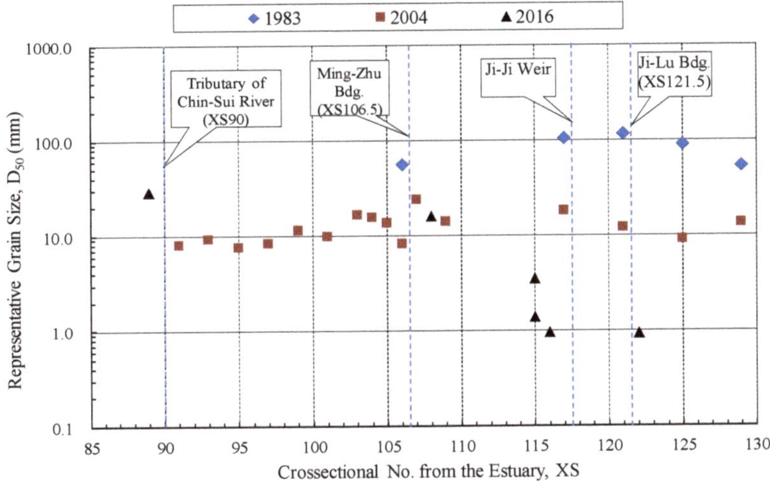

Figure 3. Variations of the bed material size in three different years.

2.4. Sediment Supply

Sediment supply from the watershed and the subsequent transport downstream are also important to understand the channel evolution processes. Unfortunately, only suspended sediment is monitored by the hydrological stations in Taiwan through the Water Resources Agency [20,21]; data was available only during low to medium flows. Two river sediment monitoring stations are selected and analyzed to understand the upstream sediment supply for the study area. One is located about 5 km upstream of the study reach, named Yu-Fong Bridge station (XS133.5); the other is immediately downstream of the study reach at the Chang-Yuan Bridge (XS86.5). After the bedrock exposure in 2004, the sediment concentration data may be divided into three periods according to the geological disturbance and channel erosion behavior: The pre-1999 earthquake period (period 1, 1995–1999), the post-1999 earthquake period (period 2, 2000–2004), and the exposed-bedrock period (period 3, 2005–2018). They are then analyzed to compare the variation of the sediment supply and transport through the study reach. The measured data and the regression curves are plotted in Figure 4. It can be seen for both stations that the sediment supply and transport rate were altered significantly by the 1999 earthquake. At the upstream inlet (Figure 4a), the period 1 regression curve has a higher exponent (1.18) than the period 2 and period 3 lines (1.05 and 1.06, respectively). Two intersections exist, one is in the medium flow discharge range (about 2000 m^3/s) for period 1 and period 2; the other is in the low flow discharge range (about 20 m^3/s) for period 2 and period 3. In period 2, the sediment rate increased initially after the earthquake when the flow was lower than 2000 m^3/s; then, in period 3, it decreased when the flow exceeded 20 m^3/s. At the downstream outlet (Figure 4b), however, the regression curve exponent decreased (1.01, 0.95, and 0.82, respectively) with an overlap in the very high flow (approximately 5000 m^3/s); the decreased exponent shows the increased sediment concentration in the downstream outlet. The data indicates that the downstream suspended sediment concentration increased in period 2 and period 3 under most flows although the upstream suspended sediment supply initially increased in the post-1999 period (period 2) and then decreased in the exposed-bedrock period (period 3). The increasing suspended sediment transport suggests a sediment transport imbalance and channel degradation in the study reach.

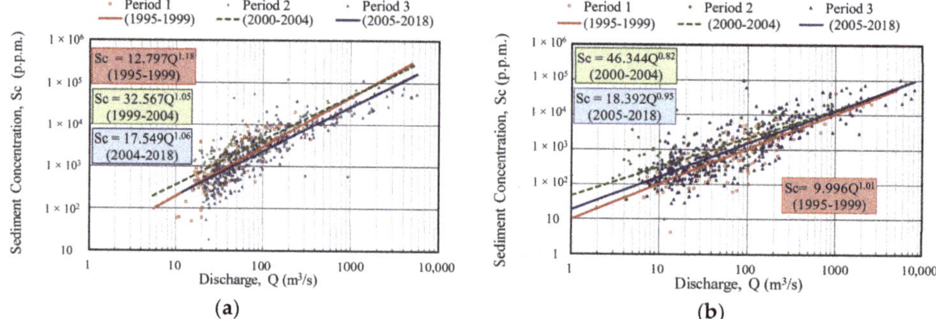

Figure 4. The variation of measured suspended sediment discharge rating curve for pre-1999 earthquake period (period 1, 1995–1999), post-1999 earthquake period (period 2, 2000–2004), and exposed-bedrock period (period 3, 2005–2018): (**a**) Upstream suspended sediment rating curve, 5 km upstream of the study reach, Yu-Fong Bridge (Station No. WRA1510H063); (**b**) Downstream suspended sediment rating curve, immediately downstream of the study reach, Chang-Yuan Bridge (Station No. WRA1510H057).

3. Numerical Modeling Approach

Numerical modeling is carried out to understand the past channel evolution processes so that a transient stable longitudinal channel profile for the future may be determined. The modeling tool, once validated, may be used to predict future channel evolution trends and assist in the design of engineering stability schemes at the study site. In the following, the numerical model is first described; it is then used to determine the dominant flow discharge for channel slope estimation. Next, it is employed to simulate the channel longitudinal profiles in the past decades with different spatial scales. The model is finally applied to predict the future evolution of the study reach.

Traditionally, channel slope is computed using the measured channel cross-section data and a straight line between two cross sections is assumed to represent the longitudinal distance. This approach usually underestimates the streamwise channel distance and overestimates the slope for continuously migrating channels [12,22]. Our approach is more accurate than the traditional method as continuous longitudinal and lateral channel profiles are computed and used to evaluate the channel slope evolution. With the numerical simulation, measured historical channel topography is used to represent the initial riverbed. The computed water depth versus discharge rating curve is used to obtain the bankfull properties at every cross section. The longitudinal profiles are determined and stable channel slopes are finally computed from the numerical results. Once the historical channel profile and slope analysis is completed, the future channel evolution trend is predicted in the next five years. The predicted channel profiles are compared with the historical data to develop transient stable channel profiles.

3.1. SRH-2D Model

The two-dimensional (2D), depth-averaged flow and mobile-bed model, SRH-2D, is adopted [23,24] and used in this study. The model was extended to simulate soft bedrock erosion and bank erosion through a collaboration between the U.S. Bureau of Reclamation and Taiwan WRA: The soft bedrock module was verified using data at the Ji-Ji Weir, Taiwan [10] and the bank erosion module was verified at a site near the river mouth of the Cho-Shui River [25]. The chosen model, therefore, is adequate for the present study objectives. Simulation processes include unsteady flow, granular sediment transport, mixed alluvial and soft rock erosion, and the combined vertical and lateral channel changes. The alluvial channel evolution and soft bedrock incision are simultaneously simulated to predict the longitudinal channel profile evolutions.

The flow solver was documented in [24] and the sediment transport theory was documented by [26]. Detailed theories of the model were described by the two above papers and are not repeated herein. Only the key processes are discussed below.

The total load is divided into a number of size classes and each size class is governed by the non-equilibrium transport equation. The multi-size approach may be compared with the alternative of the single-size method in which all sediments are represented by a representative size (D_{50} is usually used). Non-equilibrium transport refers to the situation in which the sediment concentration does not equal the transport capacity. Each sediment size class, say class k, is governed by the following equation:

$$\begin{aligned}\frac{\partial hC_k}{\partial t} + \frac{\partial \cos(\alpha_k)\beta_k V_t h C_k}{\partial x} + \frac{\partial \sin(\alpha_k)\beta_k V_t h C_k}{\partial y} \\ = \frac{\partial}{\partial x}\left(f_k D_x \frac{\partial hC_k}{\partial x}\right) + \frac{\partial}{\partial y}\left(f_k D_y \frac{\partial hC_k}{\partial y}\right) + \dot{V}_k\end{aligned} \quad (1)$$

In the above, the subscript, k, denotes a variable for size class k, h is water depth, t is time, x and y are two horizontal Cartesian coordinates, respectively, C_k is the depth averaged sediment concentration by volume, $\beta_k = V_{sed,k}/V_t$ represents the sediment-to-flow velocity ratio, V_t is the depth-averaged flow velocity (m/s), $V_{sed,k}$ is the concentration weighted, α_k is the sediment transport angle that represents the angle between the sediment moving direction and the x-axis direction, f_k is the "load" parameter representing the percentage of the sediments in suspension, D_x and D_y are the mixing coefficients of sediments in the x- and y-directions, respectively, and $\dot{V}_k$ represents the sediment exchange between the water column and the active layer. The active layer is the top bed surface layer, where sediment exchange between the water column and the bed occurs.

3.2. Soft Bedrock Erosion Equation

The bedrock erosion rate is computed by combining the stream power based hydraulic scour model, using the excess shear stress, with the modified abrasive scour model of Sklar and Dietrich [17]. The erosion rate is computed by:

$$E = k_h V_t\left(\frac{\tau_b}{\tau_{ch}} - 1\right)F_e + 0.08k_a(\gamma - 1)gq_s\left(\frac{\tau_b}{\tau_{ci}} - 1\right)^{-0.5}\left[1 - \left(\frac{u_*}{\omega_f}\right)^2\right]^{1.5} F_e \quad (2)$$

In the above, E is the bedrock erosion rate (m/s), k_h is the nondimensional hydraulic erodibility, τ_b and τ_{ch} are the bed shear stress and critical shear stress for hydraulic scour (Pa), k_a is the abrasive erodibility parameter (ms^2/kg), γ is the specific density of sediment, g is the gravitational acceleration (m/s^2), q_s is the sediment supply rate by mass per unit width (kg/ms), τ_{ci} is the critical shear stress for sediment incipient motion (P_a), ω_f is the sediment fall velocity (m/s), and F_e represents the effect of sediment cover over the bedrock.

The first term on the right hand side of Equation (2) accounts for the hydraulic scour. It is a function of the stream power ($\tau_b V_t$). The adopted form has the advantage that the hydraulic erodibility coefficient is dimensionless; it is also consistent in form to the erosion of the cohesive sediment transport in the mobile-bed models. Two model parameters are needed for the hydraulic scour model: k_h and τ_{ch}. The hydraulic erodibility, k_h, may be measured or be used as a calibration parameter. The critical stress, τ_{ch}, may be estimated using the approach of Annandale (2006), field tests, or numerical model calibration. The second term on the right side of Equation (2) models the abrasive scour according to Sklar and Dietrich [17]. Only one model parameter, k_a, is needed.

3.3. Numerical Model Domains

Two numerical models are developed: The primary and the assistance model. The primary model is developed for the transient stable longitudinal channel profile analysis. The assistance model is a larger model than the primary model so that the transient sediment supply at the upstream inlet of the primary model may be obtained. The two model domains are shown in Figure 5. The primary model domain is about 18 km in the longitudinal direction (Figure 5a). Its upstream boundary is at the Ji-Lu Bridge, about 2 km upstream of the Ji-Ji weir, so that the sediment trap by the Ji-Ji Weir is taken into account. The downstream boundary is located at the junction with the tributary of Chin-Shui River, about 14 km downstream of the weir, so that both alluvial and soft bedrock sections are included. The lateral extent of the solution domain is constrained by the levees and embankments and has an average channel width of about 850 m. The domain is wide enough to fully contain the highest possible flows simulated. The 2D mesh used for the solution domain has a total of 10,392 hybrid quadrilateral and triangular cells, which is sufficient according to the study of [10]. The assistance model solution domain is in Figure 5b. The upstream boundary of the assistance model is extended far away from the Ji-Lu Bridge and located at the Syuan-Long Bridge, about 20 km upstream of the Ji-Ji weir. A total of 17,955 hybrid quadrilateral and triangular cells are used in the assistance model.

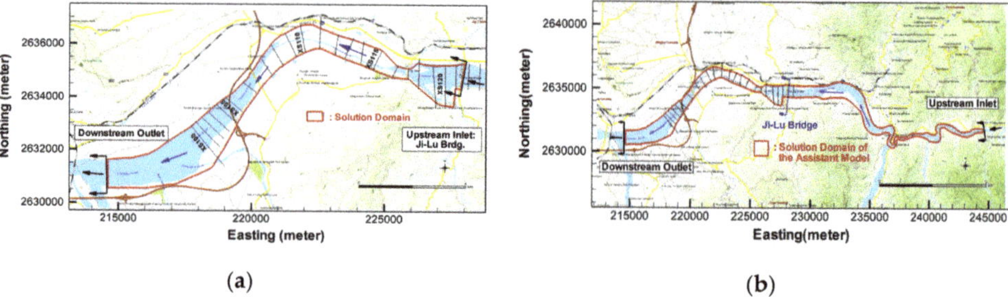

Figure 5. The solution domains of the primary model and the assistance model (background map source: maps.nlsc.gov.tw web map service). (**a**) The primary model; (**b**) The assistance model.

3.4. Input Parameters

The primary calibration parameter for the flow simulation is the Manning's roughness coefficient [23]. In this study, a constant Manning's coefficient of 0.04 is used in the alluvial area while it is 0.03 in the soft bedrock area, which was first estimated from an empirical formula (such as Lane, Einstein, and Strickler) [19] and then calibrated from measured water surface elevation and physical modeling results [11,12,15,22]. The use of a uniform roughness is based on several field visits by the authors; there was no appreciable spatial variation of the granular materials (D_{50} is about 14 mm) in the alluvial area. In addition, an extensive field study determined that the roughness was about uniform in the soft bedrock area also. The roughness values of 0.04 and 0.03 were based on the field data and supported by the full-scale model study. No calibration is carried out to alter these values.

Sediment gradation on the river bed is an important input for morphological modeling. Fourteen zones are used to represent the different bed gradations for the primary model (Figure 6a). The exposed soft bedrock is delineated first and represented by four zones using the bedrock property field test and a previous model calibration study [12,15]. Three erosion parameters are the key inputs for the soft bedrock modeling related to both the hydraulic and abrasive scours: Critical shear stress (τ_{ch}) and hydraulic erodibility (k_h) for the hydraulic scour and abrasive erodibility parameter (k_a) for the abrasive scour. Past modeling studies showed that the hydraulic scour model dominated the first 1.5 km downstream of the weir, where the knickpoint migration dominated, while the combined hydraulic and abrasive erosion was needed in the remaining reach [4,12,15]. In this study, the hydraulic scour parameters adopted the same values from these previous studies; however, the abrasive erodibility parameter is calibrated. The adopted soft bedrock input parameters (zone 11 through 14) are: τ_{ch} = 200 Pa, $k_h = 5.0 \times 10^{-7}$; $k_a = 3.3 \times 10^{-7}$ (zone 11 to zone 12); and $k_a = 5.9 \times 10^{-7}$ (zone 13 to zone 14), respectively.

The remaining 10 granular material zones are created using the surveyed gradation data gathered in [11]. The sediment size distributions for zone 1 through to 9 are plotted in Figure 6b; zone 10 is specified as non-erodible to represent the concrete structure of the weir.

The bed gradation partition for the assistance model is similarly carried out. The extended alluvial material zones are based on the 2004 survey data [12,15].

The granular sediments are divided into eight sediment size classes, the lower bound of the diameter from size class 1 to size class 8 are 0.001 mm, 0.074 mm, 0.6 mm, 2.0 mm, 8.0 mm, 24.0 mm, 72.0 mm, and 216 mm, respectively. The non-equilibrium adaptation length is set to be 500 m, close to the channel width, while the active layer thickness is 0.25 m. Both parameters are estimated based on previous 2D modeling studies of [10–12,15,22].

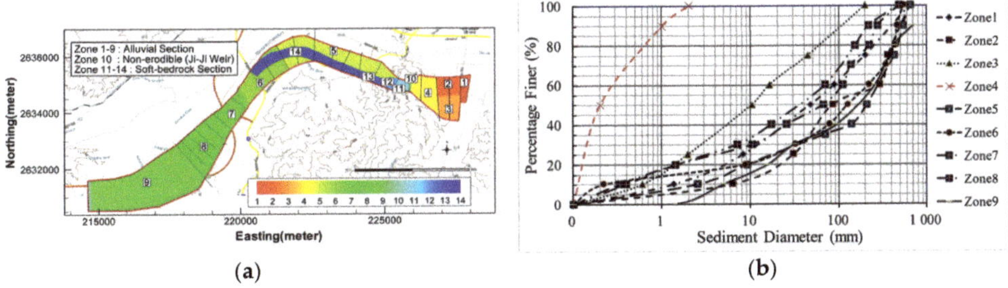

(a) (b)

Figure 6. Bed representation for the primary model: (**a**) Bed gradation zones to specify the sediment gradations on the bed (background map source: maps.nlsc.gov.tw web map service). (**b**) Cumulative sediment size distribution for various bed zones.

3.5. Boundary Conditions

At the upstream boundaries of both models, a time series flow hydrograph and a sediment supply rate are specified. The recorded hourly flow discharge in the 18-year period of 1998 to 2015 at Ji-Lu Bridge and Ji-Ji Weir are used to develop the 1-year, 3-year, and 5-year events for the primary model. The design hydrographs of 1-year, 3-years, and 5-year events are plotted in Figure 7a. The simulation time period is 156, 470, and 782 hours, respectively, for the three designed events.

The simulated sediment rates of the assistance model at the Ji-Lu Bridge are processed to produce the sediment supply condition of all size classes for the primary model. The generated data are used to develop the sediment rating curves according to the bed material size classes, which are plotted in Figure 7b.

At the downstream boundary, a water surface elevation is needed as the boundary condition. A stage-discharge rating curve is developed using the measured data at the Chang-Yuan Bridge, immediately downstream of the study reach. The data were documented in the technical report of [11,12]. No sediment boundary condition is needed.

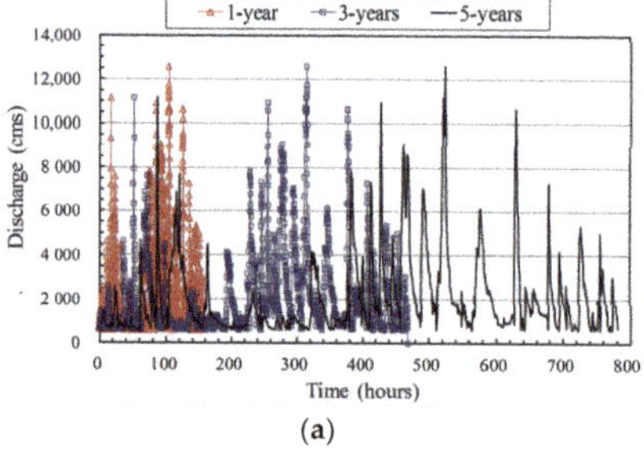

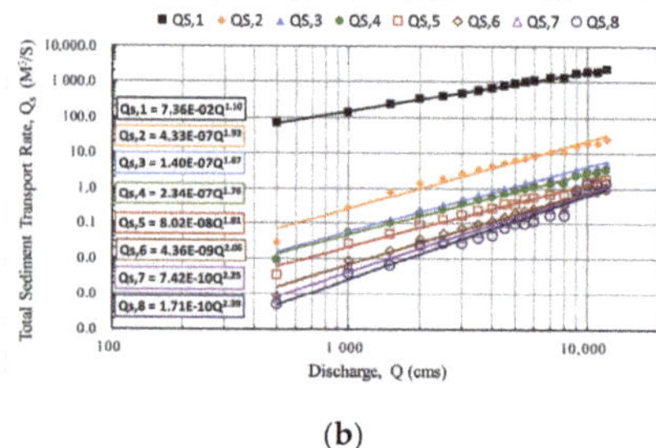

Figure 7. The upstream boundary conditions for the primary model: (**a**) The designed inlet flow hydrographs at the Ji-Lu Bridge (XS121.5). (**b**) The sediment supply rating curves of each size class, computed by the assistance model. Abbreviations of the suffix: $Q_{s,1}$ = size class 1, $Q_{s,2}$ = size class 2, $Q_{s,3}$ = size class 3, $Q_{s,4}$ = size class 4, $Q_{s,5}$ = size class 5, $Q_{s,6}$ = size class 6, $Q_{s,7}$ = size class 7, and $Q_{s,8}$ = size class 8.

4. Empirical Equations Assessment

In this section, the empirical approach is introduced as a comparison with the numerical modelling results. Two empirical equations are select to evaluate the stable slope first; the hydraulic geometries and characteristic discharges are then determined by numerical simulation. Finally, the stable channel slopes based on the empirical formula are calculated.

Many studies exist with regard to the stable alluvial channel slope under the equilibrium sediment transport, which is a function of the sediment, such as the representative size and the channel flow conveyance. Two empirical formulas, Schoklitsh [27] and Meyer-Peter-Muller (MPM) [28], are selected to evaluate the stable slope of the study reach; they are expressed as:

$$i = 0.000293 \times \left(\frac{D_m B}{Q}\right)^{\frac{3}{4}} \tag{3}$$

$$i = \frac{0.058 \times \left(\frac{n}{D_{90}^{\frac{1}{6}}}\right) \times D_m}{H} \tag{4}$$

In the above, i is the stable channel slope; D_m and D_{90} are the mean and 90% finer grain size of the bed materials; Q is the dominant flow discharge; B and H are the channel width and depth of the dominant flow; n is the Manning's roughness coefficient.

The determination of the dominant flow discharge is important in the application of the above equations. The river bankfull discharge is widely used as the dominant flow to characterize the reach-scale channel morphology [4,29,30]. There are, however, many different ways of determining the bankfull discharge. A popular method is to use the stage–discharge relation. In this study, the stage–discharge relation is obtained first using the present numerical model. The bankfull discharge, Q_{bf}, is then determined from the relation as water overtops the bank and flows to the floodplain. In addition, we defined the "channel-full" discharge, Q_{cf}, as another dominate flow discharge for comparison, which is derived from the asymptote of the stage–discharge relation when water covers the entire active floodplain with the rising water surface elevation approaching the top of embankment [12]. The channel-full discharge is used to represent the possible maximum recorded discharge of the reach. In this study, three stage–discharge relations are produced at three cross sections to demonstrate the determination of the dominated flow discharges (Figure 8). The results show that the bankfull discharge ranges from 2000 m^3/s to 3000 m^3/s and the channel-full discharge is between 9000 m^3/s to 11,000 m^3/s. Such determined bankfull discharge is close to the 1-year return period flow while the channel-full discharge is about the 5-year return period flow in the study reach [12].

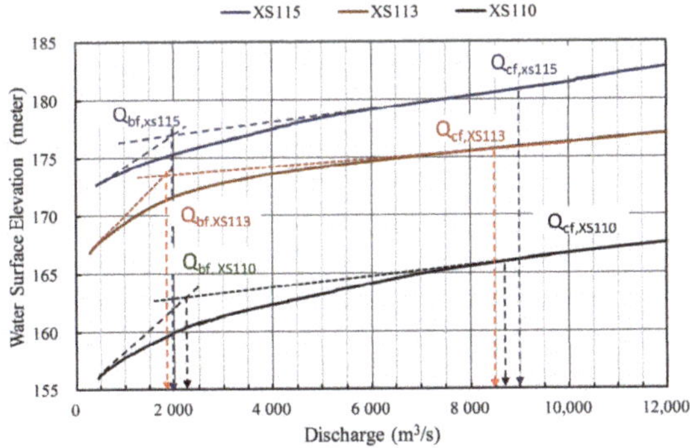

Figure 8. The stage–discharge relations computed by SRH-2D at three cross sections of the bedrock-exposed reach.

The necessary parameters for stable channel slope estimation are collected and tabulated in Table 1 at different cross sections. The representative bed material grain sizes are based on the available field measured data in recent years and the cross-sectional averaged channel geometries of both the bankfull discharge (W_{bf}, H_{bf}) and channel-full discharge (W_{cf}, H_{cf}) are computed through the 2D flow modeling.

Table 1. The measured representative bed material grain size and calculated cross-sectional averaged channel geometries at the bankfull and channel-full discharges.

XS No.	Representative Grain Size		W_{bf}	W_{cf}	H_{bf}	H_{cf}
	D_m (mm)	D_{90} (mm)	(meter)	(meter)	(meter)	(meter)
95	92.9	251.8	1328.3	219.1	2.2	2.58
105	48.5	168.4	671.9	512.1	1.6	2.93
107	54.1	190.3	273.6	136.8	4.4	5.13
109	71.3	222.9	315.0	93.5	3.8	3.82
111	94.9	263.3	222.4	146.2	4.0	6.13
113	107.2	300.0	155.0	84.2	5.6	6.05
115	33.6	90.0	104.6	83.7	4.0	7.87

Equations (3) and (4) are used to estimate the stable channel slope using the variables determined above. The computed stable channel is listed in Table 2. It is seen that the channel slopes range from 0.0004 to 0.0044 with an average value of 0.0011. This value is very close to the longitudinal slope in the lower part of most Rivers in Taiwan (near the river mouth/estuary), but different from the slopes of the middle and upper reaches of mountainous rivers. The result is not surprising as the adopted stable channel equations were developed for alluvial rivers and are not suitable for the purely soft-bedrock or mixed channels.

Table 2. The estimated stable channel slopes at different cross-sections in the study reach.

XS No.	Schoklitsch (Equation (3))		MPM (Equation (4))	
	Q_{bf}	Q_{cf}	Q_{bf}	Q_{cf}
95	0.0017	0.0044	0.0040	0.0034
105	0.0019	0.0020	0.0033	0.0017
107	0.0008	0.0011	0.0012	0.0011
109	0.0008	0.0014	0.0011	0.0018
111	0.0013	0.0012	0.0014	0.0015
113	0.0009	0.0010	0.0018	0.0016
115	0.0004	0.0003	0.0010	0.0005
Average	**0.0009**	**0.0010**	**0.0013**	**0.0014**

5. Numerical Modeling Results

The numerical modeling results of the transient stable channel slope and channel evolution trend are presented below.

5.1. Channel Slope Evolution within Three Spatial Scales

2D flow modeling is used to calculate the channel slopes in the past decades. With the measured topography since 1976, the longitudinal channel slope of the study reach is analyzed at three spatial scales: Large scale (the entire study reach), medium scale (four sub-reaches), and small scale (cross-sections {XS}).

First, the large scale longitudinal channel slope is computed; this has been widely used for planning channel stabilizing projects in Taiwan. The two stations near the boundaries of the study reach are XS117 to XS90 and the reach between the two stations is about 14 km. The measured channel slope evolution is shown in Figure 9a. It shows that the large-scale longitudinal channel profile maintained a transient stable slope of 0.0068 before the earthquake and weir operation events. After these events, the average channel slope decreased at a rate of about 0.00008 per year. The sharp decrease started in 2004, which coincided with the exposure of the soft bedrock due to the weir operation.

Next, the medium scale channel slopes are computed using the measured data. After 2004, severe soft bedrock incision occurred. The study reach is divided into four sub-reaches based on the channel morphology (from upstream to downstream): The head-cutting (XS117-XS115), the soft bedrock (XS115-XS108), the transition (mixed alluvium and bedrock, XS108-XS105), and the alluvial (XS105-XS90) sub-reaches. These sub-reaches have different bed materials and are delineated from the aerial-photos and field observations. In general, the alluvial channel sub-reach has a larger channel width-to-depth ratio than the bedrock channel; the transition and head-cutting sub-reaches have a medium width-to-depth ratio, and the bedrock section has the smallest ratio [12]. The longitudinal slope evolution for the four sub-reaches is shown in Figure 9b. It is seen that the channel slopes of the four sub-reaches maintained a dynamically stable average of about 0.0068 before the weir operation and earthquake. After these events, however, the channel slope evolved differently in different sub-reaches depending on the bed material characteristics. The head-cutting sub-reach experienced the most incision while the bedrock sub-reach showed a channel geometry self-regulating feature [31]. The channel slope evolution was observed to occur as follows. The alluvial and soft bedrock sub-reaches progressed with a predictable trend: The soft bedrock sub-reach developed a constant average slope of about 0.0050 while the alluvial channel maintained a transient stable slope of about 0.0068. The transition sub-reach decreased its slope in response to the upstream bedrock incision and downstream fluvial process. The head-cut sub-reach increased its slope as the knickpoint moved upstream towards the Ji-Ji weir.

Finally, the small scale channel slopes are computed at all river cross sections of the study reach. The evolution of local channel slopes is plotted in Figure 9c. The results show that severe channel incision occurred in the head-cutting, bedrock, and transition sub-reach. The most violent slope developed in the upper part of the head-cutting sub-reach, where the local channel experienced a slope variation up to eight folders, ranging from 0.005 to 0.040. Then a minor violent local slope change is found in the mixed transition sub-reach, the mutation of the local channel slope ranges from 0.002 to 0.014. Followed by the soft bedrock sub-reach, the variation of the local channel slope varies between the lower limit of 0.002 and the upper limit of 0.008. However, the alluvial channel retained a relative steady progress of the local channel slope.

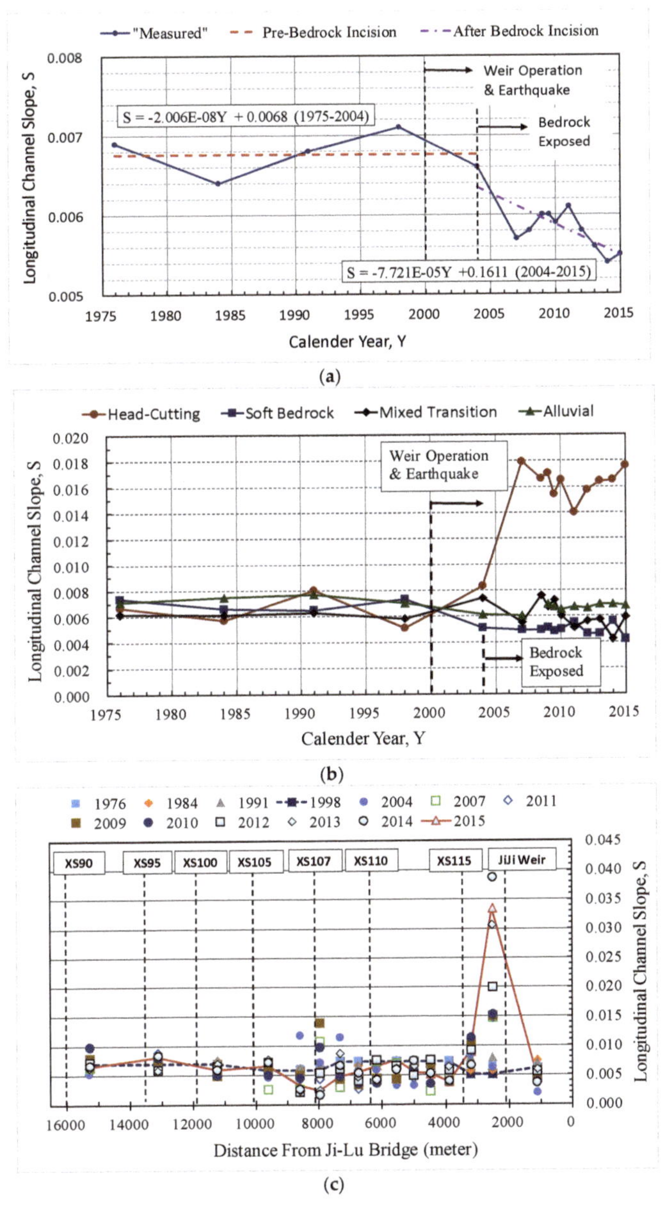

Figure 9. Longitudinal slope evolution with three different spatial scale: (**a**) Large-scale: The entire study reach; (**b**) medium-scale: Divided the study reach into four sub-reaches; and (**c**) local scale: Calculated between each cross-section.

5.2. Channel Slope Evolution and Prediction at Medium-Scale

2D mobile-bed modeling is carried out to understand the medium-scale channel slope evolution in the near future. The measured DEM in November 2015 is used as the initial bed topography. The predicted channel slope evolution is shown in Figure 10. As a comparison, the measured data in 2016 and 2017 are also displayed.

Figure 10a displays the predicted and measured longitudinal channel slopes for the head-cutting sub-reach. The channel slope is predicted reasonably well in comparison with the measured data in 2016 and 2017. The results show that the slope of this reach may continue to increase for the next few years with a rate of about 0.00035 per year. The slope increase will slow down and approach to a constant slope of 0.018 in 2020 when the knight point migration is near the Ji-Ji Weir.

Figure 10b shows the results for the soft bedrock sub-reach. The slope in this sub-reach is predicted to maintain at a transient stable value of about 0.0058. Despite this, the channel incision will continue in the next few years. The predicted channel slope agrees well with the measured data in 2016 and 2017. Both the predicted and measured slopes after 2015 indicate a slight slope increase above 0.000056 up to 2020. This slope increase may be owing to the cover effect of the fine sediments on the bedrock during the low-flow period.

The transition sub-reach is between the upstream bedrock and the downstream alluvial sections. Figure 10c shows the predicted and the measured channel slope evolution. The results show that the slope of the sub-reach is decreasing at a rate of about 0.00034 per year. This is similar to the rate in the upstream head-cutting sub-reach. The predicted channel slope matches well with the measured data in 2016 and 2017. The declining slope may be due to the dominance of the fine sediment deposition in the mixed zone. The numerical model predicts that the channel longitudinal slope will approach to the historical channel slope at the river mouth of Cho-Shui River in 2020.

Finally, Figure 10d shows the predicted and measured channel slopes in the downstream alluvial sub-reach. At the early stage before 2009, the slope of the sub-reach is increasing at a rate of about 0.000036 per year. Then, the slope fluctuates with an average value of 0.0068, which is the same as the transient stable slope over the past few decades. The slope maintaining the historical slope suggests that the sub-reach is influenced primarily by the natural river condition. The impacts of the upstream soft bedrock incision and knickpoint propagation are minimal. Again, the predicted slopes in 2016 and 2017 match the measured data well, although the slope in 2017 is overestimated. The mismatch in 2017 may be explained by the upstream channel stabilization work in early 2017. The discrepancy, however, is relatively small in comparison with the fluctuations at the beginning of the bedrock incision. The channel evolution in this sub-reach has attained a transient equilibrium state.

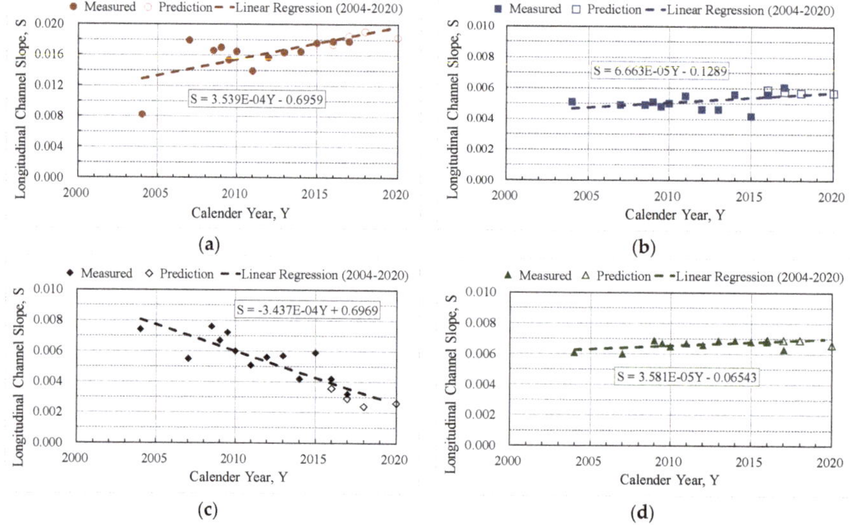

Figure 10. Measured and predicted longitudinal channel slopes for the four sub-reaches with medium-scale: (**a**) The head-cutting sub-reach: XS117-XS115; (**b**) the soft bedrock sub-reach: XS115-XS108; (**c**) the transition sub-reach: XS108-XS105; and (**d**) the alluvial sub-reach: XS105-XS90.

6. Discussion and Conclusions

A new approach is developed to assess the longitudinal channel slope of a mixed alluvial-soft-bedrock stream. The approach adopts both the historical field data and the 2D numerical model, and is applied the procedure to a 14 km reach downstream of the Ji-Ji Weir on the Chuo-Shui River, Taiwan.

6.1. Discussion

Three spatial scales are used to assess the temporal variation of channel slope in the study site. From the large spatial scale shown, it is difficult to characterize the mixed alluvial-bedrock streams, which may be adequate for purely alluvial rivers. The local spatial scale shows the widely fluctuating slope in the soft-bedrock incision reach, reflecting the compound environmental forcing and complex riverbed setting. With the medium spatial scale, the channel slopes follow a predictive trend if the reach is partitioned into four distinctive sub-reaches. The above observation and simulation results show that the temporal variation of the channel slope turns into a predictable trend when using the medium spatial scale, which could be primarily used in formulating river regulations and river stabilization projects in mixed alluvial-bedrock stretches.

For most alluvial rivers with uniform bed materials, it is easy to obtain the reach-averaged hydraulic properties, which do not change at different spatial scales. For the mixed alluvial-bedrock reaches, channel slopes are discontinuous and bed material compositions are non-uniform. Channel slopes and the evolution are different at different scales, which often complicates the planning and design process for river stabilization projects.

The transient stable longitudinal channel slope and evolution of the study reach is schematized in Figure 11, based on the above results. Four sub-reaches may be identified and represented in the figure by the "rigid bars" with components of "roller", "pin", and "hinge". The use of the rigid bar and connection points between sub-reaches follow the terminology in the structural engineering. Each rigid bar may move independently according to the fluvial behavior of their bed material characteristics. The "roller" may move freely in both vertical and horizontal direction, the "hinge" can move only in the vertical or horizontal direction, and the "pin" has fixed elevation and longitudinal position. This figure shows clearly that the longitudinal channel profile is dominated by the fluvial process in the soft bedrock and alluvium sub-reaches. The slope of the two sub-reaches evolves independently. The slope of the transient sub-reach depends on the evolution results of the bedrock and alluvial sub-reaches; and the slope of the head-cutting sub-reach is controlled by the Ji-Ji Weir.

In the head-cutting sub-reach, channel slope will increase owing to the bedrock incision and knickpoint migration. The channel will act like a rigid bar: The upstream end is fixed at the Ji-Ji Weir while the bar rotates in the counter-clockwise direction. The length of the sub-reach will decrease due to the knickpoint migration upstream. Without proper protection measures, such as energy dissipation and grade control structures, the knickpoint will continue to move upstream and endanger the weir.

For the bedrock sub-reach, the channel incision will continue but the slope remains at a constant of 0.0058 which is smaller than the alluvial channel downstream. The two end points of the sub-reach are the two knickpoints, which will continue to migrate upstream. The sub-reach length will increase as the incision rate of this sub-reach is smaller than the head-cutting zone upstream, but higher than the alluvial zone downstream.

In the transition sub-reach, erosion and deposition processes will occur simultaneously in response to the upstream soft bedrock incision and the downstream alluvial sedimentation. A higher incision rate in the bedrock zone and lower degradation rate in the alluvium zone will decrease the slope in this sub-reach to about 0.0030. Due to the differences in the movement rates of the two boundaries, this sub-reach will act like a rigid bar rotating clockwise. The longitudinal length of the bar will decrease when the downstream aggregation rate is greater than the upstream erosion rate. The evolution of the alluvial sub-reach is relatively simple: It will remain at a constant slope of 0.0068 although the sub-reach will experience both degradation and aggregation. The rigid bar representing the alluvial sub-reach may move both upward and downward. Movement of the upstream boundary may alter the length of the alluvial channel: Aggregation extends the channel into the transition sub-reach, but degradation does the opposite.

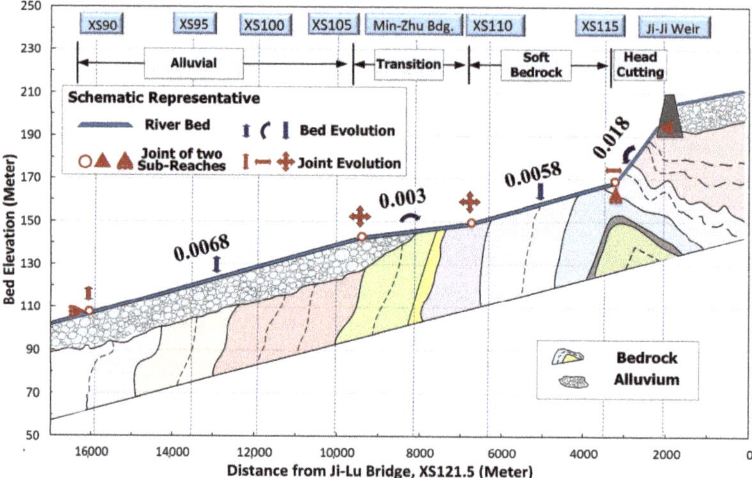

Figure 11. Schematic sketch of the transient stable channel slope evolution for the four sub-reaches of the study site.

6.2. Conclusions

In this study, a procedure was developed and demonstrated. First, the channel slope was assessed numerically at three spatial scales: The large, the medium, and the local scale. The large scale analysis may be adequate for purely alluvial streams, but is difficult to use to characterize mixed alluvial-bedrock streams. The local scale analysis showed that the soft-bedrock incision has a widely fluctuating slope, reflecting that the soft-bedrock incision is influenced by both unbalanced sediment transportation and knickpoint migration. With the medium scale analysis, the channel slope evolution follows a predictable trend and was the most appropriate for characterizing the channel evolution processes of the study site.

The medium scale analysis showed that the study site may be divided into four sub-reaches: The head-cutting, the soft bedrock, the transition, and the alluvial sub-reaches. Each sub-reach follows a predictable slope evolution trend. The numerical model results were shown to agree with the field data well. It was found that: (1) A 2D mobile-bed model may be useful in assessing the evolution of the mixed alluvial-bedrock stream; in particular, the evolution of the soft bedrock channel slope is predicted well in comparison with the available data; the numerical model provides future morphological trends that are unavailable from the field data alone. (2) The channel slope of the study reach maintains a transient stable slope of 0.0068 before the weir operation and co-seismic fault movement. (3) After river channel incision into soft bedrock, the channel slope of the head-cutting sub-reach increases in response to the manmade cross-stream structure; soft-bedrock channel develops an average slope of 0.0058 while the downstream alluvial channel evolves to an average slope of 0.0068. The transition sub-reach is found to have its channel slope decreased.

From the engineering viewpoint, it is important to predict river channel adjustments to different control or regulation schemes in order to avoid certain potential problems [32]. However, most river management in dynamic rivers are mostly based on design guidelines derived from alluvial rivers. The lack of knowledge of complex interacting geomorphological processes in the mixed alluvial-soft-bedrock reach leads to continuous maintenance for river stability. Such a study for these kinds of dynamic rivers is unique and has not been analyzed in depth. The study results can be used to select the proper locations and types of engineering stabilizing structures. Additionally, the proposed approach could also be applied to other stretches with the same geological conditions and river setting.

Author Contributions: Conceptualization, K.W.; methodology, K.W. and K.-C.Y.; software, Y.G.L.; validation, K.W.; investigation, K.W.; resources, K.-C.Y., and Y.G.L.; writing—original draft preparation, K.W.; writing—review and editing, K.W., K.-C.Y., and Y.G.L.

Funding: This research received no external funding.

Acknowledgments: The writers would like to acknowledge that the Water Resources Agency, Ministry of Economic Affairs, Taiwan has supported this study. Colleagues from Water Resources Planning Institute, Central Region Water Resources Office and Fourth River Management Office, WRA, have provided valuable field data and technical assistant. Peer review provided by Chung-Ta Liao at the National Chiao-Tung University is greatly appreciated.

Conflicts of Interest: The authors declare no conflict of interest.

References

1. Lee, Y.H.; Lu, S.T.; Shih, T.S.; Hsieh, M.L.; Wu, W.Y. Structures associated with the northern end of the 1999 Chi-Chi earthquake rupture, Central Taiwan: Implications for seismic-hazard assessment. *Bull. Seismol. Soc. Am.* **2005**, *95*, 471–485. [CrossRef]
2. Yanites, B.J.; Tucker, G.E.; Mueller, K.J.; Chen, Y.G. How rivers react to large earthquakes: Evidence from central Taiwan. *Geology* **2010**, *38*, 639–642. [CrossRef]
3. Huang, M.W.; Pan, Y.W.; Liao, J.J. A case of rapid rock riverbed incision in a coseismic uplift reach and its implications. *Geomorphology* **2013**, *184*, 98–110. [CrossRef]
4. Leopold, L.B.; Maddock, T., Jr. The hydraulic geometry of stream channels and some physiographic implications, U.S. *Geol. Surv. Prof. Pap.* **1953**, *252*.
5. Wohl, E.E. Bedrock channel morphology in relation to erosional processes, in rivers over rock: Fluvial processes in bedrock channels. *Geophys. Monogr. Ser.* **1998**, *107*, 133–152.
6. Montgomery, D.R.; Gran, K.B. Downstream variations in the width of bedrock channels. *Water Resour. Res.* **2001**, *37*, 1841–1846. [CrossRef]
7. Whipple, K.X. Bedrock rivers and the geomorphology of active orogens. *Annu. Rev. Earth Planet. Sci.* **2004**, *32*, 151–185. [CrossRef]
8. Howard, A.D. Long profile development of bedrock channels: Interaction of weathering, mass wasting, bed erosion, and sediment transport, in Rivers Over Rock: Fluvial Processes in Bedrock Channels. *Geophys. Monogr. Ser.* **1998**, *107*, 297–319.
9. Heritage, G.; Entwistle, N.; Milan, D.; Tooth, S. Quantifying and contextualising cyclone-driven, extreme flood magnitudes in bedrock-influenced dryland rivers. *Adv. Water Resour.* **2019**, *123*, 145–159. [CrossRef]
10. Lai, Y.G.; Greimann, B.P.; Wu, K. Soft bedrock erosion modeling with a two-dimensional depth-averaged model. *J. Hydraul. Eng.* **2011**, *137*, 804–814. [CrossRef]
11. Water Resources Agency (WRA). *Study on Riverbed Stability Engineering Scheme of Chou-Shi River*; Project Rep.; Water Resource Planning Institute, WRA, MOEA, Taiwan (ROC): Taichung, Taiwan, 2015.
12. Water Resources Agency (WRA). *Dynamic Stabile Channel Study for a Severe Soft Bedrock Erosion Reach of Cho-Shui River*; Project Report; Water Resource Planning Institute, WRA, MOEA, Taiwan (ROC): Taichung, Taiwan, 2017.
13. Bieniawski, Z.T. *Engineering Rock Mass Classifications: A Complete Manual for Engineers and Geologists in Mining, Civil and Petroleum Engineering*; Wiley: New York, NY, USA, 1989.
14. Kanji, M.A. Critical issues in soft rocks. *J. Rock Mech. Geotech. Eng.* **2014**, *6*, 186–195. [CrossRef]
15. Wu, K.W.; Lai, Y.G.; Yeh, K.C.; Liao, C.T. Coupled geo-fluvial channel evolution and bedrock erosion modeling on a river in Taiwan. In Proceedings of the World Environmental and Water Resources Congress 2015, Austin, TX, USA, 17–21 May 2015; pp. 1724–1735.
16. Sklar, L.S.; Dietrich, W.E. River longitudinal profiles and bedrock incision models: Stream power and the influence of sediment supply, in rivers over Rock: Fluvial processes in bedrock channels. *Geophys. Monogr. Ser.* **1998**, *107*, 237–260.

17. Sklar, L.S.; Dietrich, W.E. A mechanistic model for river incision into bedrock by saltating bed load. *Water Resour. Res.* **2004**, *40*, W06301. [CrossRef]
18. Chen, S.C.; An, S.P. The framework and application of river morphology five level classification. *J. Chin. Soil Water Conserv.* **2012**, *43*, 21–40. (In Chinese)
19. Water Resources Agency (WRA). *The River Regulation Master Plan of Cho-Shui River*; Project Rep.; Water Resource Planning Institute, WRA, MOEA, Taiwan (ROC): Taichung, Taiwan, 2007. (In Chinese)
20. Water Resources Agency (WRA). *The Study for Bedload Monitoring and Sediment Transport of Chuo-Shui River*; Project Rep.; Water Resource Planning Institute, WRA, MOEA, Taiwan (ROC): Taichung, Taiwan, 2012. (In Chinese)
21. Su, C.C.; Lu, J.Y. Comparison of sediment load and riverbed scour during floods for gravel-bed and sand-bed reaches of intermittent rivers: Case study. *J. Hydraul. Eng.* **2016**, *142*, 05016001. [CrossRef]
22. Wu, K.W.; Yeh, K.C.; Liao, C.T.; Lai, Y.G. Bedrock channel morphological modeling on the river in Taiwan. In Proceedings of the 13th International Symposium (ISRS) 2016 on River Sedimentation, Stuttgart, Germany, 19–22 September 2016.
23. Lai, Y.G. *SRH-2D Theory and User's Manual Version 2*; Technical Service Center, Bureau of Reclamation: Denver, CO, USA, 2008.
24. Lai, Y.G. Two-dimensional depth-averaged flow modeling with an unstructured hybrid mesh. *J. Hydraul. Eng.* **2010**, *136*, 12–23. [CrossRef]
25. Lai, Y.G.; Thomas, R.; Ozeren, Y.; Simon, A.; Greimann, B.P.; Wu, K. Modeling of multi-layer cohesive bank erosion with a coupled bank stability and mobile-bed model. *Geomorphology* **2015**, *243*, 116–129. [CrossRef]
26. Greimann, B.P.; Lai, Y.G.; Huang, J. Two-dimensional total sediment load model equations. *J. Hydraul. Eng.* **2008**, *134*, 1142–1146. [CrossRef]
27. Schoklitsh, A. Der Geschiebetrieb und die Geschiebefacht. *Wasserkraft Wasserwirtsch.* **1934**, *29*, 37–43.
28. Meyer-Peter, E.; Muller, R. *Formula for Bed-Load Transport*; International Association for Hydraulic Structure: Stockholm, Sweden, 1948.
29. Leopold, L.B.; Wolman, M.G.; Miller, J.P. *Fluvial Processes in Geomorphology*; W.H. Freeman: London, UK, 1964.
30. Parker, G.; Wilcock, P.R.; Paola, C.; Dietrich, W.E.; Pitlick, J. Physical basis for quasi-universal relations describing bankfull hydraulic geometry of single-thread gravel bed rivers. *J. Geophys. Res.* **2007**, *112*, F04005. [CrossRef]
31. Stark, C.P. A self-regulating model of bedrock river channel geometry. *Geophys. Res. Lett.* **2006**, *33*, L04402. [CrossRef]
32. Chang, H.H. *Fluvial Processes in River Engineering*; John Wiley & Sons: New York, NY, USA, 1988.

Article

Analysis of Shear Stress and Stream Power Spatial Distributions for Detection of Operational Problems in the Stare Miasto Reservoir

Joanna Wicher-Dysarz

Department of Hydraulic and Sanitary Engineering, Poznan University of Life Sciences, ul. Wojska Polskiego 28, 60-637 Poznan, Poland; joanna.wicher@up.poznan.pl; Tel.: +48 61 848 7730

Received: 8 March 2019; Accepted: 1 April 2019; Published: 4 April 2019

Abstract: In this paper an analysis of the lowland reservoir operation in atypical conditions is presented. The chosen study object is the Stare Miasto reservoir in the Powa river (Poland), which has been in operation since 2006. It is a two-stage reservoir, consisting of an upper sedimentation part and a lower main reservoir. The upper part is separated from the main part by an internal dam with a sluice gate. Such a construction enabled better control of sediment deposits and their removal. The atypical conditions were caused by flood wave propagation in the Powa river and the reservoir in 2014. In the research, three reservoir bathymetries are analyzed—from 2006, 2013, and 2018. Two-dimensional (2D) hydrodynamic modeling is applied to analyze spatial variability of investigated hydraulic parameters. Such an approach enabled better recognition of the changes observed in the reservoir during 2006–2018. In the research, the spatial distributions of the velocities, the shear stresses, and the stream power are the basis for the analyses and comparisons. The simulations enabled identification of the main elements prone to collapse during flood wave propagation. The presented results and approach may be applied for improvement of reservoir design, with special emphasis on specific structures located in a reservoir basin.

Keywords: 2D model flow; two-stage reservoir; shear stress; stream power; operational problem

---

1. Introduction

This paper presents the problem of reservoir operation in the specific conditions of lowland areas. The paper focuses on special type of reservoir, whose construction has become rather popular. These are two-stage reservoirs, consisting of a preliminary sedimentation zone and a main zone. Only the main zone stores water for basic purposes of reservoir operation. The popularity of the two-stage reservoir is related to the fact that its construction reflects the nature of the reservoir sedimentation process. This feature simplifies removal of sediments without raising additional problems in the reservoir operation. Because in the past such solutions were not very popular and have become quite popular recently, the functioning of two-stage reservoirs is not understood at a satisfactory level.

The water resources in Poland compared to other European countries are relatively limited. Hence, it is important to build water reservoirs improving water balance, environmental water conditions, as well as microclimate. Typical reservoirs may be used for different purposes including drinking water supply, industrial water supply, and flood protection. Building reservoirs is economically profitable due to the production of electricity in water power plants. Reservoirs also may play an important role in tourism or as elements of the inland waterway.

The Wielkopolska (Great Poland) district is the region with the lowest rainfalls level in Poland. This is the reason why small retention is intensively developed in this area. There are 38 reservoirs classified as small retention objects. According to the Polish regulations, these are the reservoirs with storage smaller than 5 hm^3. In Wielkopolska, there are also two large reservoirs with storage exceeding

5 hm^3 [1]. The number of reservoirs indicates the significance of the discussed problem. Additionally, it is worth noting the economic importance of the reservoirs.

Operation of the water reservoir may cause several problems. In general, processes which are difficult to control occur in the reservoir and around it. Two problems are common for small and large reservoirs. These are, (1) erosion downstream of the reservoir dam, and (2) sediment deposition in the backward part of the inflowing river [2–4]. These processes occur relatively fast and cause a direct threat of reservoir catastrophe [3,5]. Although the processes occurring in the entire reservoir volume are also very important, they are slower and less visible in general [4–7]. The sediment transport in equilibrium conditions, occurring in the inflowing river, changes when water flows into the reservoir. The sediment transport potential rapidly decreases and the stream is not able to transport the same amount of sediments. Suspended and bed-loaded grains are deposited, creating alluvial fans and deltas in the inlet part of the reservoir. The density currents influence the water circulation in the reservoir. Finally, the fine sediments deposited in the deeper parts of the reservoir reduce the available volume and change the operational conditions in the longer time horizon [8–12]. In the inlet part of the reservoir, the deposition causes a decrease in the local depths and expansion of the riparian vegetation [6,13]. In two-stage reservoirs, these processes occur in the preliminary sedimentation zone. Water pollutants are also inhibited there, which improves water quality in the main part of the reservoir. However, pollutants are frequently deposited in the sediments (e.g., heavy metals) [14–17].

The object analyzed is the Stare Miasto reservoir, located in the Powa river in central Poland. This is a lowland reservoir with specific construction of the bottom. There is a preliminary zone separated from the rest of the reservoir by an internal dam. The flow between the upper and lower parts of the reservoir is limited. The upper part is usually smaller and plays the role of the initial sediment container, whereas the lower part is greater and it is designed to provide water for the main purpose of the reservoir operation. Hence, the lower part is usually called the main reservoir. There are basic reservoir capacities (i.e., active conservation storage, flood storage capacity, flood surcharge capacity, as well as dead storage). The described solution is one of the methods applied for prevention of sedimentation and water quality degradation in reservoirs [11,18,19]. Such an approach is not prohibitively expensive, though there are also some drawbacks. From the economic point of view, separation of the upper part means the loss of its conservation or flood storage. For this reason, broad research on such reservoirs, describing their operational conditions, seems worthwhile.

Taking into account the simplicity of the described concept, it is expected that such reservoirs will be built more frequently. In literature such objects have been rarely analyzed. Two other problems are discussed: (a) water and sediment quality [15–17,19,20]; (b) reservoir sedimentation [4,5,11,21–23]. These two processes make the described reservoir construction so attractive. Today a number of two-stage reservoirs in Poland can be found, for example, the Poraj reservoir in the Warta river, the Rydzyny reservoir in the Sama river, Kowalskie Lake in the Główna river, or the new concept of the Wielowieś Klasztorna reservoir in the Prosna river.

The basic feature of lowland reservoirs is very good vertical mixing of heat and dissolved substances. It means the lack of temperature stratification and uniform distribution of solutes along the depth. Due to the small depths, density currents caused by temperature and salinity gradients should not be present in such reservoirs. As reported by Krenkel et al. [24], the processes of heat and dissolved mass exchange in reservoirs without stratification vary in horizontal dimensions as in rivers.

Measurements and field surveys provide limited information on the spatial as well as temporal scale. The sampling techniques provide point information. Detailed description of the spatial distributions of investigated parameters (e.g., velocities, temperature, etc.) requires a huge number of such measurements and, obviously, it is time-consuming and expensive. A similar problem arises when the temporal variability of the selected parameters is formulated. The duration of the analyzed processes is important. The description of their dynamics requires continuous measurements or very frequent momentary measurements. Hence, the application of mathematical modeling in the analysis may help. Different types of models for description of river and reservoir dynamics have been applied for many years. One-dimensional (1D) unsteady flow models are the simplest models applied for reservoirs [25,26]. However, models of this kind are able to reconstruct properly only the variability of the depth in the reservoir. Other hydraulic parameters (e.g., velocities, shear stress, etc.) are determined approximately. Modeling of other processes in 1D mode for reservoirs is burdened with significant inaccuracies. Because of this, 2D and 3D modeling is more frequently used for analysis of reservoirs. In the lowland reservoir, 2D models are preferred due to relatively small depths [25,27–30]. Such an approach enables determination of quite accurate spatial distributions of basic hydraulic variables (e.g., depth and velocity) and parameters dependent on their values (e.g., shear stress, stream power, and others) [29,30]. More accurate reconstruction of the hydraulic variables range is also the basis for modeling of other processes (e.g., transport of pollutants or sediment transport). The field measurements impact the construction of the model during the procedure of identification and verification. The final product is a model giving results (e.g., depth, velocities, etc.), consistent with observations in the field. It may be applied for test computations and forecasting. Such a model may support analysis of spatial and temporal variability of the investigated variables.

The purpose of the present research is to analyze two-stage reservoir operation, with its hydraulic and operational problems. The presented approach is based on the 2D simulations of a hydrodynamic model made for a flood scenario selected from historical data. The flood wave observed between 10 April and 19 July of 2014 is used. The attention is focused on three variables identified as the measures of potential hazards: (1) magnitude of flow velocity, (2) shear stress, and (3) stream power. The role of the flow velocity in morphodynamical changes in rivers and reservoirs is well known. Such processes as sediment deposition and erosion are observed in locations of significant velocity changes (e.g., reservoir inlets, contractions in channels, etc.). The shear stress describes these processes better with the so-called incipient motion criteria. In such objects as reservoirs, the shear stresses have specific distributions, and they may be concentrated near hydraulic obstacles (e.g., dams and bridge piers). The stream power is a combination of velocity and shear stress, with potential to link the features of both variables. Hence the stream power is a basis of many sediment transport formulae applied for modeling of sediment deposition and erosion in rivers and reservoirs. The spatial distributions of these variables are confronted with the known locations of threats (e.g., internal dam almost crashed in 2014). The main assumption is that the careful analysis of such spatial distributions may help to prevent undesirable threats in the stage of reservoir design. The analyses are made for three bathymetries reconstructed for the years 2006, 2013, and 2018. The comparison of velocity, shear stress, and stream power distributions in these three moments of time could help to understand better the nature of processes responsible for morphodynamic changes in the reservoir.

2. Case Study System

Taking into account the purpose of the research, quite a specific object, the Stare Miasto reservoir in the Powa river, was chosen as a case study. The Powa is a moderately sized river flowing in the lowland of Great Poland province in the central part of Poland. The Stare Miasto reservoir was built as an element of the small retention program [31]. The purposes of this program are the improvement of watershed capacity, flood and drought protection, and prevention of a decrease in the groundwater table. The analyzed reservoir has been in operation since 2006. It is divided into two parts (Figure 1), with an internal dam located at km 12 + 000 of the Powa river. There is a regulated sluice applied to control flow through the dam. The upper zone works as a sediment trap.

The inundation area of the upper part is 27 ha and its storage equals 0.294 hm^3. The lower part is the so-called main reservoir. This part is additionally split into two internal parts by highway A2 (Figure 1), running from Poznań to Warsaw [32]. The watershed area of the Stare Miasto reservoir in the cross-section of the main dam is 299.7 km^2. The average depth of the reservoir is 2.4 m. The estimated length of this object is 4.5 km. The inundation area for the minimum headwater level (MinPP = 92.0 m a.s.l.) is 75.77 ha. For the normal headwater level (NPP = 93.5 m a.s.l.) the inundation area is 90.68 ha. The maximum headwater level (MaxPP) is 94.0 m a.s.l. The total reservoir capacity is

2.159 hm^3. The active conservation storage is 1.216 hm^3. The inundation area of the upper zone is 13.62% of the total water surface area. The land use in the neighborhood of the reservoir agriculture. Because the usefulness of the terrain is limited, crop production has been stopped in this region [32,33]. The reservoir is working in an annual compensation cycle, which may cause annual variation in the water level, from MinPP to NPP. The operational water level is from 92.00 m a.s.l. to 94.00 m a.s.l. The reservoir is filled in March and the water surface is kept at the level of NPP until the end of October. After October, the main reservoir is emptied to the level of MinPP. To protect the inlet part from degradation and sediment accumulation, the water surface in the upper part is kept at the NPP level for the entire year.

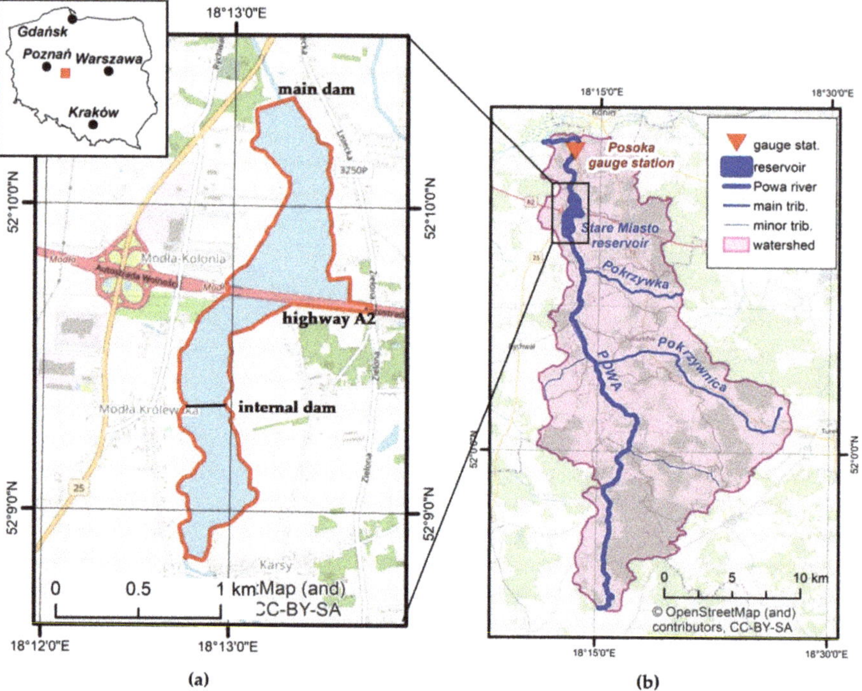

Figure 1. Chosen case study—Stare Miasto reservoir on the Powa river: (**a**) The reservoir and its main elements—internal dam, highway bridge, and main dam; (**b**) the watershed with main elements—river, reservoirs, and gauge station.

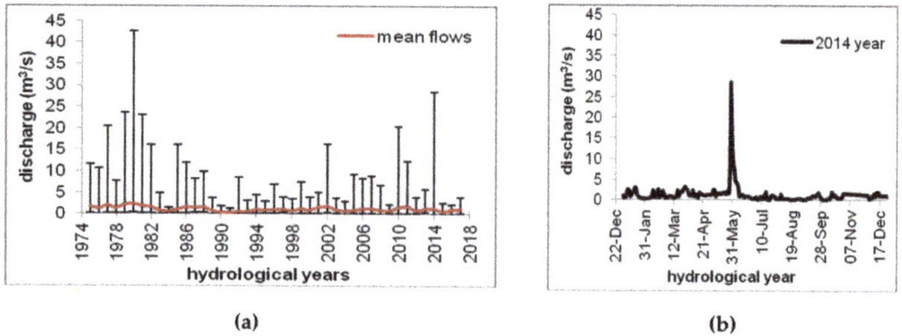

Figure 2. Variability of discharge at the Posoka gauge station, (**a**) for the period 1971–2017; (**b**) for the year 2014.

In the Powa river there is one gauge station called Posoka (Figure 1b). This is located downstream of the reservoir at km 3 + 900 of the Powa course. The watershed area in the cross-section of the gauge is 332 km^2. The characteristic flows were determined on the basis of recorded observations at this gauge station during the period 1975–2017 [34]. The results are presented in Figure 2. The total minimum observed was 0.006 m^3/s, while the total maximum was 42.6 m^3/s. The mean flow was about 1.15 m^3/s. During the time of reservoir operation, from 2006 to 2017, the maximum flow of 28.5 m^3/s (Figure 2b) occurred in 2014. The minimum and mean flows did not differ much from those estimated for the entire period. In the analyzed period, the average annual outflow from the reservoir was 36.9 hm^3.

Operational Problems in the Reservoir

Operational problems have been present in the Stare Miasto reservoir since its construction. The first of them is caused by the internal dam splitting the reservoir into two parts. As mentioned above, the headwater level should be constant over the entire year (Figure 3a–c). The flow through the internal dam from the preliminary sedimentation zone to the main part of the reservoir is regulated by the sluice installed in the internal dam (Figure 4b,c). Due to the internal dam instability and the too small span of the sluice, equal to 3 m, the sluice was totally opened almost all the year. The problems with the internal dam increased in 2014, when huge flows in the Powa river occurred (Figure 4b) after heavy rainfalls in the upper watershed, which caused local flooding.

Figure 3. Stare Miasto reservoir in the Powa river: (**a**) The internal dam in 2011; (**b**) the internal dam destroyed after the flood wave propagation in 2014; (**c**) the uncovered reservoir bottom near the internal dam (author: J. Wicher-Dysarz).

High water stages and a fast-flowing flood wave reached the reservoir and met the internal dam. Due to the huge force acting on the dam during the flood wave propagation in 2014, the structure was damaged (Figure 3b).

Figure 4. Stare Miasto in the Powa river: (**a**) The bridge with highway A2; (**b**) the sluice in the internal dam in 2011; (**c**) the sluice in the internal dam in 2014 (author J. Wicher-Dysarz).

In 2014, it was decided that the internal dam would be reconstructed and the sediments would be removed from the upper part of the reservoir. The construction works were carried out in 2015. In the upper part the layer of sediments from the reservoir bottom was removed. The depth of the layer was 40 cm. The area of removal covered about 2 ha and the estimated volume of the removed sediments was 8000 m^3. The soil removed was distributed more or less uniformly along the banks. The sluice gate was reconstructed in such a way that inverted elevation was decreased.

3. Materials and Methods

3.1. Available Data

In the research four types of data were applied: (1) Hydrologic data for the Posoka gauge station; (2) the measurements of reservoir bathymetry made in 2006, 2013, and 2018; (3) a digital elevation model (DEM); and (4) sediment samples collected in 2011 and 2018.

The hydrological data for the Posoka gauge station were obtained from the Institute of Meteorology and Water Management (IMGW—Polish: Instytut Meteorologii i Gospodarki Wodnej). Although the available daily data was for the period 1971–2017 (Figure 2), the single flood wave observed between 10 April and 19 July of 2014 was used for simulations. The DEM applied in this research was obtained from the Head Office of Geodesy and Cartography (GUGiK—Polish Główny Urząd Geodezji i Kartografii). Its spatial resolution was 1 × 1 m. The vertical accuracy was 15 cm.

Examples of the data applied to create a computer model of bathymetries are presented in Figure 5. Figure 5a shows the reservoir shape and the location of the area for presentation of details. This area was chosen as a part of the reservoir near the internal dam. The bathymetry reflecting the year 2006 was reconstructed on the basis of topographic maps used for the reservoir design purposes (Figure 5b). The digital elevation model was built on the basis of elevations and isolines presented in the maps. In 2013, the bathymetry of the reservoir was measured in cross-sections with two devices: (1) GPS; and (2) an Echotrac CVM depth sounder. The total number of cross-sections was 98 and the distances between them varied from 50 to 100 m. Examples of these data is shown in Figure 5c. The last bathymetry was measured in 2018 as a part of the ISOK project (Polish: ISOK = Informatyczny System Osłony Kraju przed nadzwyczajnymi zagrożeniami, English: IT system of the Country's Protection Against Extreme Hazards). The second part of the project started in 2017 and the data from this part were applied for the presented research. The total number of cross-sections was 14 and the distances between them were in the range 200–300 m. Measurement points locations are shown in Figure 5d.

The last set of data applied in the research included sediment samples measured in 2011 and 2018. In 2011, 36 sediment samples were collected. In 2018 the measurements were repeated and 30 samples were taken from the reservoir bottom. The granulation based on 66 samples measured in 2011 and 2018 was determined as a combination of sieve and aerometer analysis, according to Polish norm PN-R-04032:1998 "Soils and minerals—samples and analysis of granulation." The sieve analysis was performed in wet conditions with a normalized set of sieves. The aerometric analysis was performed with a set of aerometric devices produced by Eijkelkamp. According to the above-mentioned Polish norm, the soil fractions taken into account were as follows: sand (2–0.05 mm), silt (0.05–0.002 mm), and clay (0.002 mm).

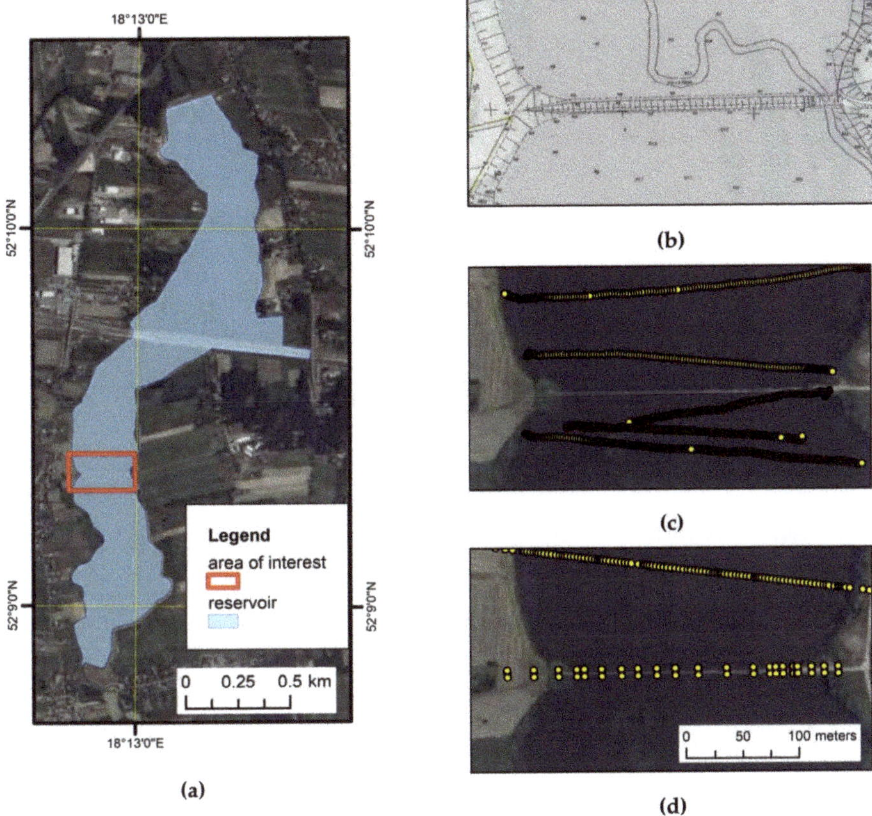

Figure 5. Examples of data used for the reconstruction of the reservoir bathymetries: (**a**) Location of the selected area; (**b**) design map from 2006; (**c**) measurement points from 2013; (**d**) measurement points from 2018.

3.2. Applied Methods

The ArcGIS 10.5.1 software developed by Esri Inc. was applied which is described in e.g., Docan [35]. It enables quite easy processing of GIS data, such as vector and raster layers. An integral part of ArcGIS is the ArcToolbox. It is a module including the main external tools and methods. Some of them are concurrent to the methods available in the basic ArcGIS interface, while others extend the capabilities of the standard interface. Extension of ArcGIS is also possible with specific plug-ins installed as ArcGIS toolbars. One such plug-in is HEC-GeoRAS [36], which is a toolbar with methods designed to support preparation of the river flow model. The tools available in HEC-GeoRAS may also be useful in preparation of a 2D model for simulation of flow in a reservoir.

The second plug-in applied is RiverBox, which is a set of geo-processing tools developed in Python by Dysarz [37]. Although the software was designed primarily for reconstruction of a river bed, after preliminary data processing it has been also successfully applied for this reservoir. The basic idea of interpolation in the channel-oriented coordinates is shown in Figure 6a. The bottom elevations are reconstructed from measurement cross-sections (green lines) through linear interpolation in two directions: the longitudinal and the transversal. The plug-in includes a number of algorithms for measurement data processing and three algorithms for reconstruction of the bed. These are described in more detail in the quoted publication [37]. The scheme of the third algorithm applied here is presented in Figure 6b. The main idea was to create a number of random points over the interpolated area, channel-oriented interpolation of the elevations in the points, and finally application of "typical" spatial interpolation. In the third algorithm of the RiverBox, the linear interpolation in the two dimensions was applied for this purpose. It wa implemented by adoption of two tools from ArcToollbox: creation of TIN (Triangular Irregular Network) and transformation of the TIN into standard raster. The choice of the third algorithm was determined by its effectiveness analyzed in Dysarz [37].

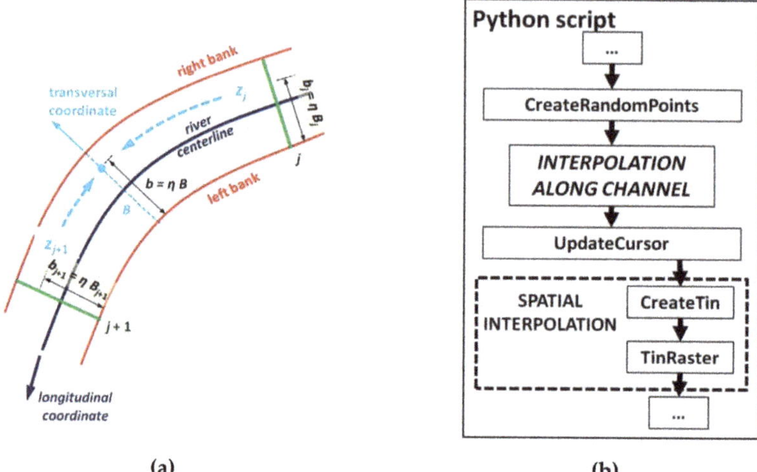

Figure 6. Main ideas of the bathymetry reconstruction with RiverBox [37]: (**a**) Scheme of interpolation; (**b**) applied algorithm.

Three bathymetries of the analyzed reservoir were reconstructed from available data, they represent the reservoir bed from 2006, 2013, and 2018. The first was composed from detailed design maps of the reservoir, with application of geo-referencing and digitization. The next two were reconstructed with the help of the RiverBox plug-in for ArcGIS Desktop software.

The HEC-RAS is a well-known hydrodynamic model for rivers and water reservoirs. This program was designed at the Hydrologic Engineering Center (HEC). The acronym RAS means River Analysis System, which well defines the application area. The concepts applied in the package are described by [38]. The main features of HEC-RAS are 1D simulations of flow and transport processes in river networks, including floodplains and reservoirs, as well as 2D simulations of pure flow. The HEC-RAS package also includes several useful tools for data preparation and results processing. These tools include the module for GIS data processing called RAS Mapper [39].

The 2D flow module was applied in this research. The basic model for such simulations implemented in HEC-RAS is the full dynamic wave [38] presented below

$$\frac{\partial H}{\partial t} + \frac{\partial (hu)}{\partial x} + \frac{\partial (hv)}{\partial y} = 0 \tag{1}$$

$$\frac{\partial u}{\partial t} + u\frac{\partial u}{\partial x} + v\frac{\partial u}{\partial y} = -g\frac{\partial H}{\partial x} + \vartheta_t\left(\frac{\partial^2 u}{\partial x^2} + \frac{\partial^2 u}{\partial y^2}\right) - c_f u \tag{2}$$

$$\frac{\partial v}{\partial t} + u\frac{\partial v}{\partial x} + v\frac{\partial v}{\partial y} = -g\frac{\partial H}{\partial v} + \vartheta_t\left(\frac{\partial^2 v}{\partial x^2} + \frac{\partial^2 v}{\partial y^2}\right) - c_f v \tag{3}$$

where the Equation (1) describes mass balance and the next two, Equations (2) and (3), are momentum balance equations. The independent variables are t—time and x, y—spatial dimensions. There are three dependent variables, namely h—the water depth, and u, v—the depth-averaged components of flow velocity. The water surface elevation H depends on the depth and g is acceleration of gravity. There are also two coefficients of more complex nature. The first is kinematic eddy viscosity coefficient ϑ_t defined as follows

$$\vartheta_t = Dhu_* \tag{4}$$

where D is non-dimensional empirical constant describing mixing intensity and u_* is well known shear velocity. The second coefficient is bottom friction coefficient c_f derived from Chezy or Manning formulae as follows

$$c_f = \frac{g|V|}{C^2R} \quad c_f = \frac{n^2g|V|}{R^{4/3}} \tag{5}$$

where R is hydraulic radius and $|V|$ is magnitude of velocity vector. There are also velocity coefficient for Chezy equation C and roughness coefficient for Manning equation n.

The interpretation of mass balance in Equation (1) is relatively simple. In arbitrary control volume, the inflow and outflow fluxes described by the second and third terms of the equation should be in balance with the increase or decrease in the water volume stored, which is expressed in terms of water surface elevation changes in the first term. The left sides of Equations (2) and (3) describe the change in the momentum stored in the control volume, including the momentum stored locally and the momentum fluxes. The first terms of the right sides represent the pressure and gravity forces, while the second is responsible for internal eddy stresses and mixing. The last terms of the right sides describe the friction force generated on the contact surface of the liquid and bottom.

In HEC-RAS, the simulation may be performed with a simplified version of the model (1)–(3). If the left side of momentum equations and the eddy viscosity terms are neglected, the model (1)–(3) becomes the so-called diffusive wave model [38].

$$\frac{\partial H}{\partial t} - \frac{\partial}{\partial x}\left\{\beta\frac{\partial H}{\partial x}\right\} - \frac{\partial}{\partial y}\left\{\beta\frac{\partial H}{\partial y}\right\} = 0 \tag{6}$$

In the above equation β plays a role of diffusion coefficient and it is defined as

$$\beta = \frac{R^{5/3}}{n|\nabla H|^{1/2}} \tag{7}$$

where $|\nabla H|$ is the magnitude of water surface gradient or, in other words, the water surface slope. The model (6) with (7) is the default choice due to stability and efficiency. The simulations presented here were performed with the diffusive wave model.

The equations are approximated with a hybrid scheme based on a combination of finite difference and finite volume methods. Such an approach is applied due to the fact that the computations mesh is composed of rectangular elements inside the flow domain and multi-edge irregular elements near the boundaries [38]. The applied boundary conditions are a flow hydrograph in the reservoir inlet and a constant stage hydrograph in the reservoir outlet.

The obtained results were transformed into the cell-oriented shear stress τ and stream power SP values. These two variables are defined as follows

$$\tau = \rho g R S_f \quad SP = \tau|V| \tag{8}$$

where S_f is energy grade line slope (friction slope).

For the purposes of the present research, three models were developed. All of them were 2D models of the entire reservoir. Elements such as the bridge, internal dam, and main dam were reconstructed by modification of elevations in a digital terrain model representing the reservoir bottom. The basis for the 2D simulations was the numerical mesh covering the entire reservoir area, approximately equaling 1 km^2. The imposed cell size of rectangular cells was 2 × 2 m, but the cells near the boundaries may have different shapes. Hence, the minimum cell area was 1.74 m^2 and maximum was 7.87 m^2, when the total average was 4.01 m^2. The total number of cells covering the reservoir was 241,699. Three bottoms were tested, which were the reconstructions of the real reservoir for 2006, 2013, and 2018. The reconstruction process is described above. In all cases the flood wave observed in 2014 was simulated with constant head water kept in the reservoir outlet. The elevation of the head water simulated was 93.50 m a.s.l. This is the so-called normal head water level in the Stare Miasto reservoir. The simulation time step was chosen due to the stability requirements and it equaled 30 s. Because the semi-implicit scheme was used for time discretization, the weighting factor was set to 0.75.

Table 1. The dependence between absolute roughness and characteristics of sediment grain sizes.

Methods	Absolute Roughness (k_s)
Nikuradse	$k_s = 2.3d_{80}$
Einstein	$k_s = d_{65}$
Engelung-Hansen	$k_s = 2d_{65}$
van Rijn	$k_s = 3d_{90}$
Kamphuis	$k_s = 2d_{90}$

Besides the geometry reconstruction and flow boundary conditions, the important elements determining the flow process in the analyzed area were roughness coefficients. A simplified approach was used for this purpose. The sediment samples collected from the reservoir bottom were the basis for determination of the equivalent roughness coefficient for the entire reservoir bottom. Two step calculations were applied. In the first step, the dimensionless resistance coefficient λ from the Darcy–Weisbach formula was determined for full turbulent flow conditions:

$$\lambda = \left[-2\log\left(\frac{k_s}{14.84\ R_h}\right)\right] \tag{9}$$

where R_h is the hydraulic radius and k_s is the absolute roughness. The second coefficient was defined theoretically as the average height of bottom irregularities. In practice its value was calculated on the basis of sediment characteristics, but a number of formulae have been proposed in literature for this purpose. The chosen and applied approaches are presented in Table 1 [40]. The symbol d_x in this table means specific diameter of the grain sample determined on the basis of the sieve curve. The next step was calculation of Manning's roughness coefficient n from the formula:

$$n = R_h^{\frac{1}{6}}\sqrt{\frac{\lambda}{8g}}\ \left[\mathrm{sm}^{-1/3}\right] \tag{10}$$

In Figure 7a, the box-and-whiskers plot, representing the relationship between the five methods listed in Table 1 and the roughness coefficients, is presented. The sediment characteristics were determined on the basis of the 66 sieve curves obtained from sediment samples, presented in Figure 7b, in which dots represent the mean values. The boxes show the values of the mean ± standard deviation. The whiskers denote the values between mean ± 1.96 standard deviations, which define the 95% confidence level. The plot presented in Figure 7a summarizes 330 calculations of roughness coefficients. The obtained average value was 0.038 sm$^{-1/3}$, while the minimum and maximum values were 0.026 and 0.059 sm$^{-1/3}$, respectively.

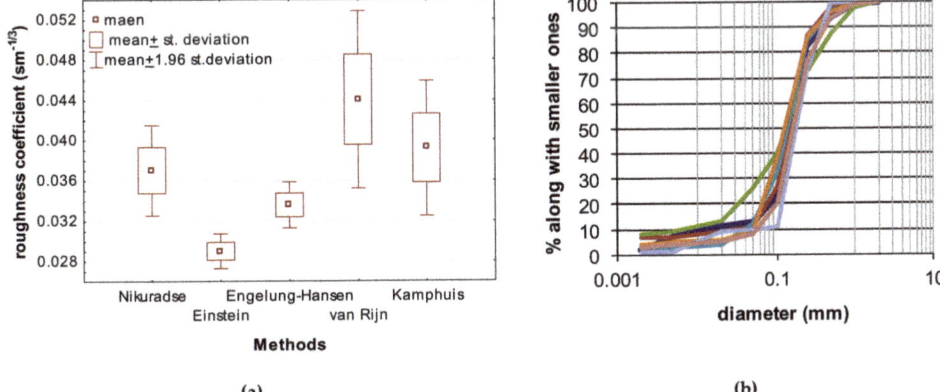

Figure 7. Estimation of roughness coefficients: (**a**) Statistics of roughness coefficients calculated on the basis of approaches presented in Table 1; (**b**) sieve curves applied for the calculation of roughness coefficients.

4. Results and Discussion

On the basis of 2D simulations, the depths, flow velocities, shear stresses, and stream power were determined for the entire reservoir with each of the three different bathymetries. Due to the huge amount of data, the results were kept in three groups: maps, graphs of selected cross-sections, and graphs with statistical analyses. In Figure 8, the locations of the main elements applied to analyze the results is shown. The reservoir was divided into five parts as presented in Figure 8a. They were, (1) the main reservoir 1, (2) the main reservoir 2, (3) pre-reservoir, (4) area under bridge, and (5) the internal dam. The four cross-sections were located in characteristic places of the greatest parts. There was 241,699 computational cells in the numerical grid. Because this number of cells was difficult for processing, the sample points (Figure 8b) were used for monitoring the obtained results. The total number of sample points was 1670, which was about 0.67% of the available locations of computed values. Although this number was much smaller than the number of computational cells, the sample points covered the entire reservoir with sufficient density. Because the reservoir parts differed in size, their distribution over the parts was not uniform. In the main reservoir 1, the main reservoir 2, and pre-reservoir there were 500 points, while 20 points were generated in the area under the bridge and 150 points generated over the internal dam.

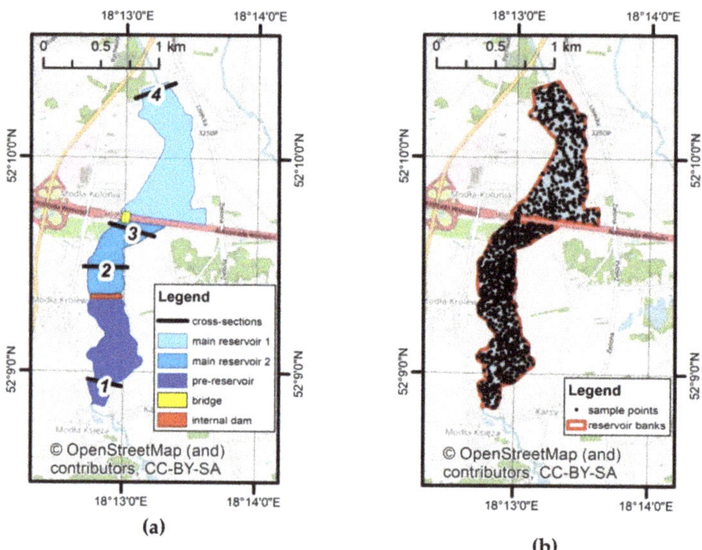

Figure 8. Elements used in the analyses: (**a**) Parts of the reservoir and location of cross-sections; (**b**) location of sample points.

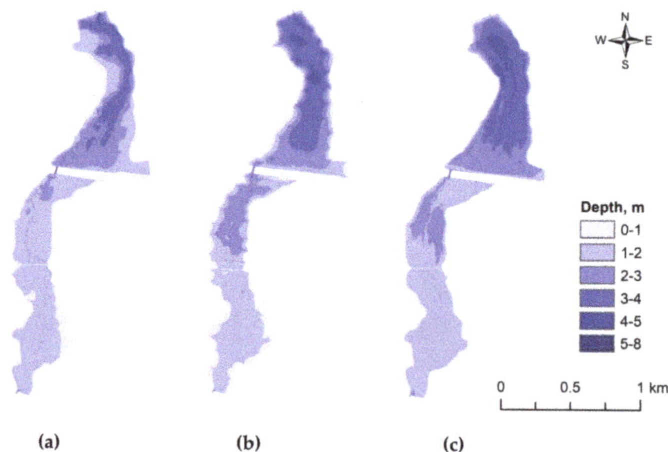

Figure 9. Analysis of depth distributions over the reservoir bottom in the Stare Miasto reservoir: (**a**) Bathymetry 2006; (**b**) bathymetry 2013; (**c**) bathymetry 2018.

The maps presented in Figure 9, Figure 11, and Figure 12 show the spatial distribution of depth, velocity, and shear stress over the three reservoir bathymetries, namely 2006, 2013, and 2018. As may be observed, the reservoir bottom changed during 12 years. Figure 9 presents the maps of maximum depths obtained as the results of simulations. The following maps (Figure 9a–c) show the results for bathymetries of 2006, 2013, and 2018, respectively. The comparison of bathymetries is made in Figure 10—part A (Figure 10(a1–a4)), where the geometry of cross-sections marked in Figure 8 is shown. In part B of Figure 10 (Figure 10(b1–b4)), the differences of the bottom elevations related to the initial bathymetry 2006 were analyzed. The blue line denotes increase in the elevations, from the initial bathymetry to the bathymetry measurement in 2013. The red lines represent differences between measurements of 2018 and the initial data.

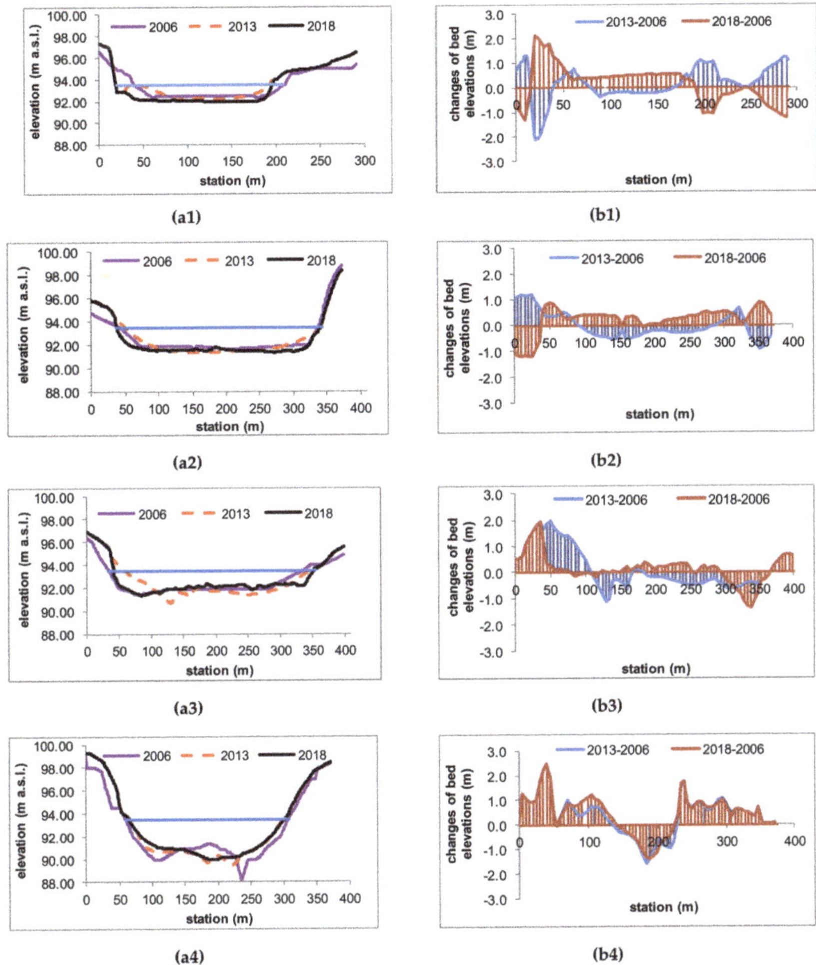

Figure 10. Selected cross-sections: (**a**) Bottom and water surface elevations for three bathymetries; (**b**) difference in bottom elevations between 2006 and 2013, as well as between 2006 and 2018.

As it can be seen, during the life of the reservoir, since 2006, the reservoir bottom was considerably transformed. Figure 10, graph b1 shows the cross-section located in the inlet of the Stare Miasto reservoir. The differences in the bottom elevations varied from 2 m of accumulated sediments to 2 m of removed sediments. The comparison with the bathymetry of 2006 shows that the greatest differences occurred in 2018 as a result of natural processes, as well as artificial removal of sediments in the end of 2014 and the beginning of 2015. At this time, a 0.4-m layer of sediments was removed from the bottom of the upper part. Figure 10(b2,b3) show similar bottom increases for the bathymetries of 2018 and 2006. In the last cross-section located near the main dam, the sediments were accumulated during 12 years of reservoir life. The only place where erosion occurred was located near the valve tower and spillway.

The graphs presented in Figure 11 illustrate the spatial distribution of maximum magnitudes of velocity in the reservoir. According to the maps, the maximum velocities occurred near structures such as the internal dam, the sluice in the internal dam, the flow below the bridge, and the reservoir outflow. There were also greater velocities in the reservoir inlet, though the area is so small that it was not visible well in the scale of Figure 11. The internal dam separates the preliminary part of the reservoir from its main part. The single and very small sluice caused local increase in the velocity magnitude and erosion of the alluvial bed. The maximum velocities over 3 m/s occurred also below the bridge. The reason was the too narrow bridge span.

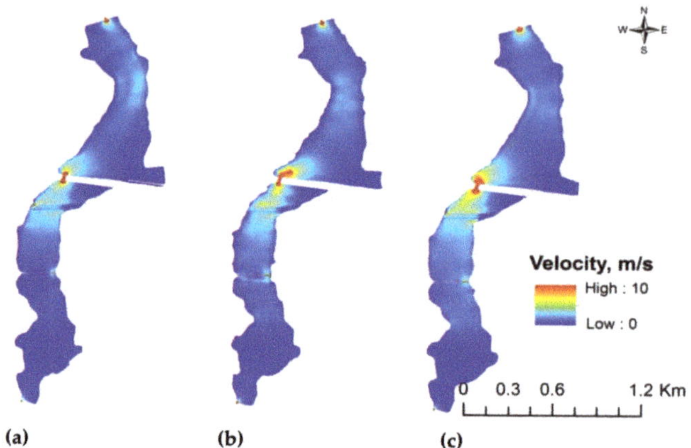

Figure 11. Spatial distributions of velocities in the Stare Miasto reservoir for three bathymetries: (**a**) Bathymetry 2006; (**b**) bathymetry 2013; (**c**) bathymetry 2018.

The results presented ealier were confirmed by the results shown in Figure 12a–c, illustrating that the distribution of the shear stresses in the reservoir was similar to that of velocity. The maximum values occurred at the reservoir inlet, near the internal dam, and in the bridge span. Additionally, the reservoir inlet was more prone to shear stress increase. The maximum shear stresses occurred for the bathymetry of 2013 (Figure 12b). In the bathymetry of 2018, the corresponding values were lower, especially, in the pre-reservoir. Supposedly, the artificial change in the bottom configuration as a result of dredging in 2014/2015 was the reason for such a change. Additionally, the very specific cross-sections marked in Figure 12 are cross-sections near the inlet, in the internal dam, and under the bridge. They were applied for analyses presented in Figure 13.

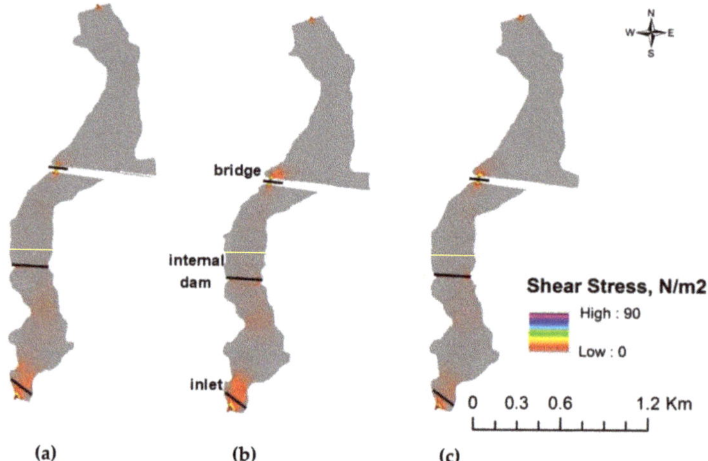

Figure 12. Shear stress calculated for the Stare Miasto reservoir and three bathymetries: (**a**) Bathymetry 2006; (**b**) bathymetry 2013; (**c**) bathymetry 2018.

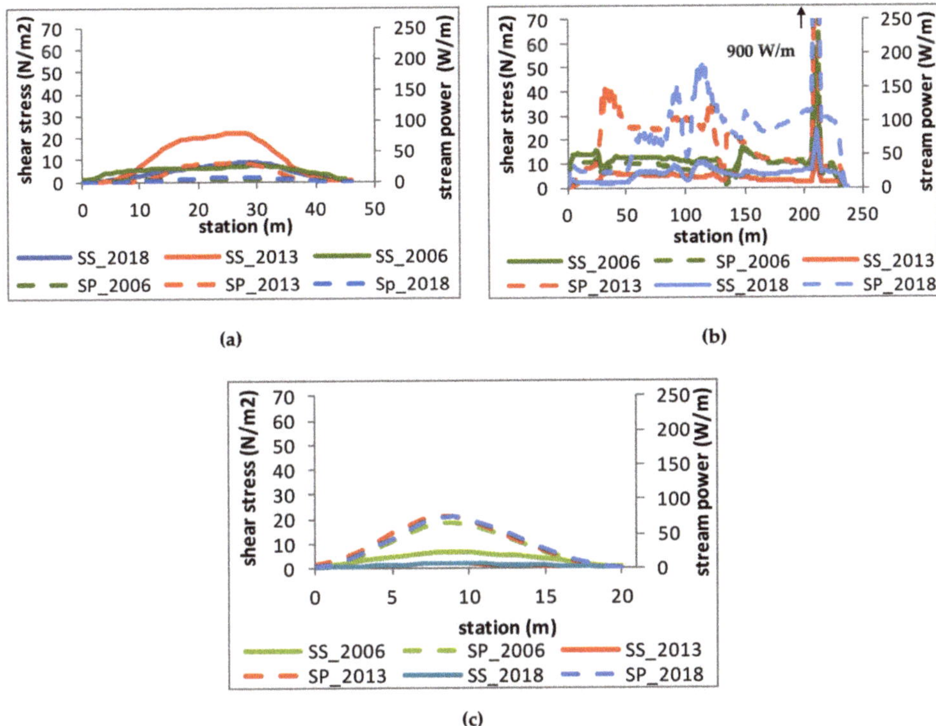

Figure 13. Shear stress and stream power in chosen cross-section: (**a**) inlet; (**b**) internal dam; (**c**) bridge.

The next variable analyzed was the stream power [41–43], but it is shown only in the graphs created with one of two methods: (1) as cross-sections with results extracted from results in raster format (Figures 8a and 12) (2) as statistics of values in the random sample points (Figure 8b). In Figure 13a–c, the graphs of shear stresses (SS) and stream power (SP) in selected cross-sections are shown for comparison. The cross-sections applied were those shown in Figure 12. In Figure 13a, the inlet cross-section is presented. The shear stresses (SS) simulated there were over 20 N/m^2, while the stream power (SP) values reached about 100 W/m. The highest values of shear stress as well as stream power were observed for the reservoir in 2013. Supposedly, the accumulation of sediments in the period 2006–2013 changed the configuration of the bottom in such a way that there was an increase in these two factors. The values of shear stress and stream power were lower for the conditions in 2018. The removal of sediments in the pre-reservoir in 2014/2015 induced another kind of change. The next specific cross-section is shown in Figure 13b, at the internal dam with its sluice gate. The shear stresses as well as stream power in the sluice were increasing rapidly, independently of the bottom configuration tested. However, the sluice gate is made of concrete and it should be resistant to such forces. The huge values of the analyzed variable are more dangerous for the rest of the internal dam. The maximum values of shear stress were about 40 N/m^2 there, while the stream power reached about 200 W/m. The values of these two factors were relatively small for the conditions in 2006, but in other configurations of the bottom, 2013 and 2018, high shear stresses and stream power were noticed. These values were quite high, especially in the lowland reservoirs where fine sand is deposited with grain sizes smaller than 0.5 mm. It means that, irrespective of the dredging and rebuild of the sluice gate, the internal dam is still exposed to huge forces and is prone to break. In the cross-section of the bridge, the shear stresses were smaller, about 7 N/m^2 and seem to be stable. The same was noticed about the stream power values obtained there. The hydraulic conditions under the bridge seemed to be less dependent on the bottom configuration.

The next stage of the research was statistical analysis performed with Statistica 13. The values of maximum magnitudes of velocity, maximum shear stresses, and maximum stream power were read in the locations of the randomly generated points shown in Figure 8b, and then processed over the set of points and subsets included in particular reservoir parts (Figure 8a). Such results are presented as standard box-and-whisker plots in Figure 14a–c. The graph in Figure 14a shows the maximum flow velocities for different parts of the reservoir indicated in Figure 8a. The total medians were 0.32, 0.41, and 0.44 for bathymetries 2006, 2013, and 2018, respectively, while the mean values were 0.53, 0.67, 0.73 m/s. The significant differences between median and mean values for all bathymetries were caused by greater velocities calculated for the bridge and the internal dam. In the first case, the medians were 4.69, 4.81, and 6.38 m/s for the same bathymetries as previously. In the internal dam these factors reach values 2.33, 3.65, and 4.11 m/s. Such great irregularities indicate the importance of these two locations. The greatest values for all bathymetries were noticed under the bridge and over the internal dam. The velocities simulated with the initial bathymetry were slightly smaller than the others. The similarity of the internal dam median velocities for 2013 and 2018 shows that the dredging in 2014/2015 did not improve the safety of this dam much. Fortunately, the maximum velocities decreased from about 10 m/s in 2013 to 6–7 m/s in 2018. In the other parts of the reservoir, the velocities obtained were much smaller irrespective of the bathymetry tested.

The trends observed in the spatial distributions of shear stresses (Figure 14b) and stream power (Figure 14c) were different. Definitely the highest values of these two variables were observed in the internal dam. The discrepancy of the velocity magnitudes with shear stress and stream power observed under the bridge may be explained by relatively great narrowing of the flow area but small hydraulic slope governed by the head in the main dam. The internal dam influenced from the upstream and downstream was in different conditions. Hence, the shear stress and stream power confirm the observations of velocities.

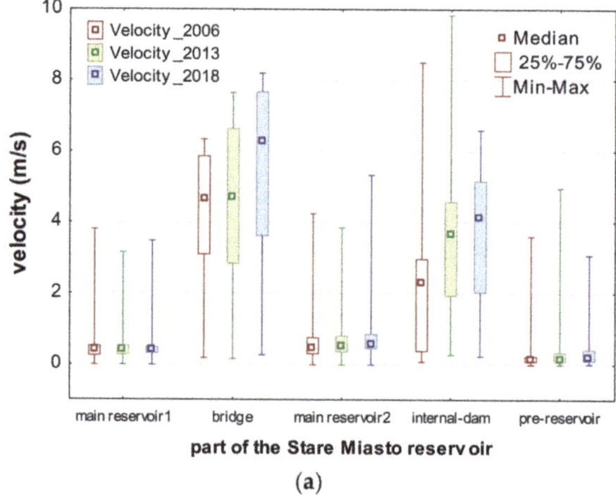

(a)

Figure 14. *Cont.*

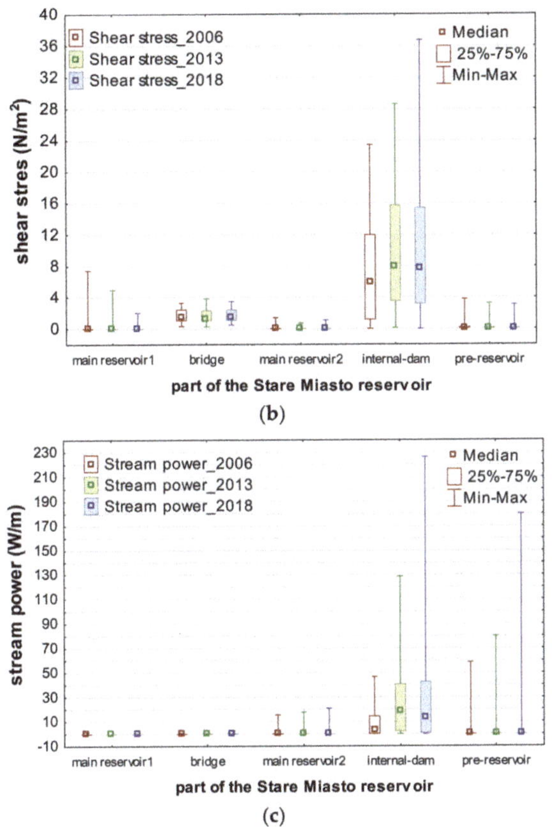

Figure 14. Characteristic values of analyzed hydraulic parameters simulated for bathymetries 2006, 2013, and 2018, and recorded at random points: (**a**) Velocity; (**b**) shear stress; (**c**) stream power.

It is also important that there is a difference in trend between the median and maximum values. The median values of the shear stress as well as stream power more or less decreased between 2013 and 2018. The simulation over the bottom for 2013 provided median shear stress equal to 7.95 N/m^2 and median stream power of 27.97 W/m over the internal dam. The median results obtained for bathymetry 2018 were 7.77 N/m^2 and 19.27 W/m, respectively. Hence, the decreases were 2.2% and 31.1% for shear stress and stream power, respectively. The same results indicate significant increases of the maximum values in the case of these two factors. The maximum shear stress and stream power values for bottom for 2013 were 28.63 N/m2 and 653.78 W/m, respectively. The results for the bathymetry 2018 gave maximum values of the same variables equaling 36.75 N/m^2 and 815.12 W/m. Hence, the increase of shear stress was 28.4% and the increase of the stream power was 24.7%. It confirms the previous conclusion that the rebuild of the internal dam and sluice gate did not improve the safety of this object.

The analysis of bed changes and shear stress or stream power in reservoirs is relatively difficult for modeling. In many papers there is a description of the shear stresses acting on the river bed in mountainous as well as lowland conditions [30,44,45]. The majority of such concepts is based on the simulation in 1D or calculated with empirical formulae [30,45]. In results provided by Glock et al. [30], who simulated flow in Austrian rivers and a laboratory flume, shear stresses varied from 0.4 to 93 N/m^2. The course of the channel, meandering or straight, influenced the obtained values greatly. The results were also dependent on the type of model applied. They compared results of 1D, 2D, and 3D models. It is obvious that shear stresses are impacted by the roughness of the channel bed and flow velocities. In the Stare Miasto reservoir maximum velocities up to 10 m/s (Figure 14a) generated huge shear stresses, reaching values of 90 N/m^2 (Figure 13b). Such magnitudes were noted in the internal dam and near the sluice installed there. Huge velocities and shear stresses caused huge values of stream power. In the sluice it was 900 W/m (Figure 13b). During the flood wave propagation in the reservoir, the most critical location was the reservoir inlet. The velocities simulated there equaled 5 m/s (Figure 11), which was related to huge shear stresses reaching values of 25 N/m^2 (Figure 13a).

5. Conclusions

The impact of hydraulic parameters, such as velocity, shear stress, and stream power, on the operational conditions in a two-stage reservoir was analyzed. The basis for the investigations was simulations assuming the 2D hydrodynamic model. Such an approach enabled better illustration of the processes and changes in the reservoir for different moments of time. The method applied was different from the 1D models usually applied, as it enabled better recognition of the advantages and drawbacks of two-stage constructions. The shear stresses were calculated for the entire reservoir basin and especially for the structures located in it, and visualization revealed significant differences in response to impacts such as flood wave propagation. The simulations were made for the bathymetries of 2006, 2013, and 2018. The conclusions drawn from the results indicated really great stresses exerted on the internal dam, which may lead to breakage of the dam structure. In general, the analyses proved that structures in the reservoir are prone to profound stresses. The magnitudes of shear stresses in specific flood conditions are not observed frequently in lowland rivers and reservoirs. The contraction caused by the bridge span is also very dangerous. In flooding conditions, it may cause great acceleration of water flow. The highway bridge is a massive construction and no direct threat related to great stress has been detected in this case. However, it is not possible to exclude the probability of erosion below the bridge and collapse of the construction due to bridge scour in future. One thing is obvious—the velocities, the shear stress, and stream power values are not typical of lowland conditions if the flood propagates through the reservoir. The presented analyses confirm that two-stage reservoirs are a good solution of some reservoir problems, if the inflows are moderate. However, the increase in inflows up to flood magnitudes may generate new problems related to the stability of the structures located in the reservoir basin.

The applied 2D hydrodynamic modeling enabled more detailed analysis focused on spatial distributions of important factors: the velocity, the shear stress, and the stream power. The obtained results are satisfactory and they may be applied in several ways. The approach presented enabled identification of potential threats, which is well shown on the example of the internal dam. However, such an approach required greater computational costs than in other cases—more simulation time, faster processors, more effective models, more memory on disks, etc. Although, the use of three different bathymetries and comparisons between them increased the complexity of the problem, it also enabled simplified verification of the object vulnerability to flood hazard. As indicated, the rebuild of the internal dam did not improve the safety of this structure and future breakage could be expected. Some application areas of the results presented seem to be obvious—improvement of reservoir design, optimization of structure shapes, invention of protective elements reducing detected threats. Others are not so obvious (e.g., operational control of reservoir to safely propagate the flood wave). However, such an approach requires greater computational power than that applied in this research.

As can be seen, the two-stage reservoirs are quite an interesting alternative approach to prevention of reservoir sedimentation. However, their construction should be carefully analyzed and computationally tested before the main dam is built. The results of the present research could help in the design of crucial elements, such as the sluice in the internal dam, the span of the bridge, etc. All such elements could cause unexpected effects dangerous for the stability of the reservoir construction and safety of the water management in the reservoir.

Author Contributions: Conceptualization; methodology; software; validation; formal analysis; investigation; resources; data curation; writing—original draft preparation; writing—review and editing; visualization; supervision; project administration; and funding acquisition, J.W.-D.

Funding: This research received no external funding.

Acknowledgments: This research would not be possible without the access to the data provided by such institutions as IMGW, CODGiK, and GUGiK. The access to the free data available in the Internet, such as Geoportal 2, was also very helpful.

Conflicts of Interest: The author declares no conflict of interest.

References

1. RZGW. National Board for Water Management. Available online: https://warszawa.rzgw.gov.pl/the-regional-water-management-authority-in-warsaw2018 (accessed on 5 March 2019).
2. Schleiss, A.J.; Franca, M.J.; Juez, C.; De Cesare, G. Reservoir sedimentation. *J. Hydraul. Res.* **2016**, *54*, 595–614. [CrossRef]
3. Dysarz, T.; Wicher-Dysarz, J.; Przedwojski, B. Man-induced morphological processes in Warta river and their impact on the evolution of hydrological conditions. In *River Flow*; Ferreira, R.M.L., Alves, E.C.T.L., Leal, J.G.A.B., Cardosa, A.H., Eds.; Taylor & Francis Groups: London, UK, 2006; pp. 1301–1310.
4. Wicher-Dysarz, J.; Dysarz, T. Uncertainty in assessment of annual inflow of sediments to the Stare Miasto reservoir based on empirical formulae. *Acta Sci. Pol. Form. Circumiectus* **2015**, *14*, 201–212. [CrossRef]
5. Sojka, M.; Jaskuła, J.; Wicher-Dysarz, J.; Dysarz, T. Poznań Univeristy of Life Sciences Analysis of Selected Reservoirs Functioning in the Wielkopolska Region. *Acta Sci. Pol. Form. Circumiectus* **2017**, *4*, 205–215. [CrossRef]
6. Szoszkiewicz, K.; Wicher-Dysarz, J.; Sojka, M.; Dysarz, T. Assessment of hydraulic, hydrological and physicochemical factors affecting vegetation development in dam reservoir with separated inlet zone—Stare Miasto (central Poland) reservoir as a case study. *Fresenius Environ. Bull.* **2016**, *25*, 2772–2783.
7. Ptak, M.; Sojka, M.; Choiński, A.; Nowak, B. Effect of environmental conditions and morphometric parameters on surface water temperature in Polish lakes. *Water* **2018**, *10*, 580. [CrossRef]
8. Michalec, B. Appraisal of methods of determination of sediment quantity supplied to a small water reservoir. *Arch. Environ. Prot.* **2008**, *34*, 49–62.
9. Michalec, B.; Tarnawski, M. The influence of small water reservoir operational changes on capacity reduction. *Environ. Prot. Eng.* **2008**, *34*, 117–124.
10. Michalec, B. Evaluation of an empirical reservoir shape function to define sediment distributions in small reservoirs. *Water* **2015**, *7*, 4409–4426. [CrossRef]
11. Dysarz, T.; Wicher-Dysarz, J. Analysis of flow conditions in the Stare Miasto Reservoir taking into account sediment settling properties. *Annu. Set Environ. Prot.* **2013**, *15*, 584–605.
12. Dong, N.; Yang, M.; Meng, X.; Liu, X.; Wang, Z.; Wang, H.; Yang, C. CMADS-driven simulation and analysis of reservoir impacts on the streamflow with a simple statistical approach. *Water* **2019**, *11*, 178. [CrossRef]
13. Jaskuła, J.; Sojka, M.; Wicher-Dysarz, J. Analysis of the vegetation process in a two-stage reservoir on the basis of satellite imagery—A case study: Radzyny Reservoir on the Sama River. *Annu. Set Environ. Prot.* **2018**, *20*, 203–220.
14. Sojka, M.; Jaskuła, J.; Siepak, M. Heavy metals in bottom sediments of reservoirs in the lowland area of western Poland: Concentrations, distribution, sources and ecological risk. *Water* **2019**, *11*, 56. [CrossRef]
15. Sojka, M.; Siepak, M.; Jaskuła, J.; Wicher-Dysarz, J. Heavy metal transport in a river-reservoir system: A case study from central Poland. *Pol. J. Environ. Stud.* **2018**, *27*, 1725–1734. [CrossRef]
16. Sojka, M.; Jaskuła, J.; Wróżynski, R.; Waligórski, B. Application of sentinel-2 satellite imagery to assessment of spatio-temporal changes in the reservoir overgrowth process—A case study: Przebedowo, West Poland. *Carpathian J. Earth Environ. Sci.* **2019**, *14*, 39–50. [CrossRef]
17. Sojka, M.; Siepak, M.; Gnojska, E. Assessment of heavy metal concentration in bottom sediments of Stare Miasto pre-dam reservoir on the Powa River. *Annu. Set Environ. Prot.* **2013**, *15*, 1916–1928.
18. Wicher-Dysarz, J.; Kanclerz, J. Functioning of small lowland reservoirs with pre-dam zone on the example of Kowalskie and Stare Miasto Lakes. *Rocznik Ochrona Srodowiska* **2012**, *14*, 885–897.
19. Kasperek, R.; Wiatkowski, M. Research on the Mściwojów reservoir. *Sci. Rev. Eng. Environ. Sci.* **2008**, *40*, 194–201. (In Polish)
20. Paul, L. Nutrient elimination in pre-dams: Results of long term studies. *Hydrobiologia* **2003**, *504*, 289–295. [CrossRef]

21. Pikul, K.; Mokwa, M. Influence of pre-dams on the main reservoir silting process. *Sci. Rev. Eng. Environ. Sci.* **2008**, *17*, 185–193. (In Polish)
22. Paul, L.; Pütz, K. Suspended matter elimination in a pre-dam with discharge dependent storage level regulation. *Limnol. Ecol. Manag. Inland Waters* **2008**, *38*, 388–399. [CrossRef]
23. Dysarz, T.; Wicher-Dysarz, J.; Sojka, M. Two approaches to forecasting of sedimentation in the Stare Miasto reservoir, Poland. In *Reservoir Sedimentation*; Taylor & Francis Group: London, UK, 2014; ISBN 978-1-138-02675-9.
24. Krenkel, P.A.; Novotny, V. *Water Quality Management*; Academic Press: Cambridge, MA, USA, 1980.
25. Szymkiewicz, R. *Numerical Modeling in Open Channel Hydraulics*; Springer: Berlin, Germany, 2010.
26. Chow, V.T. *Open-Channel Hydraulic*; McGraw-Hill: New York, NY, USA, 1959.
27. Wu, W. *Computational River Dynamics*; CRC Press: Boca Raton, FL, USA, 2007.
28. Radecki-Pawlik, A.; Pagliara, S.; Hradecky, J. *Basics of Open Channel Hydraulics, River Training and Fluvial Geomorphology*; CRC Press: Boca Raton, FL, USA, 2016.
29. Juárez, A.; Adeva-Bustos, A.; Alfredsen, K.; Dønnum, B.O. Performance of a two-dimensional hydraulic model for the evaluation of stranding areas and characterization of rapid fluctuations in hydropeaking rivers. *Water* **2019**, *11*, 201. [CrossRef]
30. Glock, K.; Tritthart, M.; Habersack, H.; Hauer, C. Comparison of hydrodynamics simulated by 1D, 2D and 3D models focusing on bed shear stresses. *Water* **2019**, *11*, 226. [CrossRef]
31. Marshal Office of the Wielkopolska Region in Poznan. Small Retention Program. Available online: https://umww.pl/o-programie-malej-retencji (accessed on 2018). (In Polish)
32. Woliński, J.; Zgrabczyński, J. *The Stare Miasto Reservoir in the Powa River: Water Management Rules*; BIPROWODMEL Sp. z o.o.: Poznań, Poland, 2008. (In Polish)
33. Preibisz, J.; Matan, J.; Woliński, J. *Studium zabudowy retencyjnej rzeki Powy, Dokumentacja*; BIPROWODMEL Sp. z o.o.: Poznań, Poland, 2000. (In Polish)
34. IMGW. Public Data IMGW-PIB. Available online: https://dane.imgw.pl/ (accessed on 5 March 2019).
35. Docan, D.C. *Learning ArcGIS for Desktop*; Packt Publishing: Birmingham, UK, 2016; Available online: https://www.packtpub.com/application-development/learning-arcgis-desktop (accessed on 7 July 2018).
36. Cameron, T.; Ackerman, P.E. *HEC-GeoRAS GIS Tools for Support of HEC-RAS using ArcGIS User's Manual*; US Army Corps of Engineers, Institute for Water Resources, Hydrologic Engineering Center (HEC): Davis, CA, USA, 2012. Available online: http://www.hec.usace.army.mil/software/hec-georas/documentation/HEC-GeoRAS_43_Users_Manual.pdf (accessed on 7 July 2018).
37. Dysarz, T. Development of RiverBox—An ArcGIS toolbox for river bathymetry reconstruction. *Water* **2018**, *10*, 1266. [CrossRef]
38. Brunner, G.W. *HEC-RAS River Analysis System Hydraulic Reference Manual; US Army Corps of Engineers*; Report No. CPD-69; Hydrologic Engineering Center (HEC): Davis, CA, USA, 2016.
39. Brunner, G.W. *HEC-RAS River Analysis System User's Manual Version 5.0; US Army Corps of Engineers*; Report No. CPD-68; Hydrologic Engineering Center (HEC): Davis, CA, USA, 2016.
40. Van Rijn, L.C. Sediment transport. Part I: Bed load transport. *J. Hydraul. Eng.* **1984**, *110*, 1431–1456. [CrossRef]
41. Bagnold, R. *An Approach to the Sediment Transport Problem from General Physics*; United States Government Printing Office: Washington, DC, USA, 1966.
42. Yang, C.T. Unit stream power equation for gravel. *J. Hydraul. Divis.* **1984**, *110*, 1783–1797. [CrossRef]
43. Yang, C.T. *Sediment Transport: Theory and Practice*; McGraw-Hill Companies, Inc.: New York, NY, USA, 1996.
44. Radecki-Pawlik, A.; Kuboń, P.; Radecki-Pawlik, B.; Plesiński, K. Bed-load transport in two different-sized mountain catchments: Mlynne and Lososina streams, Polish Carpathians. *Water* **2019**, *11*, 272. [CrossRef]
45. Radecki-Pawlik, A. Determination of nondimensionalization shear stress shields criterion used to calculate the initiationof motion of sediment using different formulae. *Acta Sci. Pol. Form. Circumiectus* **2015**, *14*, 75–107. (In Polish)

Article

Hydraulic Modeling and Evaluation Equations for the Incipient Motion of Sandbags for Levee Breach Closure Operations

Ahmed M. A. Sattar [1,2], Hossein Bonakdari [3,*], Bahram Gharabaghi [3,*] and Artur Radecki-Pawlik [4]

1. Irrigation and Hydraulics Department, Faculty of Engineering, Cairo University, Giza 12613, Egypt; ahmoudy77@yahoo.com
2. Civil department, German University in Cairo, New Cairo City 13611, Egypt
3. School of Engineering, University of Guelph, Guelph, ON N1G 2W1, Canada
4. Institute of Structural Mechanics, Faculty of Civil Engineering, Cracow University of Technology, 31-155 Krakow, Poland; rmradeck@cyf-kr.edu.pl

* Correspondence: hbonakda@uoguelph.ca (H.B.); bgharaba@uoguelph.ca (B.G.)

Received: 27 November 2018; Accepted: 1 February 2019; Published: 6 February 2019

Abstract: Open channel levees are used extensively in hydraulic and environmental engineering applications to protect the surrounding area from inundation. However, levees may fail to produce an unsteady flow that is inherently three dimensional. Such a failure may lead to a destructive change in morphology of the river channel and valley. To avoid such a situation arising, hydraulic laboratory modeling was performed on an open channel levee breach model capturing velocity, in x, y and z plans, at selected locations in the breach. Sandbags of various shapes and sizes are tested for incipient motion by the breach flow. We found that a prism sandbag has a better hydrodynamic characteristic and more stability than spherical bags with the same weight. Experimental results are then used to evaluate existing empirical equations and to develop more accurate equations for predicting critical flow velocity at the initial stage of sandbag motion. Results showed the superior predictions a few of the equations could be considered with an uncertainty range of ±10%. These equations explained the initial failed attempts of the United States Army Corps of Engineers (USACE) for breach closure of the case study, and confirmed the experimental results are simulating the case study of breach closure.

Keywords: breach closure; levee breach; channel side flow; sandbags; open channel flow; critical velocity; incipient motion

1. Introduction

Over seven million people have lost their lives in floods between 1990 and 2017 in Asia alone [1,2]. Predicting flood water levels, the design of stable alluvial channels, and protecting lives and properties during major flood events is one of the greatest challenges of the 21st century [3,4]. Open channel levees are used extensively in hydraulic and environmental engineering applications. They are constructed for the purpose of water constriction within the channel and act as an efficient structural method of floodwater control. However, breaches are very likely to occur in levees during high flood events and in locations where levees are not constructed properly. These breaches release flood waves that inundate the surrounding area, causing property damage, social interruptions and can threaten human lives. A severe example is the 2005 levee breach in New Orleans, LA [5]. In an open-channel levee breach, flow is diverted laterally in a hydraulic manner similar to side weir flow and dividing flow [6,7]. The flow in the main channel approaching the breach starts to divide as the flow enters the breach. A separation zone develops in the main channel after the breach, and a stagnation zone is formed near the downstream corner of the breach. At the breach outflow, a contracting flow region is observed with free over-fall conditions, similar to dam-break flow.

For side weir flow, most of the earlier studies were devoted to the evaluation of the discharge coefficient for various weir geometries, with few studies discussing hydraulic behaviors of flow at the weir exit [8–17]. However, for dividing flows in the open channel, more attention was given by earlier studies to the hydraulic characteristics of dividing flow (e.g., [18–25]). Research in levee breach hydraulics has fallen into two categories: (1) Studying the impact of breach flow on the hydraulics of the main upstream channel, and (2) studying the impact of breach flood waves propagation on downstream urban areas.

Within the first category, Yen [26] studied the impact of levee breach on water levels in the main channel upstream of the breach, and discussed various methods for estimating the channel capacity. Jaffe and Sanders [27] studied engineered levee breaches using a shallow water model for an optimal design to reduce flood stages. Apel et al. [28] investigated the effect of levee breaches on the flood frequency distribution in the main channel. In the second category, Sattar [29] used a 1:50 scale model to investigate flood waves generated by the 17th Street Canal breach in New Orleans following Hurricane Katrina. Using the same scale model, Sylvie et al. [30] compared measurements of urban flooding in the breach neighborhood with the 2D numerical model. LaRocque et al. [31] studied the flow pattern of the New Orleans flooded the neighborhood. The flow characteristics of a side channel breach were studied by Sattar [32]. Others have investigated the interaction of general flood waves with city-like blocks, either experimentally [33,34] or numerically [35–39].

Mitigating the adverse effects of breach floods on urban areas can be achieved using inundation plans, based on which evacuation plans will be devised, or by breach closure and stopping flood wave releases as early as possible. The most efficient way for closure of levee breaches is the utilization of large sand bags. These bags are dropped in flowing water to decrease flooding and eventually close the breach. Following the extreme events of Hurricane Katrina, levees protecting the city of New Orleans breached at several locations, with the 17th Street Canal levee breach being the biggest, threatening the whole Metropolitan Orleans East Bank. The United States Army Corps of Engineers [5] tried to drop large sandbags to close the breach and stop flood waves from inundating the east bank, but the sandbags were washed away with the breach strong flow. This failure to close the breach with sandbags led to the inundation of the entire Orleans east basin.

It is very important to know the hydrodynamic characteristics of the sand bags, especially critical velocity at incipient motion. Although many researchers have studied the subject of particle incipient motion for more than two centuries after Brahms in 1753, the effect of particle shape has not been examined, except in very few studies. The effect of particle shape on the threshold of motion has been investigated using particles with different shapes and sizes and constant density under subcritical, uniform flow conditions in a titling flume [40,41]. The effect of particle shape on the threshold of motion using constant specific weight for all shapes was studied by Gogus and Defne [41].

Gulcu [42] studied the effect of shape and size of individual particles on the initiation of motion over a smooth sloping channel bed when the particles are resting behind an obstruction of known height. In spite of such importance of sandbags for breach closures, relevant studies are still lacking, and hydrodynamic characteristics of sandbags in open channel flow are not reported in the literature [43]. An equation was proposed by Izbash [44] for incipient motion of rocks deposited in running water for the purpose of cofferdam construction. The stability of sandbags placed on non-uniform slopes has been studied by Kobayashi and Jacobs [45]. Additionally, Zhu et al. [43] investigated the hydrodynamic characteristics of sand-filled geosynthetic bags used for the construction of submerged dikes in rivers.

Neill et al. [46] conducted experiments using sandbags to protect the bank from erosion, using a previously developed formulation [5]. In Korkut et al. [47] study, Geobag were used to check the stability of the sandbags. To study the protection of the bridge abutment using sandbags, lab experiments were conducted [47], and the size of the bags and ripraps were determined based on methods in Pilarczyk [48] and Richardson and Davis [49]. El-Kholy and Chaudhry [50] studied the path of sandbags as they are washed away by flood waves and incipient motion for spherical bags. Juez et al. [51] carried out an extensive study on numerical, hydrodynamic, and morphological models of dam break flow over mobile beds. Their study showed that the interactions between dam break flow and mobile bed have a great impact on obtained results.

The experimental hydraulic laboratory study in this paper is intended to provide more understanding for the hydraulics of breach flows and incipient motion of sandbags used for closure. This study has four main objectives: (1) To provide 3D velocity measurements at selected locations in an open channel levee breach, (2) to investigate the effect of shape and size of sandbags on the threshold of motion, (3) to calibrate existing formulae and develop new ones for incipient motion of sandbags used for breach closure, and finally, (4) to apply developed equations to the case of the 17th Street Canal levee breach closure.

The aim of the study is to perform laboratory modelling for evaluation of the empirical equations for the critical velocity of incipient motion of sandbags for a levee breach closure. It is especially crucial in urbanized areas: In such places, unexpected fluvial processes, against which we use river engineering methods, destroy urban systems that are difficult to re-establish.

The results of this research may be used for further investigations to develop universal optimum open channel breach closure procedures for breaches in other locations and with both erodible and intact levees. In this paper, first a description of the experimental setup used to simulate an open channel levee breach is presented. Sandbags with various shapes and sizes are prepared and tested to measure their velocity for incipient motion under the impact of levee breach waves. The hydrodynamic characteristics of sandbags are discussed with relation to flow hydraulics. Based on experimental hydraulics measurements, two equations for sandbag incipient motions are calibrated, and new equations are developed. Statistical analysis is performed on the developed equations for error and uncertainty bands. Finally, the developed equations are applied to the 17th Street Canal levee breach.

2. Equations for Sandbag Incipient Motion

The size of the sandbags can be determined by relating the critical flow conditions at the onset of bag instability to the bag size that is not washed away. The sandbags could then be sized so that they settle at the channel bed to close the breach. Stability conditions can be evaluated using either the critical bed shear stress or the critical velocity at the initiation of sandbag motion. The latter approach is followed herein. The earliest research was conducted by Brahms in 1753, and resulted in the following equation for spherical particles:

$$V_{cr} = C_1 W^{1/6} \tag{1}$$

where V_{cr} = depth averaged velocity at the location of the particle at which the particle starts moving, W = weight of the particle, and C_1 = an empirical constant.

Using the particle diameter instead of the weight, Yu et al. [37] suggested the following power function:

$$V_{cr} = 2.5 D^{0.44} \tag{2}$$

The following expression for spherical rocks used in toe dumping in flowing water [42]:

$$V_{cr} = C_2 \sqrt{2gD\left(\frac{\rho_s - \rho_w}{\rho_w}\right)} \tag{3}$$

where D = diameter of the spherical particle, g = gravitational constant, ρ_s = particle density, ρ_w = water density, and C_2 = an empirical constant. Novak and Nalluri [52] proposed an equation for the critical motion of single particles on smooth and rough beds in the form

$$V_{cr} = C_3 \sqrt{gD\left(\frac{\rho_s - \rho_w}{\rho_w}\right)} \left(\frac{D}{R}\right)^{C_4} \tag{4}$$

where R is the channel hydraulic radius and was included in the empirical equation to generalize their approach, which has been done using circular shape channels; and C_3 and C_4 are empirical constants. This relation is stated to be valid for $D/R < 0.3$. On the other hand, for prism sandbags, Gulcu [42] used force balance on single solid particles and the proposed relation between critical velocity and particle dimensions as follow:

$$\frac{V_{cr}}{\sqrt{gb\left(\frac{\rho_s-\rho_w}{\rho_w}\right)}} = \sqrt{\mp C_5\left(\frac{b}{c}\right)^2 + C_6\,\frac{b}{c} + C_7} \tag{5}$$

where b is sandbag height, c is sandbag length parallel to the flow direction, and C represents empirical constants. Zhu et al. [43] worked on sandbags and assumed that the sandbag would start to slide at the incipient condition and to apply a simple force balance equation to maintain static equilibrium, they proposed the following equation for prism sandbags:

$$V_{cr} = C_8\left(\frac{H}{b}\right)^{1/6}\sqrt{\frac{c}{b}}\sqrt{gb\left(\frac{\rho_s-\rho_w}{\rho_w}\right)} \tag{6}$$

where H is water depth above the sandbag, and C_8 is a constant to be calibrated from experimental data.

3. Experimental Setup

To model a general case of levee breach failure, experiments were performed in a straight flume, which is constructed in 11.6 × 6.7 m basin (Figure 1) at the Department of Irrigation & Hydraulics, Cairo University, Egypt. The experiments were designed to simulate breach flow after full formation of the breach and did not account for breach formation. The flume was 6.4 m long, 0.6 m deep and 0.4 m wide. The channel bed and sides were covered with rough concrete with an average Manning roughness coefficient of 0.01. All experiments were performed on fixed beds, in the channel, and in breach. The discharge was measured by the electromagnetic flow meter installed on the discharge pipe.

The current study was on a single lateral breach located 3 m from the channel entrance to have developed flow in the channel upstream from the breach. The floor of the flume was horizontal, and the breach discharged on a horizontal platform constructed at the same level as the channel. Breach flow was allowed to flow freely without any impact of this boundary on breach flow. The downstream end of the flume was left open with an adjustable tailgate for controlling the water level and discharge in the downstream channel. Two 0.16 m^3/s (2500 gallons per minutes) axial pumps were used for flow supply from an underground sump to an overhead tank that supplies the flume through a 0.305 m supply pipe. To ensure uniformly distributed flow at the breach location, perforated screens and a 60 mm long honeycomb were placed at the flume inlet.

A point gauge was used to measure water surface in the channel with an accuracy of 0.1 mm on the Vernier scale. A 10 MHz 3D Sontek Acoustic Doppler Velocimeter (ADV) (SonTek, San Diego, CA, USA) was used to measure 3D velocity and turbulent components. To verify the repeatability of experiments, the tests were conducted several times, and the velocity components at a selected location were recorded. It was found that experiments are repeatable and the difference in velocity was less than 1%.

A coordinate system was chosen such that the positive x-axis was in the flow direction of the main channel. The positive y-axis was pointing toward the channel side, while the breach and the positive z-axis was upward in the vertical direction, as shown in Figure 1a. The origin of this coordinate system was at the bed of the flume at the upstream end of the breach opening. The experiments were performed with the flow rate and the depth downstream of the flume held constant. Measurements were taken after conditions became steady in the channel and flow was fully turbulent for all tests. This experimental setup has been used previously to perform hydraulic analyses for channel levee breach flow [29]. The authors of that study used the grid shown in Figure 1b to measure flow velocity at various locations in the channel and through the breach.

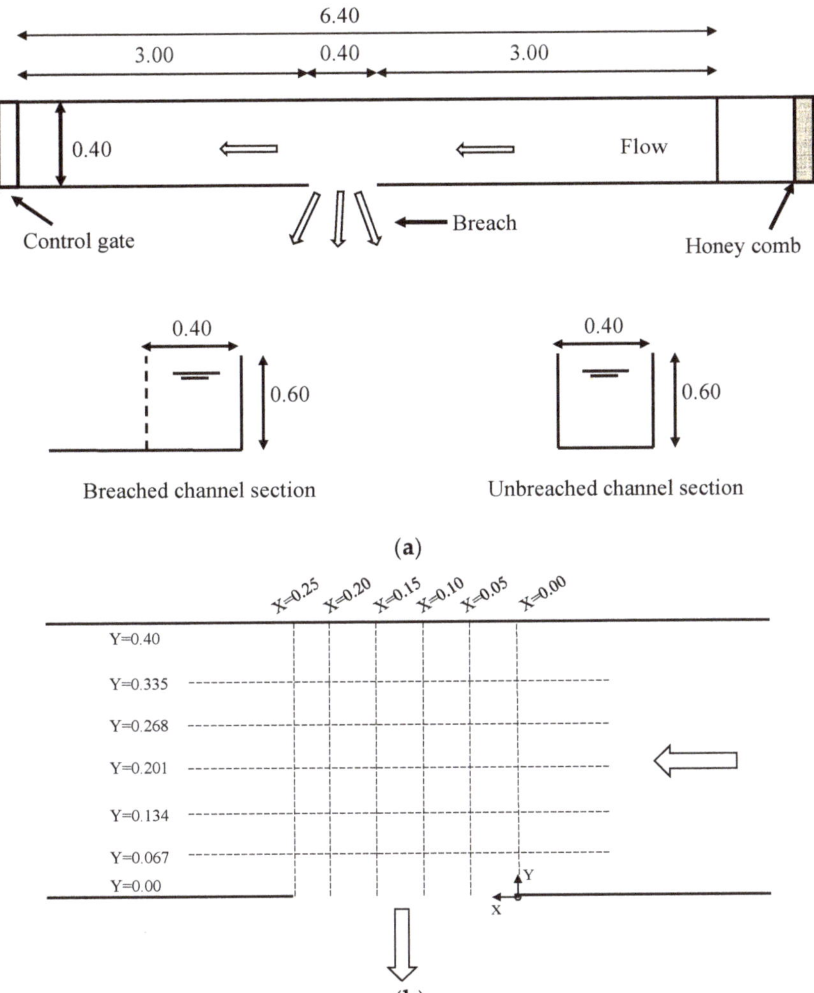

Figure 1. Schematic description of the experimental setup. (**a**) The experimental levee breach model, (**b**) the grid used for velocity measurements.

Sandbags of different shapes and sizes were made of sewed cloth bags filled with sand. Densities of material were determined by weighing the bags on a digital balance of 0.01 g accuracy and by measuring their dimensions to calculate the volume. The dimensions of each sandbag were measured using a compass with an accuracy of 0.1 cm. The sandbags had an average density of 1.85 g/cm^3. Rectangular prisms were divided into two groups according to their orientation to the flow direction. Geometrical properties and dimensional details of the sandbags used in this investigation are shown in Table 1. It should be noted that *a*, *b*, and *c* are the width, height and length of an imaginary rectangular prism enfolding sandbag, and $\forall$ and S are the volume and surface area, respectively. All sandbags were placed such that the shortest dimension was always the height of the sandbag.

Table 1. Geometric and dimensional characteristics of sandbags used in this study.

Sandbag Shape	Statistical Parameter	*a* (cm)	*b* (cm)	*c* (cm)	∀ (cm³)	S (cm²)
Prism (29 cases)	Minimum	3	2.5	4	40	72
	Maximum	18	7	18	378	384
	Average.	6.9	3.82	9.1	214	236
	Standard deviation	4.2	1.15	4.3	103	93
Sphere (13 cases)	Minimum		5		58	72
	Maximum		22		5575	1521
	Average.		13		1739	618
	Standard deviation		6		1788	475

4. Experimental Procedure

Two breach widths L_b were considered (0.25 m and 0.40 m), and for each the downstream channel gate was altered to change flow through the breach and thus the breach flow ratio, $Q_r = Q_b/Q_u$, was changed from 1 to 0.80, where Q_b is the breach discharge and Q_u is the main channel discharge upstream of the breach. Fr_u is the main channel Froude number upstream of the breach, and Fr_d is the main channel Froude number downstream of the breach. Table 2 presents the hydraulic parameters which were recorded in the experimental study.

Table 2. Breach hydraulic parameters.

Breach Width	Q_u (m³/s)	Q_b (m³/s)	Q_r	Fr_u	Fr_d
0.25 m	0.16	0.16	1	0.62	-
	0.16	0.128	0.8	0.62	0.11
0.4 m	0.16	0.16	1	0.62	-
	0.16	0.128	0.8	0.62	0.11

In each experiment, discharge was kept constant until steady flow conditions had developed in the channel upstream of the breach section. A sandbag with a selected dimension was then placed at the bottom of the flume at the mid-breach section. Sattar [32] extensively studied various procedures for closure of levee breaches and found that staring by dumping sandbags in mid-breach sections is the most effective way for closure. The flow was increased in small increments, and this procedure was repeated until the motion of sandbag was observed. Then the flow was decreased in three steps, and the test was repeated to make sure that it reflected real critical conditions for the incipient motion of the sandbag. This procedure for determining the critical conditions at the incipient motion is defined in detail by Gogus and Defne [41] and Gulcu [42].

Once the criterion for sandbag motion was satisfied, the depth of water above the sandbag was measured; the vertical velocity profile over the flow depth was recorded with ADV. Due to the non-uniform distribution of vertical velocity along the breach section (velocity increases along the breach to reach its maximum at the end of the contraction zone), and the presence of velocity components in both x- and y-directions, the critical velocity for incipient motion was taken as the velocity magnitude averaged over the height of the sandbag at the test location.

5. Results and Discussions

This section first presents some aspects of breach hydraulics, including water surface profiles and flow velocity in the x-, y- and z-directions around the breach section. This is followed by presenting measurements of critical velocity for various shapes and sizes of sandbag. These measurements are then used to calibrate exiting equations and develop new equations that are tested statistically and then applied to the 17th Street Canal levee breach in New Orleans.

5.1. Water Surface Mapping

The flow in the main channel approaching the breach started to divide as the flow enters the breach. A separation zone developed in the main channel after the breach, and a stagnation zone formed near the downstream corner of the breach. At the breach outflow, a contracting flow region is observed with free overfall conditions similar to dam-break flow. For L_b = 40 cm, and Q_r = 1, water depth in the main channel upstream of the breach was 38 cm, which increased longitudinally across the main channel to 42 cm downstream of the breach, and decreased rapidly in the transverse direction at the breach location to 29 cm at the breach. Figure 2 shows the mapping of water depth for L_b = 40 cm.

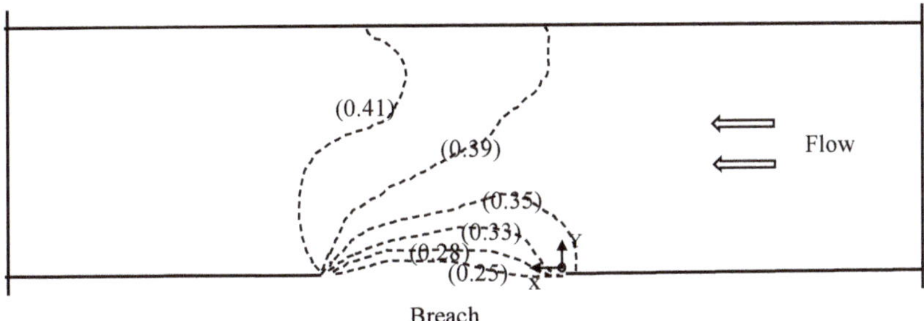

Figure 2. Water depth along the channel for L_b = 40 cm, and Q_r = 1 m^3/s.

At the breach section, the variations in water depth were more pronounced than in the case of open channel dividing flow. There was a rapid drop in the water surface from the outer channel bank to the breach location and a corresponding rapid increase in the flow velocity, which was highest at the breach location. Flow depth in the stagnation zone was the highest along the breach section. Due to flow circulation and a significant decrease in the flow velocity in the separation zone, the water depth in this zone was higher than the main channel and was equal to that of the stagnation zone, downstream of the breach.

The highest water depth opposite to the breach line occurred in the separation zone. At Q_r = 1 m^3/s, the recirculation effects and the separation zone were highest when compared to those in the other case of Q_r < 1. Unlike for open channel dividing flow, the breach flow separation occurred at the near side of the breach in the main channel, causing a flow contraction. This is obvious from the velocity vectors across the breach. However, for the same Q_r, the larger the breach opening, the larger the angle of entry of flow, and the extent of flow separation in the main channel just before the flow exits from the breach. Also, the larger the breach opening, the smaller the contraction zone. Michelazzo et al. [7,53] discussed more details on the influence of breach width on channel flow. These characteristics are similar to those of the dividing open channel flows [20]. Figure 3 shows the hydraulics of an open channel levee breach for the two considered cases, showing clearly the separation and stagnation zones.

5.2. Velocity Fields

The following section presents the findings related to velocity measurements in the open channel and side breach. Michelazzo et al. [7,53] have presented 3D velocity measurements for open channel side flow similar to those discussed herein. Figure 4 shows the flow velocity in the x- and y-direction at the breach section in the main channel at z = 0.04 m and z = 0.11 m for Q_r = 1 m^3/s. The x-velocity decreases at the mid–section of the breach and reaches zero by the end of the breach due to the complete curvature of the vectors exiting the breach. Experimental results showed that the vectors at the tip of the stagnation zone at the breach end took a steep curvilinear path to exit the breach and were forced by the flume side to exit in the negative x-direction.

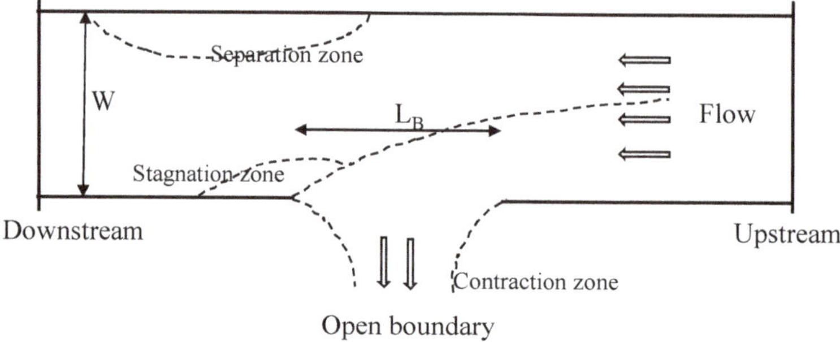

Figure 3. Schematics of the hydraulics of channel levee breach (L_B and W are breach width and channel width respectively).

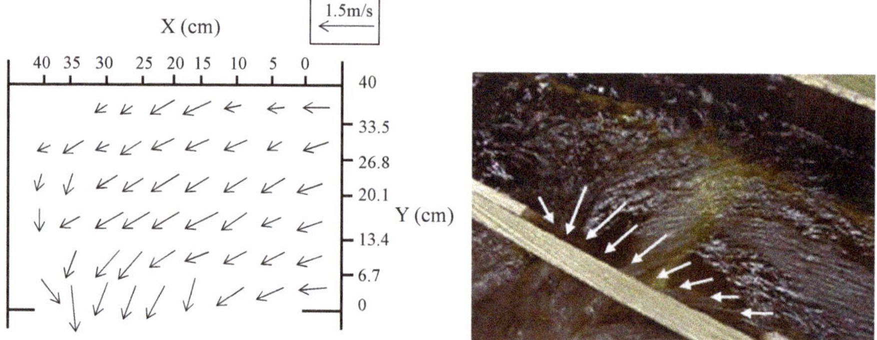

Figure 4. x- and y-velocity field at the breach section in the main flume. Q_u = 0.16 m^3/s, Q_r = 1 m^3/s, z = 0.15 m, for L_b = 0.40 m.

The increase in x-velocity can be clearly observed with the increase in distance from the flume bottom. It can also be observed that the distribution of x-velocity in the main channel around the breach is almost the same at different distances from the bottom, where the changes is only in the values. At the far end of the breach in the main channel x-velocity decreased significantly, and the direction changed at some locations at z = 0.04 m.

The velocity in the y-direction clearly showed that the exit flow velocities were perpendicular to the breach section for Q_r = 1 (m^3/s). These velocities gave a better view of the separation and contraction zones. In the separation zone that started to form in the main channel before the breach flow exit, the exit velocities in the y-direction were the lowest along the breach section.

The highest values of velocity in the y-direction indicated the zone where flow contraction occurred in the far section of the breach after the separation zone. In general, the distribution of both velocity components along the breach section is almost the opposite, that is, the maximum values of one velocity component correspond to the minimum values of the other component. This is caused by the curvature in the vectors to exit the breach section. Flow tends to exit at the far point of the breach section with a very high velocity in the opposite x-direction for flow entering the breach. This is adjacent to the stagnation point.

The y-velocity decreased with an increase in distance from the flume bottom. Similar distributions were observed at z = 0.04 m, 0.11 m, and 0.20 m. At the far end of the breach in the main channel, y-velocity decreased significantly, similar to x-velocity. Thus, the vectors curved at the exit of the breach section, starting a small distance upstream of the breach end. This caused a change in the distribution of velocities along the channel width.

Both of the velocity components decreased away from the breach and increased near its location. The vector curvature and its impact on the velocity distribution occurred at all levels above the flume bottom. However, results showed that the y-velocities at the breach exit are higher at z = 0.04 m than those at z = 0.11 m. Thus, the curvature of vectors at the breach section was observed to be greater at lower depths of water at the breach section.

For $Q_r = 1\ m^3/s$ and $L_b = 0.25$ m, Figure 5 shows the velocity field for y- and z-velocity components at two locations: x = 0.10 m and x = 0.20 m. Both sections were after the region of the flow separation at the breach, and they showed clearly the impact of the third vertical component of velocity (z-velocity) in the flow pattern near the breach section. This secondary downward current is very strong and obvious at section x = 0.20 m; however, near the end of the breach, its intensity at the other sections in the contraction zone is low. This secondary current caused the vector to drop rapidly, leading to a rapid decrease in the water level in the vicinity of the breach, until the mid-width of the channel, where it started to diminish along the channel width. The z-velocity component decreased with a decrease in z distance at the contraction zone, at x = 0.10 m and 0.20 m.

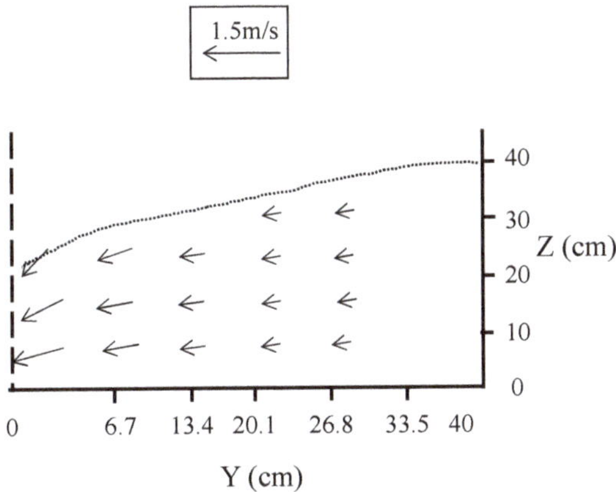

Figure 5. y-z velocity vector profile along the breach, $Q_u = 0.16\ m^3/s$, $Q_r = 1\ m^3/s$, for $L_b = 0.25$ m, at x = 0.2 m.

5.3. *Critical Velocity for Sandbags Incipient Motion*

Table 3 shows the sandbag dimensions used in the experimental study versus the critical depth averaged velocity required for the onset of sandbag instability. Where non-dimensional critical velocity is defined as $V_{cr}/\sqrt{\frac{(\rho_s-\rho_w)}{\rho_w}gb}$, and V_{cr} is the depth averaged velocity (at mid-section of breach) averaged over the whole flow depth. For spherical sandbags, $a = b = c = D$.

Figure 6 shows the relationship between critical velocity for incipient motion versus the ratio of *D/R* for the Shields equation, Novak and Nalluri [52] experiments (diameters were from 0.6 mm to 50 mm), and results measured in the present study for spherical sandbags with diameters from 48 to 200 mm. The Shields equation has been plotted using Stricklers' equation for Manning roughness. It can be observed that the critical velocity required for initiation of motion for a single particle is lower than that given by Shields for a given *D/R* value.

Table 3. Sandbag geometric and dimensional properties versus the critical velocity $V_{cr}/\sqrt{\frac{(\rho_s-\rho_w)}{\rho_w}gb}$.

Experiment	b/c	$\frac{V_{cr}}{\sqrt{\frac{(\rho_s-\rho_w)}{\rho_w}gb}}$	Exp.	b/c	$\frac{V_{cr}}{\sqrt{\frac{(\rho_s-\rho_w)}{\rho_w}gb}}$
1	0.63	0.76	27	1	0.49
2	0.60	0.69	28	1	0.51
3	0.50	0.69	29	1	0.52
4	0.58	0.79	30	1	0.61
5	0.50	0.79	31	1	0.52
6	0.42	0.93	32	1	0.61
7	0.35	0.81	33	1	0.57
8	0.44	0.81	34	1	0.53
9	0.44	0.81	35	1	0.50
10	0.56	0.68	36	1	0.55
11	0.50	0.68	37	1	0.53
12	0.45	0.72	38	1	0.58
13	0.32	1.15	39	1	0.56
14	0.30	1.25	40	1	0.55
15	0.23	1.25	41	1	0.52
16	0.19	1.35	42	1	0.53
17	0.18	1.41			
18	0.21	1.35			
19	0.22	1.35			
20	0.18	1.46			
21	0.70	0.65			
22	0.64	0.70			
23	0.50	0.70			
24	0.58	0.67			
25	0.44	0.67			
26	1.00	0.49			

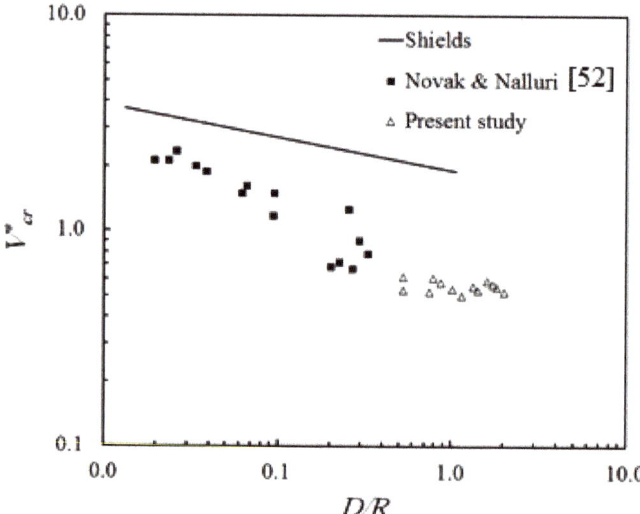

Figure 6. Relationship between the non-dimensional critical velocity (V^*_{cr}) at the initiation of sandbag motion and D/R for circular sandbags.

Experiments in the current study covered a higher range of D/R ($0.5 < D/R < 2$) and followed a similar decreasing trend as the Shields equation and experiments of Novak and Nalluri [52]. However, the rate of change of critical velocity with respect to increasing in D/R is more pronounced in the

Novak and Nalluri [52] experiments for a D/R lower than 0.3. Using both datasets for critical velocity on spherical particles on smooth and rough beds, and for spherical sandbags, the following relation has been obtained with a coefficient of determination, R^2 of 0.90:

$$\frac{V_{cr}}{\sqrt{\frac{(\rho_s-\rho_w)}{\rho_w}gD}} = 0.569\left(\frac{D}{R}\right)^{-0.34} \tag{7}$$

Introducing the channel hydraulic radius is not common for critical velocity equations describing particle incipient motion, although it has been used by Novak and Nalluri [52] to decrease the effects of flume shape, having used triangular and circular flumes in their experiments. Usually, the main influencing parameters on the critical velocity of a spherical particle are the particle weight and the diameter. The critical velocity measurements for sandbags followed the 1/6th power equation suggested by Stelczer [54], with R^2 of 0.95 as shown in Figure 7, where $V_{cr} = 0.381W^{1/6}$ using metric units, and W is the sandbag weight in Newton.

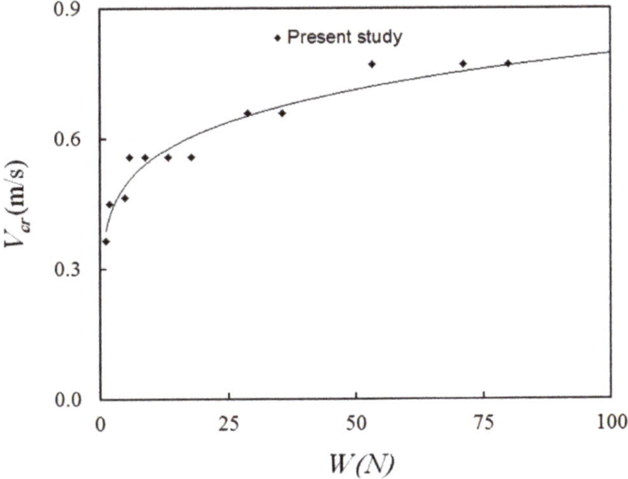

Figure 7. Variation of V_{cr} with spherical sandbag weight.

Using the sandbag diameter as a main influencing parameter on its incipient motion has been suggested by numerous studies. Figure 8 shows the results of some of the relevant studies in addition to the experimental findings of the present study on sandbags. It should be noted that the maximum grain diameter used in these studies is 10 mm for the Gulcu [40] and 100 mm for Defne [41]. The critical motion for sandbags followed a very similar trend to those of solitary grains in previous studies, with differences attributed to the conditions and assumptions used in every study (e.g., mixed size grains with equivalent diameter, very small size grains, using big rocks, and rough and smooth beds). Data in Figure 8 suggest that expressions in the form of power function can be used to define the relationship (using metric units) between the sandbag critical velocity and the bag diameter as follows:

$$V_{cr} = 1.633D^{0.47} \tag{8}$$

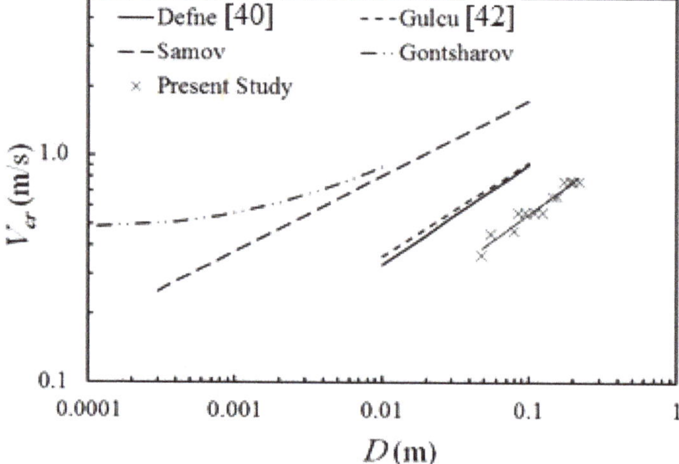

Figure 8. Variation of V_{cr} with spherical grain diameter.

Figure 9 shows the relationship between the critical velocity for prism sandbags and the height to length ratio *b/c*, using data from previous studies and this study [41,43].

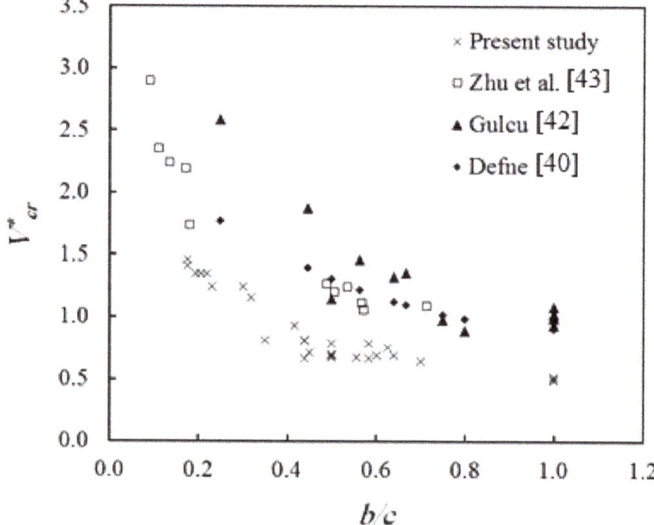

Figure 9. $V_{cr}/\sqrt{\frac{(\rho_s-\rho_w)}{\rho_w}gb}$ versus the height to length ratio (*b/c*) for prism sandbags.

The critical velocity required for initiation of particle motion decreases rapidly with the increase of *b/c* for $b/c \leq 0.7$. For $b/c > 0.7$, the critical velocity remained almost constant until $b/c = 1$ for a cube sandbag. Cube sandbags had the lowest critical velocity and thus were more easily washed away by flowing water than prism sandbags. Gogus and Defne [41] found that the critical velocity required for moving a cube particle is three times lower than for a prism particle, while results of this study found it to be around two times lower. Zhu et al. [43] did not use cube sandbags, and thus no results are reported for $b/c = 1$.

Using the experimental results of Zhu et al. [43] and those measured in the current study, a power function can be used to develop a relationship between critical velocity for sandbag threshold motion and the sandbag height to length ratio with R^2 of 0.83 as follows:

$$\frac{V_{cr}}{\sqrt{\frac{(\rho_s-\rho_w)}{\rho_w}gb}} = 0.522\left(\frac{b}{c}\right)^{-0.67} \quad (9)$$

The database of incipient motion of solitary particles measured in this study and measured previously [43] were used to calibrate some traditional equations and fine tune numerical constants in them by using multiple regression analysis. Table 4 shows these equations with original numeric constants and the same equations after being calibrated with the database.

Table 4. Original and calibrated traditional critical velocity equations.

Study		Model Equation
Novak and Nalluri [52]	Original	$V_{cr} = C_3\left(\frac{D}{R}\right)^{C_4}\sqrt{\frac{(\rho_s-\rho_w)}{\rho_w}gD}$
	Calibrated	$V_{cr} = 0.569\left(\frac{D}{R}\right)^{-0.34}\sqrt{\frac{(\rho_s-\rho_w)}{\rho_w}gD}$
Brahms	Original	$V_{cr} = C_1W^{1/6}$
	Calibrated	$V_{cr} = 0.381W^{1/6}$
Izbash [44]	Original	$V_{cr} = 3.76\sqrt{D\left(\frac{\rho_s-\rho_w}{\rho_w}\right)}$
	Calibrated	$V_{cr} = 1.77\sqrt{D\left(\frac{\rho_s-\rho_w}{\rho_w}\right)}$

Uncertainty Analysis

In the levee breach closure risk assessment study, the uncertainties of many influencing parameters have to be included. Various choices of sandbag dimensions can influence the outcome in a drastic way, which could lead to the sandbag being swept away by flowing water. Using the calibrated and developed equations for sandbag incipient motion together with the experimental measurements, a quantitative assessment for the uncertainty in the prediction of critical velocity can be presented. The uncertainty analysis defines the prediction error in log cycles as [55–65]

$$e_i = log_{10}(P_i) - log_{10}(T_i) \quad (10)$$

where e_i is the prediction error, P_i is the predicted value of parameter, and T_i is the measured value of the parameter. Data were then used to calculate main indicators defined as mean prediction error $\bar{e} = \sum_{i=1}^{n} e_i$, the width of uncertainty band, $B_{ub} = \pm 1.96S_e$ and the confidence band around the predicted value:

$$\left\{P_i \times 10^{\bar{e}+2S_e},\ P_i \times 10^{\bar{e}-2S_e}\right\} \quad (11)$$

where S_e is the standard deviation of prediction errors and P_i is the predicted value and is taken as unity. Table 5 summarizes the results of the uncertainty analysis performed on both calibrated and developed models for prediction of critical velocity for sandbag motion in flowing water. All models were tested on experimental data in this study and Zhu et al. [43] data.

Table 5. Results of prediction uncertainty analysis for sandbag incipient motion models.

Sandbags Shape	Study	Model	The Width of Uncertainty Band	Prediction Interval Around Hypothetical Predicted V_{cr}^{*} of Unity
Sphere	Novak and Nalluri [52] (Calibrated)	$\frac{V_{cr}}{\sqrt{\frac{(\rho_s-\rho_w)}{\rho_w}gD}} = 0.569\left(\frac{D}{R}\right)^{-0.34}$	±0.14	(0.78–1.51)
	Brahms (Calibrated)	$V_{cr} = 0.381W^{1/6}$	±0.05	(0.91–1.14)
	Present study—Sphere	$V_{cr} = 1.633D^{0.47}$	±0.04	(0.89–1.13)
	Samov (Original)	$V_{cr} = 2.5D^{0.44}$	±0.05	(1.45–1.85)
	Izbash [44] (Original)	$V_{cr} = 3.76\sqrt{D\left(\frac{\rho_s-\rho_w}{\rho_w}\right)}$	±0.05	(1.87–2.39)
	Izbash [44] (Calibrated)	$V_{cr} = 1.77\sqrt{D\left(\frac{\rho_s-\rho_w}{\rho_w}\right)}$	±0.05	(0.88–1.12)
Prism	Present study—Prism1	$\frac{V_{cr}}{\sqrt{\frac{(\rho_s-\rho_w)}{\rho_w}gb}} = 0.295\left(\frac{b}{c}\right)^{-1}$	±0.19	(0.52–1.25)
	Present study—Prism2	$\frac{V_{cr}}{\sqrt{\frac{(\rho_s-\rho_w)}{\rho_w}gb}} = 0.522\left(\frac{b}{c}\right)^{-0.67}$	±0.08	(0.91–1.33)
	Zhu et al. [43] (Original)	$\frac{V_{cr}}{\sqrt{\frac{(\rho_s-\rho_w)}{\rho_w}gb}} = 0.44\left(\frac{H}{b}\right)^{1/6}\left(\frac{b}{c}\right)^{-0.5}$	±0.08	(0.90–1.30)

The lower the uncertainty in the prediction of a model and the narrower the prediction interval, the more reliable the model is. It is obvious that the widest uncertainty bands for predictions were due to the calibrated [38] proposed equation, with a value of ±0.14 versus ±0.05 for other proposed models for spherical sandbags. The Izbash [44] model, based on spherical grains and rock material, seems to significantly overestimate critical velocity. Other models for spherical sandbags had narrower prediction intervals, with an average of −10% to +12% of predicted critical velocity. On the other hand, for prism sandbags, equations seemed to behave in a similar way, with a narrow lower prediction uncertainty of −10% and slightly wider high prediction uncertainty of +30%. In the following section, it will be shown how the width of prediction intervals can affect the correct choice of the sandbag size.

5.4. 17th Street Canal Levee Breach Closure

Hurricane Katrina resulted in a breach of the levees and floodwalls in approximately 20 places. One of the major breaches was on the 17th Street Canal, approximately 305 m from the Old Hammond Highway Bridge, with a breach width of 137 m [5]. The flood protection system in the 17th Street Canal was a concrete floodwall with an I-section over a levee embankment of fill material over a marsh layer [5].

The storm surge moved the levee and floodwall horizontally for about 13.7 m, and this breach—as well as others—accounted for the flooding of the city, the destruction of infrastructure (sewers, water, phone and electricity lines), and the tripping of pumping stations due to water rise. The Army Corps dropped 10 large sandbags to close the breach [29]. They started with 1359 kg (3000 lb) and increased to 2718 kg (6000 lb) using the National Guard Helicopter (see Figure 10). However, the bags were completely washed away, and the initial attempt failed.

Figure 10. Black hawk dumping sandbags to close the 17th Street Canal levee breach [29].

Work in this section is intended to simulate the initial failed attempts of the United States Army Corps of Engineers (USACE) to close the breach using 2718 kg (6000 lb) sandbags on 31 August 2007. While there are several unknowns on the actual failure of the 17th Street Canal floodwalls and levees, Sattar et al. [29] considered a "base scenario" that is more conservative than flow conditions during initial failed attempts.

This scenario uses some of the assumptions used in the base scenario considered by the USACE [5]. The breach is assumed to have widened from 61 to 137 m by 10:00 am. The elevation of water after the breach had increased by 1.92 m. Corresponding to this water elevation in the channel before the breach, the measured flow across the breach was 580 m^3/s. Using these data, the critical velocity is 2.5 m/s (as measured experimentally by Sattar et al. [29] and calculated numerically by Jia et al. [66]) and V_{cr} is 0.78 m/s for 1359 kg (3000 lb) and 0.70 m/s for 2718 kg (6000 lb) sandbags.

Scaling up model parameters to prototype parameters results in significant differences due to model effects, scale effects, and measurement effects [67], with scale effects contributing most to the difference. Making the experimental model as close as possible to prototype dimensions would yield least scale effects. However, it would be uneconomic to build such models. Thus, Heller [67] proposed scales of 1:50 to 1:100 for open channel flows that would be a compromise between both reasonable model size and moderate scale effects. Moreover, Chanson [68] reported that scale effects would be minimized and/or eliminated in models with scales from 1:25 to 1:50. In our experiments, water depth in the flume was 40 cm, which implies a scale of less than 1:50 compared to the 17th Street Canal depth of 7 m. Therefore, it can be assumed that scale effects between the lab setup and case study are minimal and would not affect results.

Calculating the critical velocity from various calibrated and original models for the sandbags that were initially dropped by the USACE [5] during attempts to close the breach showed that some equations provide non-logical results. The critical velocity for sandbag motion as predicted by Samov and Izbash [44] is 2.5–2.8 m/s and 3.1–3.6 m/s for 3000 lb and 6000 lb sandbags, respectively. These velocities greatly exceed the actual critical velocity, suggesting that both sizes of sandbags dropped by the USACE could not have been washed away by flowing water, which is not what happened. Therefore, these equations provided false predictions for sandbag incipient motion and are not recommended to be used without calibration.

The calibrated equations of Brahms and Izbash [44], and the equation developed in the present study, predicted similar values for critical velocity required for the motion of sandbags. Critical velocities had average values from 1.6 m/s to 1.8 m/s for 3000 lb and 6000 lb sandbags, respectively. These predicted velocities were lower than the actual flow velocity when dropping the bags, confirming what really occurred in initial failed attempts where the dropped sandbags could be easily swept away by breach flood waves. These models had the narrowest prediction interval of ±10%, which implies that even when considering the uncertainty in predictions within this interval, the sandbags would still be washed away easily.

As discussed before, spherical and cube shaped sandbags had the lowest critical velocity for initiation of motion among prism sandbags with the same weights. Prism sandbags (with $b/c < 0.4$) have better hydrodynamic characteristics that give them higher stability and resistance to being moved by flowing water. This is confirmed by calculations of critical velocity for prism sandbags that could have been used for initial breach closure attempts.

The actual critical velocity V_{cr} for 1359 kg (3000 lb) and 2718 kg (6000 lb) prism sandbags are 1.10 and 0.90 m/s, respectively. The equations developed in the current study predicted values close to this for critical velocity, which were higher than those predicted by the equation in Zhu et al. [43]. V_{cr} was predicted as 0.83 m/s and 1.04 m/s for developed equations in the present study, compared with 0.37 m/s for Zhu et al. [43]. However, the uncertainty range for the first equation was wider than other equations, where the predicted critical velocity could lie within the range of 0.42–1. On the other hand, the predicted velocity—taking into account the uncertainty range—from the present study was equal to or greater than the actual critical velocity, suggesting that the sandbags would resist being washed away by the flow. The same is true for the 2718 kg (6000 lb) sandbags, where the average critical velocity was higher than or equal to the actual breach flow conditions, and thus, would resist being moved by breach water.

Using this type of prism sandbag, the available helicopters could have been used to carry more sandbags, and breach closure would have been very likely possible in a timely manner. The Zhu et al. [43] model predictions for the critical velocity were lower than those of the fitted model and did not show a clear differentiation between various orientations of sandbag with respect to flow. The Zhu et al. [43] model's critical velocity was a half less than other models, suggesting a considerable error in sandbag choice.

According to Sattar et al. [29] a sandbag of weight 6795 kg (15,000 lb) was not washed away at about 40% of the breach length and a similar ratio for a sandbag of 3171 kg (7000 lb), while only 20% of the breach required 4530 kg (10,000 lb) sandbag. An experimental simulation [29] and a numerical simulation [66] of the breach flow showed that the depth averaged velocity for the base scenario ranged from 2 to 2.5 m/s. Table 6 presents the critical depth averaged velocity required sandbag motion as predicted by the available models for large spherical sandbags. As discussed above, the Samov equation and Izbash [44] model predicted the highest critical velocity amongst all other models. All models, except for the Novak and Nalluri [52] model, confirmed experimental findings by predicting higher critical velocity for sandbag motion than actual flow conditions; thus, the chosen sandbag was not washed away. This is true when considering the prediction uncertainty interval for each model. Utilizing these results, a recommendation can be made to use the calibrated Brahms model, the model developed in the current study for spherical sandbags, and the calibrated Izbash [44] model to predict the critical velocity for sandbag motion threshold when used in levee breach closure. Considering the three models with a prediction uncertainty of ±10% can give decision makers a rough estimate for the correct choice of sandbag weight and/or diameter for breach closure.

Table 6. Critical velocity (V_{cr}) for spherical sandbag motion for breach closure in Base Scenario studied by Sattar et al. [29].

Study	Model	V_{cr} 95% Confidence Interval (m/s)	10,000 lbs Sandbag	15,000 lbs Sandbag
Novak and Nalluri [52] (Calibrated)	$V_{cr} = 0.569\left(\frac{D}{R}\right)^{-0.34}\sqrt{\frac{(\rho_s-\rho_w)}{\rho_w}gD}$	Predicted Range	0.96 (0.73–1.45)	0.98 (0.75–1.48)
Brahms (Calibrated)	$V_{cr} = 0.381W^{1/6}$	Predicted Range	2.27 (2.01–2.50)	2.43 (2.21–2.77)
Present study—Sphere	$V_{cr} = 1.633D^{0.47}$	Predicted Range	2.00 (1.78–2.26)	2.16 (1.92–2.44)
Samov (Original)	$V_{cr} = 2.5D^{0.44}$	Predicted Range	3.00 (4.35–5.55)	3.24 (4.70–6.00)
Izbash [44] (Original)	$V_{cr} = 3.76\sqrt{D\left(\frac{\rho_s-\rho_w}{\rho_w}\right)}$	Predicted Range	5.48 (10.25–13.10)	5.77 (10.79–13.79)
Izbash [44] (Calibrated)	$V_{cr} = 1.77\sqrt{D\left(\frac{\rho_s-\rho_w}{\rho_w}\right)}$	Predicted Range	2.20 (1.94–2.46)	2.37 (2.08–2.66)

6. Conclusions

Motivated by the importance of open channel levee breach and the engineering requirements for timely breach closure, this study experimentally investigated some of the hydraulic characteristics of a breach flow together with the effect of sandbag shape and size on the incipient motion of the bags. Experimental measurements during hydraulic modelling in the laboratory showed that prism sandbags have better hydrodynamic characteristics than spherical bags with the same weight. Measurements were used to calibrate existing models and develop new ones for sandbag incipient motion for breach closure. A power function linking critical velocity for sandbag threshold motion and the sandbag height to length ratio was proposed. The developed models for the incipient motion are based on the built experiment that has the following important assumptions: Breach formation and dike failure process are not accounted for, and breach and channel beds are immobile; the breach and channel beds have the same elevation; and the inundation area after a breach is assumed as the open boundary with no effect on breach outflow. Most importantly, similar to previous research, only one sandbag is placed in breach and tested for incipient motion.

Statistical analyses were performed on the prediction models to provide a range of uncertainty in the predicted critical velocity. The available calibrated and developed models were then applied to 17th Street Canal levee breach closure. The obtained results showed that Izbash [44] models predicted the highest critical velocity amongst all other models. All models, except for the Novak and Nalluri [52] model, confirmed experimental findings by predicting higher critical velocity for sandbag motion than actual flow conditions, and thus, the chosen sandbag was not washed away. Reasonable agreement was achieved in comparing the predicted critical velocity for sandbags with the actual velocities during the USACE attempts for breach closure. Results suggested that some equations can be used considering their uncertainty in prediction to provide a rough view for decision makers on the required sandbags for closing a breach flow. Moreover, through the use of a new formula, it provides engineers with applicable results for implementation in practical cases.

In this work, all experiments for incipient motion of sandbags were carried out in side breach flows to calibrate existing and develop new empirical equations for linking sandbag size and critical flow velocity. Various parameters including breach failure, downstream bed morphological changes, and the associated change in breach flow/velocity, can affect obtained results and could be considered in future research.

Author Contributions: A.M.A.S. performed the measurements, drafted the manuscript and designed the figures. H.B. and A.M.A.S. analysed the data. B.G. and A.R.-P. aided in interpreting the results and worked on the manuscript. All authors discussed the results and commented on the manuscript.

Funding: This research received no external funding.

Conflicts of Interest: The authors declare no conflict of interest.

Abbreviations

b	sandbag height (cm)
B_{ub}	the width of the uncertainty band
c	sandbag length parallel to the flow direction (cm)
$C_{1\text{–}8}$	empirical constants
D	the diameter of the spherical particle (mm)
e_i	the prediction error
$\bar{e}$	mean prediction error
H	water depth above sandbag (m)
L_b	breach width (m)
P_i	the predicted value of the parameter
Q_b	the breach discharge (m^3/s)
Q_u	main channel discharge (m^3/s)
Q_r	breach flow ratio = Q_b/Q_u
R	channel hydraulic radius (m)
R^2	the coefficient of determination
S_e	the standard deviation of prediction errors
T_i	the measured value of the parameter
V_{cr}^*	$V_{cr}/\sqrt{\frac{(\rho_s-\rho_w)}{\rho_w}gb}$;
ρ_s	sandbag density (g/cm^3)
ρ_w	water density (g/cm^3)

References

1. Zaji, A.H.; Bonakdari, H.; Gharabaghi, B. Applying Upstream Satellite Signals and a 2-D Error Minimization Algorithm to Advance Early Warning and Management of Flood Water Levels and River Discharge. *IEEE Trans. Geosci. Remote Sens.* **2019**, *57*, 902–910. [CrossRef]
2. Zaji, A.H.; Bonakdari, H.; Gharabaghi, B. Remote Sensing Satellite Data Preparation for Simulating and Forecasting River Discharge. *IEEE Trans. Geosci. Remote Sens.* **2018**, *56*, 3432–3441. [CrossRef]
3. Perdikaris, J.; Gharabaghi, B.; Rudra, R. Evaluation of the Simplified Dynamic Wave, Diffusion Wave and the Full Dynamic Wave Flood Routing Models. *Earth Sci. Res.* **2018**, *7*, 14. [CrossRef]

4. Gholami, A.; Bonakdari, H.; Ebtehaj, I.; Mohammadian, M.; Gharabaghi, B.; Khodashenas, S.R. Uncertainty analysis of an intelligent model of hybrid genetic algorithm and particle swarm optimization with ANFIS to predict threshold bank profile shape based on digital laser approach sensing. *Measurement* **2018**, *121*, 294–303. [CrossRef]
5. United States Army Corps of Engineers (USACE). Interagency Performance Evaluation Taskforce (IPET). Final Report, March–June 2007. Available online: www.ipet.wes.army.mil (accessed on 15 February 2007).
6. Kamrath, P.; Disse, M.; Hammer, M.; Koengeter, J. Assessment of discharge through a dike breach and simulation of flood wave propagation. *Nat. Hazards* **2006**, *38*, 63–78. [CrossRef]
7. Michelazzo, G.; Oumeraci, H.; Paris, E. Laboratory study on 3D flow structures induced by zero-height side weir and implications for 1D modelling. *J. Hydraul. Eng.* **2015**, *141*, 0401502. [CrossRef]
8. Agaccioglu, H.; Yüksel, Y. Side-weir flow in curved channels. *J. Irrig. Drain. Eng.* **1998**, *124*, 163–175. [CrossRef]
9. Borghei, S.M.; Jalili, M.R.; Ghodsian, M. Discharge coefficient for sharp-crested side weir in subcritical flow. *J. Hydraul. Eng.* **1999**, *125*, 1051–1056. [CrossRef]
10. Ramamurthy, A.S.; Qu, J.; Vo, D. VOF model for simulation of a free overfall in trapezoidal channels. *J. Irrig. Drain. Eng.* **2006**, *132*, 425–428. [CrossRef]
11. Emiroglu, M.E.; Kaya, N.; Agaccioglu, H. Discharge capacity of labyrinth side weir located on a straight channel. *J. Irrig. Drain. Eng.* **2010**, *136*, 37–46. [CrossRef]
12. Zaji, A.H.; Bonakdari, H. Performance evaluation of two different neural network and particle swarm optimization methods for prediction of discharge capacity of modified triangular side weirs. *Flow Meas. Instrum.* **2014**, *40*, 149–156. [CrossRef]
13. Ebtehaj, I.; Bonakdari, H.; Zaji, A.H.; Azimi, H.; Sharifi, A. Gene Expression Programming to Predict the Discharge Coefficient in Rectangular Side Weirs. *Appl. Soft Comput.* **2015**, *35*, 618–628. [CrossRef]
14. Khoshbin, F.; Bonakdari, H.; Ashraf Talesh, S.H.; Ebtehaj, I.; Zaji, A.H.; Azimi, H. ANFIS multi-objective optimization using Genetic Algorithm Singular Value Decomposition method for modeling discharge coefficient in rectangular sharp-crested side weirs. *Eng. Optim.* **2016**, *48*, 933–948. [CrossRef]
15. Azimi, H.; Bonakdari, H.; Ebtehaj, I. Sensitivity Analysis of the Factors Affecting the Discharge Capacity of Side Weirs in Trapezoidal Channels using Extreme Learning Machines. *Flow Meas. Instrum.* **2017**, *54*, 216–223. [CrossRef]
16. Bonakdari, H.; Zaji, A.H. New type side weir discharge coefficient simulation using three novel hybrid adaptive neuro-fuzzy inference systems. *Appl. Water Sci.* **2018**, *8*, 10. [CrossRef]
17. Ebtehaj, I.; Bonakdari, H.; Gharabaghi, B. Development of more accurate discharge coefficient prediction equations for rectangular side weirs using adaptive neuro-fuzzy inference system and generalized group, method of data handling. *Measurement* **2018**, *116*, 473–482. [CrossRef]
18. Neary, V.S.; Odgaard, A.J. Three-dimensional flow structure at open channel diversions. *J. Hydraul. Eng.* **1993**, *119*, 1223–1230. [CrossRef]
19. Weber, L.J.; Schumate, E.D.; Mawer, N. Experiments on flow at a 90° open-channel junction. *J. Hydraul. Eng.* **2001**, *127*, 340–350. [CrossRef]
20. Ramamurthy, A.S.; Qu, J.; Vo, D. Numerical and experimental study of dividing open-channel flow. *J. Hydraul. Eng.* **2007**, *133*, 1135–1144. [CrossRef]
21. Kesserwani, G.; Vazquez, J.; Rivière, N.; Liang, Q.; Travin, G.; Mosé, R. New approach for predicting flow bifurcation at right-angled open-channel junction. *J. Hydraul. Eng.* **2010**, *136*, 662–668. [CrossRef]
22. Nanía, L.; Gomez, M.; Dolz, J.; Comas, P.; Pomares, J. Experimental study of subcritical dividing flow in an equal-width, four-branch junction. *J. Hydraul. Eng.* **2011**, *37*, 1298–1305. [CrossRef]
23. Momplot, A.; Lipeme Kouyi, G.; Mignot, E.; Rivière, N.; Bertrand-Krajewski, J.L. Typology of the flow structures in dividing open channel flows. *J. Hydraul. Res.* **2017**, *55*, 63–71. [CrossRef]
24. Ghostine, R.; Kesserwani, G.; Vazquez, J.; Rivière, N.; Ghenaim, A.; Mosé, R. Simulation of supercritical flow in crossroads: Confrontation of a 2D and 3D numerical approaches to experimental results. *Comput. Fluids* **2009**, *38*, 425–432. [CrossRef]
25. Ghostine, R.; Vazquez, J.; Nicolas, R.; Terfous, A.; Ghenaim, A.; Mose, R. A comparative study of the 1D and 2D approaches for simulation of flows at right angled dividing junctions. *Appl. Math. Comput.* **2013**, *219*, 5070–5082.

26. Yen, C. Hydraulics and effectiveness of levees for flood control. In Proceedings of the U.S. Italy Research Workshop on the Hydrometeorology, Impacts, and Management of Extreme Floods, Perugia, Italy, 13–17 November 1995.
27. Jaffe, D.; Sanders, B. Engineered levee breaches for flood mitigation. *J. Hydraul. Eng.* **2001**, *127*, 471–477. [CrossRef]
28. Apel, H.; Merz, B.; Thieken, A. Influence of dike breaches on flood frequency estimation. *Comput. Geosci.* **2008**, *35*, 907–923. [CrossRef]
29. Sattar, A.M.A.; Kassem, A.; Chaudhry, M.H. Case study: 17th street canal breach closure procedures. *J. Hydraul. Eng.* **2008**, *134*, 1547–1558. [CrossRef]
30. Sylvie, V.E.; Sandra, S.; Cyrus, K.; Hanif Chaudhry, M.H.; Imran, J. Simulations of the new Orleans 17th street canal breach flood. *J. Hydraul. Res.* **2012**, *50*, 70–81.
31. LaRocque, L.; Elkholy, M.; Hanif Chaudhry, M.; Imran, J. Experiments on Urban Flooding Caused by a Levee Breach. *J. Hydraul. Eng.* **2013**, *139*, 960–973. [CrossRef]
32. Sattar, A.M.A. Experimental investigation of flood waves from open-channel levee breach. In *Experimental and Computational Solutions of Hydraulic Problems, GeoPlanet: Earth and Planetary Sciences*; Springer: Berlin/Heidelberg, Germany, 2013; pp. 221–235.
33. Testa, G.; Zuccal, D.; Alcrudo, F.; Mulet, J.; Soares-Frazao, S. Flashflood flow experiment in a simplified urban district. *J. Hydraul. Res.* **2007**, *45*, 37–44. [CrossRef]
34. Soares-Frazao, S.; Lhomme, J.; Guinot, V.; Zech, Y. Two dimensional shallow water model with porosity for urban flood modeling. *J. Hydraul. Res.* **2008**, *46*, 45–64. [CrossRef]
35. Liang, D.; Falconer, R.A.; Lin, B. Coupling surface and subsurface flows in a depth averaged flood wave model. *J. Hydrol.* **2007**, *337*, 147–158. [CrossRef]
36. Sanders, B.F.; Schubert, J.E.; Gallegos, H.A. Integral formulation of shallow-water equations with anisotropic porosity for urban flood modeling. *J. Hydrol.* **2008**, *362*, 19–38. [CrossRef]
37. Yu, M.; Deng, Y.; Qin, L.; Wang, D.; Chen, Y. Numerical simulation of levee breach flows under complex boundary conditions. *J. Hydrodyn.* **2009**, *21*, 633–639. [CrossRef]
38. Petaccia, G.; Soares-Frazao, S.; Natale, L.; Zech, Y. Simplified versus detailed two dimensional approaches to transient flow modeling in urban areas. *J. Hydraul. Eng.* **2010**, *136*, 262–266. [CrossRef]
39. Mignot, E.; Bonakdari, H.; Knothe, P.; Lipeme Kouyi, G.; Bessette, A.; Rivière, N.; Bertrand-Krajewski, J.L. Experiments and 3D simulations of flow structures in junctions and their influence on location of flowmeters. *Water Sci. Technol.* **2012**, *66*, 1325–1332. [CrossRef] [PubMed]
40. Defne, Z. Effect of Particle Shape and Size on Incipient Motion of Solid Particles. Master's Thesis, The Middle East Technical University, Ankara, Turkey, 2002.
41. Gogus, M.; Defne, Z. Effect of shape on incipient motion of large solitary particles. *J. Hydraul. Eng.* **2005**, *131*, 38–45. [CrossRef]
42. Gulcu, B. Incipient Motion of Coarse Solitary Particles. Master's Thesis, The Graduate School of Natural and Applied Sciences of Middle East Technical University, Ankara, Turkey, 2009.
43. Zhu, L.; Wang, J.; Cheng, N.; Ying, Q.; Zhang, D. Settling distance and incipient motion of sandbags in open channel flows. *J. Waterw. Port Coast. Ocean Eng.* **2004**, *130*, 98–103. [CrossRef]
44. Izbash, S.V. Construction of dams by depositing rock in running water. In Proceedings of the 2nd Congress on Large Dams, Washington, DC, USA, 7–12 September 1936.
45. Kobayashi, N.; Jacobs, B.K. Experimental study on sand-bag stability and runup. In *Coastal Zone '85*; Magoon, O.T., Converse, H., Miner, D., Clark, D., Tobin, L.T., Eds.; ASCE: New York, NY, USA, 1985; Volume 2, pp. 1612–1626.
46. Neill, C.; Mannerstrom, M.; Azad, A.K. Model tests on geobags for erosion protection. In Proceedings of the International Conference on Scour and Erosion, Tokyo, Japan, 5–7 November 2008.
47. Korkut, R.; Martinez, E.; Morales, R.; Ettema, R.; Barkdoll, B. Geobag performance as scour countermeasure for bridge abutments. *J. Hydraul. Eng.* **2007**, *113*, 431–439. [CrossRef]
48. Pilarczyk, K.W. Geomettessess in erosion control—An overview of design criteria. In *Filters and Drainage in Geotechnical and Environmental Engineering*; Balkema: Rotterdam, The Netherlands, 2000; pp. 331–338.
49. Richardson, E.V.; Davis, S.R. *Evaluating Scour at Bridges*, 3rd ed.; Hydraulic Engineering Circular No. 18 (HEC 18), Rep. No. FHwA-IP-90-017; Office of Technology: Washington, DC, USA, 1995.

50. El-Kholy, M.; Chaudhry, M.H. Tracking sandbags motion during levee breach closure using DPTV technique. In Proceedings of the 33rd Congress, International Association of Hydraulic Engineering and Research, Vancouver, BC, Canada, 10–14 August 2009.
51. Juez, C.; Murillo, J.; Garcia, P. A 2D weakly-coupled and efficient numerical model for transient shallow flow and movable bed. *Adv. Water Resour.* **2014**, *71*, 93–109. [CrossRef]
52. Novak, P.; Nalluri, C. Incipient motion of sediment particles over fixed beds. *J. Hydraul. Eng.* **1984**, *22*, 181–197. [CrossRef]
53. Michelazzo, G.; Oumeraci, H.; Paris, E. New hypothesis for the final equilibrium stage of a river levee breach due to overflow. *Water Resour. Res.* **2018**, *54*, 4277–4293. [CrossRef]
54. Stelczer, K. *Bed-Load Transport: Theory and Practice*; Water Resources Publications: Littleton, CO, USA, 1981.
55. Sattar, A.M.; Dickerson, J.; Chaudhry, M. A wavelet Galerkin solution to the transient flow equations. *J. Hydraul. Eng.* **2009**, *135*, 283–295. [CrossRef]
56. Sattar, A.M. Gene expression models for the prediction of longitudinal dispersion coefficients in transitional and turbulent pipe flow. *J. Pipeline Syst. Eng. Pract.* **2013**, *5*, 04013011. [CrossRef]
57. Sattar, A.M. Gene expression models for prediction of dam breach parameters. *J. Hydroinform.* **2014**, *16*, 550–571. [CrossRef]
58. Najafzadeh, M.; Sattar, A.M. Neuro-Fuzzy GMDH Approach to Predict Longitudinal Dispersion in Water Networks. *Water Resour. Manag.* **2015**, *29*, 2205–2219. [CrossRef]
59. Sattar, A.M.; Gharabaghi, B. Gene expression models for prediction of longitudinal dispersion coefficient in streams. *J. Hydrol.* **2015**, *524*, 587–596. [CrossRef]
60. Sattar, A.M. Prediction of organic micropollutant removal in soil aquifer treatment system using GEP. *J. Hydrol. Eng.* **2016**, *21*, 04016027. [CrossRef]
61. Sattar, A.M.; Gharabaghi, B.; McBean, E.A. Prediction of timing of watermain failure using gene expression models. *Water Resour. Manag.* **2016**, *30*, 1635–1651. [CrossRef]
62. Sattar, A.M. A probabilistic projection of the transient flow equations with random system parameters and internal boundary conditions. *J. Hydraul. Res.* **2016**, *54*, 342–359. [CrossRef]
63. Sattar, A.M.; Ertuğrul, Ö.F.; Gharabaghi, B.; McBean, E.A.; Cao, J. Extreme learning machine model for water network management. *Neural Comput. Appl.* **2017**, 1–13. [CrossRef]
64. Sattar, A.M.A.; Gharabaghi, B.; Sabouri, F.; Thompson, A.M. Urban stormwater thermal gene expression models for protection of sensitive receiving streams. *Hydrol. Process.* **2017**, *31*, 2330–2348. [CrossRef]
65. Sattar, A.M.; Plesiński, K.; Radecki-Pawlik, A.; Gharabaghi, B. Scour depth model for grade control structures. *J. Hydroinform.* **2018**, *20*, 117–133. [CrossRef]
66. Jia, Y.; Zhu, T.; Rihani-Nezhad, C.; Zhang, Y. Numerical modeling of flow through a breached levee and during levee closure. In Proceedings of the World Environmental and Water Resources Congress, Providence, RI, USA, 16–20 May 2010; pp. 1304–1316.
67. Heller, V. Scale effects in physical hydraulic engineering models. *J. Hydraul. Res.* **2011**, *49*, 293–306. [CrossRef]
68. Chanson, H. *The Hydraulics of Open Channel Flow*; Arnold Publisher: Wiley, NY, USA, 1999.

MDPI
St. Alban-Anlage 66
4052 Basel
Switzerland
Tel. +41 61 683 77 34
Fax +41 61 302 89 18
www.mdpi.com

Water Editorial Office
E-mail: water@mdpi.com
www.mdpi.com/journal/water

www.ingramcontent.com/pod-product-compliance
Lightning Source LLC
LaVergne TN
LVHW080458180726
843515LV00007BA/1959

* 9 7 8 3 0 3 9 3 6 4 5 1 0 *